T0191889

Principles of Applied Remote Sensing

Siamak Khorram • Cynthia F. van der Wiele
Frank H. Koch • Stacy A. C. Nelson
Matthew D. Potts

Principles of Applied Remote Sensing

Siamak Khorram
Environmental Sci. Policy & Mgmt.
University of California, Berkeley
Berkeley, California
US
and
Center for Geospatial Analytics
North Carolina State University
Raleigh, North Carolina

Cynthia F. van der Wiele
US Environmental Protection Agency
Region 4 NEPA Program Office
Research Triangle Park, North Carolina
US

Frank H. Koch
Southern Research Station
USDA Forest Service
Research Triangle Park, North Carolina
US

Stacy A. C. Nelson
North Carolina State University
Center for Geospatial Analytics
Raleigh, North Carolina
US

Matthew D. Potts
Environmental Sci Policy & Mgmt
University of California Berkeley
Berkeley, California
US

ISBN 978-3-319-79410-5
ISBN 978-3-319-22560-9 (eBook)
DOI 10.1007/978-3-319-22560-9

Springer Cham Heidelberg New York Dordrecht London

Softcover reprint of the hardcover 1st edition 2016

Printed on acid-free paper

Springer Science+Business Media LLC New York is part of Springer Science+Business Media (www.springer.com)

Acknowledgments

The authors are thankful to Steven M. Unikewicz, ASME, for his enthusiastic contributions in reviewing, critiquing, and providing suggestions on ways to present our materials to be understood by students of many disciplines. We are also thankful to Joshua Verkerke of the Department of Environmental Science, Policy, and Management (ESPM), University of California, Berkeley, for his contributions in processing certain images of Southern California for this book.

Contents

1 **Remote Sensing: Past and Present** ... 1
1.1 Introduction ... 1
1.2 A Brief History of Remote Sensing ... 2
1.3 What Is Remote Sensing, and Why Do It? ... 8
1.4 The Electromagnetic Spectrum ... 11
1.5 Photo Interpretation, Photogrammetry, and Image Processing ... 14
1.6 The Importance of Accuracy Assessment ... 15
1.7 Cost Effectiveness and Reach Versus Richness of Remote Sensing Technology ... 15
1.8 Organization of This Book ... 16
1.9 Review Questions ... 17
References ... 18
Suggested Reading ... 19
Relevant Websites ... 19

2 **Data Acquisition** ... 21
2.1 Data Resolution ... 21
2.2 Payloads and Platforms: An Overview ... 34
2.2.1 Airborne Platforms ... 35
2.2.2 Spaceborne Platforms ... 42
2.3 Review Questions ... 61
References ... 62
Suggested Reading ... 67
Relevant Websites ... 67

3 **Data Processing Tools** ... 69
3.1 Display of Multispectral Image Data ... 69
3.2 Preprocessing Image Data ... 71
3.2.1 Geometric Correction ... 71
3.2.2 Atmospheric Correction ... 73
3.2.3 Radiometric Correction ... 74
3.2.4 Band Combinations, Ratios, and Indices ... 75
3.2.5 Data Fusion ... 78

3.3 Image Processing ... 83
3.3.1 Selection of a Classification Scheme ... 85
3.3.2 Optimum Band Selection Prior to Classification ... 86
3.3.3 Unsupervised Classification ... 88
3.3.4 Supervised Classification ... 89
3.3.5 Fuzzy Logic Classification ... 93
3.3.6 Other Classification Approaches ... 95
3.4 Post-processing Image Data ... 99
3.4.1 Spatial Filters ... 99
3.4.2 Accuracy Assessment ... 101
3.4.3 Change Detection ... 102
3.4.4 Data Integration and Geospatial Modeling ... 108
3.4.5 Processing of Airborne LiDAR Data ... 114
3.5 Summary ... 116
3.6 Review Questions ... 116
References ... 117
Suggested Reading ... 124

4 Terrestrial Applications of Remote Sensing ... 125
4.1 Classifying Land Use and Land Cover ... 126
4.2 Understanding and Protecting Biodiversity Through Wildlife Tracking ... 130
4.3 Water Resources ... 132
4.4 Forest Resources ... 136
4.4.1 Forest Health ... 140
4.4.2 Biomass Estimation ... 142
4.4.3 Carbon Estimation ... 146
4.4.4 Wildland Fire Risk Assessment ... 150
4.5 Optimizing Sustainable Food and Fiber Production through Remote Sensing ... 155
4.5.1 Improving Wine Harvest and Quality ... 158
4.5.2 Using Remote Sensing to Optimize Grazing and Improve Wool Quality ... 160
4.6 Exploring and Monitoring Oil, Gas, and Mineral Resources ... 160
4.7 Using Remote Sensing for Humanitarian and Peace-Keeping Operations ... 163
4.8 Archaeology and Cultural Heritage ... 164
4.9 Summary ... 166
4.10 Review Questions ... 167
References ... 168
Additional Reading ... 175
Relevant Websites ... 176

5 Atmospheric Applications of Remote Sensing ... 177
5.1 Weather Forecasting and Extreme Weather Events ... 178
5.1.1 Measuring Precipitation from Space ... 179
5.2 Public Health ... 180
5.2.1 Measuring Air Pollution to Understand Human and Ecosystem Health Impacts ... 181
5.3 Appraising and Predicting Episodic Events ... 183
5.3.1 Monitoring and Forecasting Volcanic Activity ... 184
5.3.2 Using Remote Sensing for Early Warning of Dust Storms ... 186
5.4 Global Climate Change ... 189
5.5 Review Questions ... 196
References ... 196
Additional Reading ... 198
Relevant Websites ... 199

6 Observing Coastal and Ocean Ecosystems ... 201
6.1 Introduction ... 201
6.2 Using Remote Sensing to Map Ocean Color, Phytoplankton, and Chlorophyll Concentration ... 204
6.3 Remote Sensing of Eutrophication and Ocean Hypoxia ... 209
6.4 Using Remote Sensing to Map the Sea Surface Temperature and Circulation Patterns ... 211
6.5 Spatial Analysis of Submersed Aquatic Vegetation ... 213
6.6 Remote Sensing of Coastal Bathymetry ... 215
6.7 Remote Sensing of Coral Reefs ... 217
6.8 Achieving Sustainable Fisheries and Aquaculture Management ... 221
6.9 Ocean Observation Networks ... 222
6.9.1 Global Ocean Observing System (GOOS) ... 222
6.9.2 Australia's Integrated Marine Observing System (IMOS) ... 223
6.9.3 European Marine Observation and Data Network (EMODnet) ... 223
6.9.4 US Integrated Ocean Observing System (IOOS®) ... 223
6.10 Review Questions ... 224
References ... 225
Additional Reading ... 228
Relevant Websites ... 228

7 The Final Frontier: Building New Knowledge Through Planetary and Extrasolar Observation ... 229
7.1 Introduction ... 229
7.2 Lunar Exploration ... 232
7.3 Mercury, Venus, and Mars ... 237
7.4 Jupiter, Saturn, Uranus, and Neptune ... 242
7.5 Pluto and the Kuiper Belt ... 246

7.6 The Sun ... 247
7.7 Extrasolar Remote Sensing ... 248
7.8 Review Questions ... 253
References ... 253
Additional Reading ... 258
Relevant Websites ... 258

8 International Laws, Charters, and Policies ... 261
8.1 Introduction ... 261
8.2 Origin and Focus of International Space Law ... 262
8.3 The International Charter on Space and Major Disasters ... 265
8.4 National Policies Governing Remotely Sensed Data ... 266
8.4.1 Common Themes and Policy Solutions ... 267
8.4.2 US Laws and Policies ... 268
8.4.3 Legal Frameworks Within the European Union ... 270
8.4.4 Asian Policies ... 270
8.4.5 Australian Remote Sensing Policy ... 271
8.4.6 Remote Sensing Policies on the African Continent ... 271
8.5 The Future of Remote Sensing Laws and Policy ... 272
8.6 Review Questions ... 273
References ... 273
Suggested Reading ... 274
Relevant Websites ... 275

9 Future Trends in Remote Sensing ... 277
9.1 Future Advances in Hardware and Software ... 277
9.2 Open, Social, and Timely ... 279
9.3 Interdisciplinarity and Big Data ... 282
9.4 Concluding Thoughts ... 283
9.5 Review Questions ... 284
References ... 284
Suggested Reading ... 285

Appendix 1: Answers to Questions ... 287

Index ... 301

About the authors

Siamak Khorram has joint appointments as a professor of remote sensing and image processing at both the University of California at Berkeley and North Carolina State University. He is also the founding director of the Center for Geospatial Analytics and a professor of electrical and computer engineering at North Carolina State University and a member of the Board of Trustees at International Space University (ISU) in Strasbourg, France. Dr. Khorram was the first dean of ISU and a former vice president for academic programs as well as a former chair of the ISU's Academic Council. He has also served as the American Society for Engineering Education (ASEE) fellow at Stanford University and NASA Ames Research Center. Dr. Khorram has extensive research and teaching experience in remote sensing, image processing, and geospatial technologies and has authored well over 200 publications. He has served as the guiding professor for numerous PhD and masters graduate students. He is a member of several professional and scientific societies. His graduate degrees were awarded by the University of California at Davis and Berkeley.

Cynthia F. van der Wiele is a senior physical scientist with the US Environmental Protection Agency (USEPA), Region 4, NEPA Program Office. Previously, she was a research associate and adjunct faculty at North Carolina State University. Her research interests include the development of high accuracy land use/land cover classifications for analysis and improved land use and conservation planning and policies. Dr. van der Wiele received her BS in engineering and Masters of Landscape Architecture from North Carolina State University, a Masters in Forestry and a Masters in Environmental Economics and Policy from Duke University, and her PhD in community and environmental design from North Carolina State University. She is active in several national and international professional societies.

Frank H. Koch is a research ecologist with the US Department of Agriculture (USDA) Forest Service. Previously, he was a research assistant professor at the North Carolina State University. His primary area of research is alien forest pest invasions. Specifically, he is interested in the spatiotemporal dynamics of invasions at national and continental scales. This multidisciplinary work involves geographical information systems (GIS), remote sensing, statistics, and spatial simulation modeling. Dr. Koch regularly collaborates with other USDA Forest Service scientists as well as researchers from the Canadian Forest Service, the USDA Animal and Plant Health Inspection Service, and several universities. He has authored numerous journal articles and other publications. Dr. Koch received his BA from Duke University and MS and PhD from North Carolina State University.

Stacy A. C. Nelson is currently an associate professor and a researcher with the Center for Geospatial Analytics at North Carolina State University. Dr. Nelson received a BS from Jackson State University, an MA from The College of William & Mary, and a PhD from Michigan State University. His research centers on GIS technologies to address questions of land use and aquatic systems. He has worked with several federal and state agencies including the NASA Stennis Space Center in Mississippi, the NASA Regional Earth Science Applications Center (RESAC), the USDA Forest Service, as well as various state-level agencies. He is active in several professional societies.

Matthew D. Potts is an associate professor of forest ecosystem management at the University of California at Berkeley. He has a broad interdisciplinary background with training in mathematics, ecology, and economics, with a BS from the University of Michigan and a PhD from Harvard University. Matthew has extensive international experience conducting field research in tropical forests throughout the world. His varied research interests include spatial aspects of forest management and land use planning as well as how human actions, values, and ethics affect biodiversity conservation.

Chapter 1
Remote Sensing: Past and Present

1.1 Introduction

The use of remote sensing perhaps goes all the way back to prehistoric times when early man stood on a platform in front of his cave and glanced at the surrounding landscape (late Robert N. Colwell, UC Berkeley). These humans were remotely sensing the features in the landscape to determine the best places to gather food and water and how to avoid becoming a food for the other inhabitants of the landscape. The term "photography" is derived from two Greek words meaning "light" (phos) and "writing" (graphein) (late John E. Estes, UC Santa Barbara). All cameras and sensors utilize the same concept of light entering a camera or a sensor and being recorded on a film or on a digital media.

The term remote sensing as used for Earth observation has experienced major changes since the 1960s (Baumann 2009). It was not until the 1960s that "remote sensing" moved beyond black and white aerial photography and began to evolve towards its usage today in the twenty-first century. The past 50 years have seen the development of remote sensing platforms for high-altitude aircraft and satellites and the development of imaging sensors for collecting data from various regions of the electromagnetic spectrum. In concert with these advances in image acquisition capabilities were the developments of image display and processing techniques and algorithms. With the increase in processing power, these image processing software have migrated from mainframe computers to desktops to handheld mobile smart devices.

Along with the advances in technology, there has been a rapid and growing social acceptance of remote sensing. At first, remote sensing was a concern to the public with the "eye in the sky" and "Big Brother" concept. This perception, however, has eroded to a very large extent (but may be returning). Today, a large variety of sensors are deployed on numerous satellites and airborne platforms that collect vast amounts of remotely sensed data around the globe and around the clock. These data of various characteristics in spectral, spatial, radiometric, and temporal resolutions are commonly utilized in the areas of environmental and natural resource management, climate change, disaster management, law enforcement, military, and military

S. Khorram et al., *Principles of Applied Remote Sensing*,
DOI 10.1007/978-3-319-22560-9_1

intelligence gathering. Remote sensing has permeated our daily lives through Google Earth; global positioning systems (GPS); weather forecasting, wildland fire, hurricane, and disaster management; precision agriculture; and natural resources inventory and monitoring. We cannot live without it.

The primary goal of this book is to provide readers with a basic understanding of the principles and technologies behind remotely sensed data and their applications. In Chapter 1, we offer historic development of the remote sensing technology. Chapters 2 and 3 deal with data acquisition payloads and platforms and the techniques and tools for processing the vast amount of remotely sensed airborne and satellite data from various sensors and scanners. Chapters 4, 5, 6, and 7 include various uses of this technology including terrestrial, oceanographic, atmospheric, and planetary environments. Chapter 8 offers a discussion of political trends in the remotely sensed data acquisition and applications at international levels. In Chapter 9, the current state of the art and the future trends in remote sensing technology are reviewed.

1.2 A Brief History of Remote Sensing

According to the late John E. Estes and Jeff Hemphill, one can trace the history of photography to *Leonardo da Vinci* (1490) describing the light entering the dark room (obscura) through a pinhole on one of the walls of a dark room; *Angelo Sala* (1614) observing the darkening of silver salts when exposed to sunlight; *Sir Isaac Newton* (1666) experimenting with a prism; *Johann Christopher Sturm* (1676) discovering the lens; *Sir William Herschel* (1800) observing the visible colors and temperature; and *Thomas Young* (1802) describing the basic concepts of the Young-Von Helmholtz theory of color vision to the first photograph by *Joseph Niépce* in 1827.

The invention of photography helped to lay the groundwork for the field of remote sensing by enabling the near-instantaneous documentation of objects and events. The French inventor Joseph Niépce is generally credited with producing the first permanent photograph in 1827, which depicted the view from his upstairs workroom window (Hirsch 2008). In 1839, it was announced that Louis Daguerre—who collaborated with Niépce until his death in 1833—had invented a process for creating a fixed silver image on a copper plate, called a daguerreotype (Newhall 1982). One of his daguerreotypes, "Boulevard du Temple, Paris" (Fig. 1.1), taken in 1838 or 1839, is reputedly the oldest surviving photograph of a person. The image appears to show an empty street, but this is an artifact; the long exposure of more than 10 minutes prevented the capture of moving carriage and pedestrian traffic. However, Daguerre was able to capture a man who stopped to have his shoes shined (see lower left of Fig. 1.1). Notably, Daguerre's "Boulevard Du Temple" image has many characteristics of what is now called an *oblique* aerial photograph (i.e., an aerial photograph captured from an angle rather than vertically, or directly overhead) (Mattison 2008).

The potential cartographic applications of photography were recognized almost immediately. In 1840, François Arago, director of the French Académie des Sciences

Fig. 1.1 "Boulevard du Temple, Paris," a photograph (daguerreotype) taken by Louis Daguerre in 1838

and the man who publicly announced Daguerre's process, advocated the use of photographs to produce topographic maps (Mattison 2008; Wood 1997). Credit for the first actual aerial photograph is given to the French photographer Gaspar Felix Tournachon, who used the pseudonym "Nadar." Nadar patented the concept of using aerial photography for cartography and surveying in 1855, but experimented unsuccessfully until 1858, when he captured a photograph from a balloon tethered 80 m above the Bievre Valley (PAPA International 2011). None of Nadar's early efforts is believed to have survived. The oldest existing aerial photograph is a view of Boston, taken from a balloon by James Wallace Black in 1860 (Fig. 1.2).

During the latter part of the nineteenth and into the early twentieth century, a number of people experimented with the use of aerial photography from balloons, kites, and even birds as an effective means of mapmaking and surveying. In 1889, Canadian Dominion Lands Surveyor General E.G.D. Deville published *Photographic Surveying*, a seminal work that focused on balloon-based photography (Mattison 2008). French photographer Arthur Batut is usually credited with the first successful photograph from a kite, taken in 1887 or 1888, and published the first textbook on kite photography in 1890 (Mattison 2008). In 1908, Julius Neubronner, a German apothecary and inventor, patented a breast-mounted aerial camera for carrier pigeons (PAPA International 2011). The lightweight camera took automatic exposures at 30-s intervals. Although faster than balloons, the pigeons did not always follow their expected flight paths. When the aerial photographs taken

Fig. 1.2 The oldest surviving aerial photograph, an image of Boston taken from a balloon in 1860

by the pigeons were introduced at the 1909 Dresden International Photographic Exhibition, postcards created from them became popular with the public (PAPA International 2011). Camera-equipped pigeons have also been used for military surveillance. Figure 1.3 illustrates two examples of pigeon aerial photographs, as well as an image of a pigeon mounted with one of Neubronner's aerial cameras.

In 1880, George Eastman patented a machine for rapidly preparing a large number of "dry" photographic plates (i.e., glass plates coated with a gelatin emulsion). Searching for a lighter and less temperamental alternative to glass plates, he developed rolled paper film, but found that paper was not an ideal film base because the resulting photographs tended to be grainy and have inadequate contrast (Utterback 1995). Eastman addressed these limitations through the introduction of flexible celluloid film in 1887. In 1900, Eastman's company, Kodak, released the Brownie, an inexpensive box camera for rolled film, making photography accessible to a mass audience for the first time.

Eastman's innovations shortly preceded the Wright Brothers' first successful flight, in 1903, which took place in the shores of the Outer Banks of North Carolina. Photographic images of various kinds were followed. As an example is the oblique aerial photograph of the City of San Francisco in 1906 taken by a kite, as shown below (Fig. 1.4).

Five years later, Wilbur Wright took the first aerial photograph from an airplane (PAPA International 2011). Quickly embraced by military forces, airplane-based aerial photography was used extensively for reconnaissance during World War I (Rees 2001). Near the end of the war, US entrepreneur Sherman Fairchild began to develop what became the first true aerial camera system (PAPA International 2011).

Fig. 1.3 Examples of Julius Neubronner's pigeon aerial photography. Notably, the photograph of the Schlosshotel Kronberg *(top left)* accidentally included the pigeon's wingtips. The image on the *right* shows a pigeon with one of Neubronner's breast-mounted cameras

In 1921, Fairchild demonstrated the utility of his system for cartography, employing more than 100 overlapping aerial images to create a *photo-mosaic* of New York City's Manhattan Island (Fig. 1.5). During the period between World Wars I and II, aerial photography was also applied in other civilian contexts, including forestry, geology, and agriculture (Rees 2001).

Aerial photography experienced dramatic refinement during World War II—a period that also saw the introduction of the first infrared-sensitive instruments and radar imaging systems (Rees 2001). In fact, the basic elements of aerial photography as we know it today largely arose out of these wartime developments and related technological advances during the next two decades. False-color infrared film was

Fig. 1.4 An oblique aerial photograph of San Francisco, CA, taken in 1906 from a kite. (Courtesy of the US Geological Survey, adopted from Madry 2013)

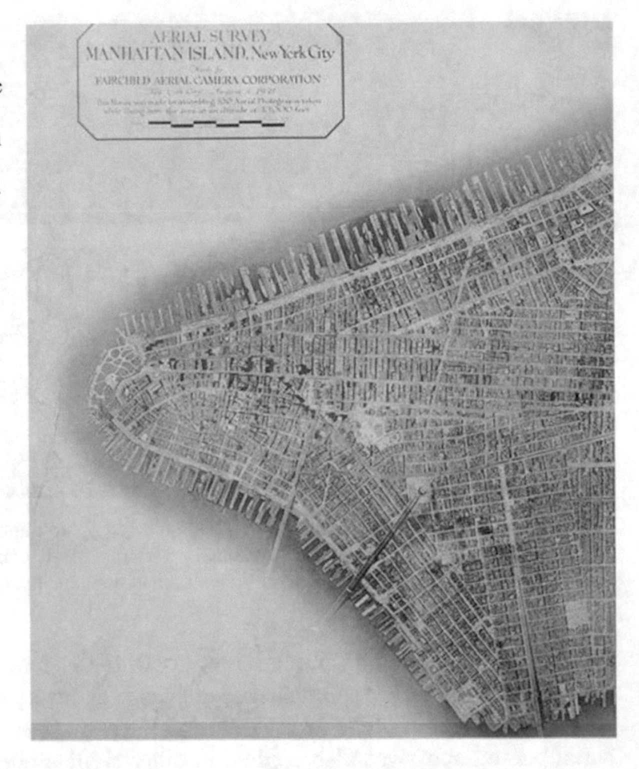

Fig. 1.5 Details from Fairchild Aerial Camera Corporation's 1921 photo-mosaic of Manhattan, New York, showing the island's southern region. (Image courtesy of Library of Congress, Geography and Map Division)

first developed during World War II for camouflage detection, and by the 1950s, it was already being applied for air-photo-based mapping of vegetation (Rees 2001). Live plants typically exhibit strong spectral reflectance in the near-infrared portion of the electromagnetic spectrum (see Gates et al. 1965.)

The Space Age was an era initiated by the Soviet Union's launch of the first man-made satellite, Sputnik-1, in 1957, from Baikonur Cosmodrome at Tyuratam (370 km southwest of the small town of Baikonur) in Kazakhstan, then part of the former Soviet Union. The Russian word "Sputnik" (http://nssdc.gsfc.nasa.gov/nmc/spacecraftDisplay.do?id=1957-001B) means "companion" ("satellite" in the astronomical sense).

The term "remote sensing" was coined in the mid-1950s by Evelyn Pruitt, a geographer with the US Office of Naval Research, allegedly because the term "aerial photography" did not sufficiently accommodate the notion of images from space (Short 2010). After the launch of Sputnik-1, the US and Soviet governments raced to design and implement new space-related technologies, including both manned spacecraft and satellites. While the first (and rather crude) satellite image of the Earth was captured by NASA's Explorer 6 in 1959 (Fig. 1.6), the US Department of Defense's CORONA (also known as "Discoverer") Reconnaissance Satellite Program, which remained classified until 1995, may be seen as a key forerunner to present-day Earth-observing satellite programs. During its period of operation (1958–1972), the CORONA Program developed an increasingly sophisticated series of high-resolution, film-based camera systems (Cloud 2001). The first photograph of the Soviet

Fig. 1.6 First satellite image of the Earth, showing a sunlit portion of the Pacific Ocean and its cloud cover. The image was captured in August 1959 by Explorer 6, a US National Aeronautics and Space Administration (NASA) satellite. (Image courtesy of NASA)

territory from space, taken in August 1960, shows an air base at Mys Shmidta, Siberia (Fig. 1.7). Within a decade, CORONA satellites had extensively mapped the United States and other parts of the world. Before the CORONA program ended, its science team had begun to experiment with color (i.e., spectral) photography, thus serving as a precursor to the sensors used by the Landsat program (discussed later and in Chapter 2) and present-day satellite imaging systems (Cloud 2001).

In the 1960s and the early 1970s, the US and Soviet Union launched an assortment of reconnaissance, meteorological, and communications satellites into the orbit. Also during this period, astronauts from NASA's Mercury, Gemini, and Apollo space missions took thousands of photographs of the Earth using handheld and automated cameras (Witze 2007). The "Blue Marble," a photograph taken in December 1972 by the crew of Apollo 17, is often cited as the most widely reproduced image of the Earth (McCarthy 2009). In 2002, NASA released a new version of the "Blue Marble," a mosaic of images from the Moderate Resolution Imaging Spectroradiometer (MODIS) instrument onboard the Earth Observing System (EOS) Terra satellite; see Fig. 1.8.

A more formative event for modern remote sensing occurred in July 1972, when NASA launched ERTS-A (Earth Resources Technology Satellite—Mission A), the first satellite dedicated to monitoring environmental conditions on the Earth's surface. Shortly after its launch, the satellite's name was changed to ERTS-1. It was followed by ERTS-2 (launched in January 1975) and ERTS-3 (launched in March 1978). Later, the names for these satellites were changed to Landsat-1, -2, and -3, respectively. The Landsat Program (along with a few other US satellite programs; see Chap. 2) served as the primary source of space-based Earth imagery until the 1980s, when a number of other countries began to develop their own Earth-observing satellite programs, particularly France and the European Union, Canada, Japan, India, Russia, China, and Brazil. More recently, a number of private companies have emerged as providers of satellite imagery, demonstrating the feasibility of commercial, space-based remote sensing.

Fig. 1.7 Photograph of a Soviet air base at Mys Shmidta, Siberia, taken in August 1960 by a camera onboard the CORONA satellite Discoverer-14. (Image courtesy of the US National Reconnaissance Office)

The European Space Agency (ESA) estimates that between the 1957 launch of Sputnik and January 1, 2008, approximately 5600 satellites were launched into the Earth's orbit (ESA 2009). The vast majority of these are no longer in service, raising concerns and the formation of an international committee regarding the high volume of space debris encircling the planet (see Fig. 1.9). Today, just fewer than 1000 operational satellites are orbiting the Earth, approximately 9 % of which are dedicated to Earth observation and remote sensing (plus roughly 4 % for meteorology and related applications) (UCS 2011). These satellites, carrying a wide range of sensors optimized for various applications, represent a rich potential source of data for remote sensing analysts. Furthermore, they also extend the nearly four-decade-old satellite image record started by Landsat-1.

1.3 What Is Remote Sensing, and Why Do It?

Today, the field of remote sensing is a well-recognized interdisciplinary field across the globe. This field is often coupled with the disciplines of image processing (IP), geographic information systems (GIS), and GPS to create the broad field of geospatial science and technologies.

Fig. 1.8 A new version of the "Blue Marble": This true-color image is a seamless mosaic of separate images, largely recorded with the Moderate Resolution Imaging Spectroradiometer (MODIS), a device mounted on NASA's Terra satellite. (Image courtesy of NASA Goddard Space Flight Center)

Remote sensing is defined as the acquisition and measurement of information about certain properties of phenomena, objects, or materials by a recording device not in physical contact with the features under surveillance. This is a rather broad definition that encompasses, for instance, medical technologies such as X-rays and magnetic resonance imaging. In an environmental context, remote sensing typically refers to technologies for recording *electromagnetic energy* that emanates from areas or objects on (or in) the Earth's land surface, oceans, or atmosphere (Short 2010). Essentially, the properties of these objects or areas, in terms of their associated levels of electromagnetic energy, provide a way to identify, delineate, and distinguish between them. Because the electromagnetic energies of these features are commonly collected by instruments mounted on aircraft or Earth-orbiting spacecraft, remote sensing also gives scientists the opportunity to capture large geographic areas with a single observation or *scene* (Fig. 1.10).

Another potential advantage of remote sensing, especially when done from satellites, is that geographic areas of interest can be revisited on a regular cycle, facilitating the acquisition of data to reveal changing conditions over time. For a given instrument, or *sensor*, onboard a satellite, the revisit time depends on the satellite's orbit and the navigational speed as well as the width of the sensor's *swath*, which

Fig. 1.9 Debris objects in the low-Earth orbit. These debris are largely composed of inactive satellites and other hardware, as well as fragments of spacecraft that have broken up over time. (Objects are not to scale.) (Image courtesy of the European Space Agency)

is the track the sensor observes as the satellite travels around the Earth (Fig. 1.11). The concept of a sensor's *temporal resolution* will be explored further in Chapter 2.

Remotely sensed data are *geospatial* in nature, meaning that the observed areas and objects are referenced according to their location in a geographic coordinate system, such that they may be located on a map (Short 2010). This allows the remotely sensed data to be analyzed in conjunction with other geospatial data sets, such as data depicting road networks or human population density. Remotely sensed data with sufficient detail can be used to characterize such things as the growth or health status of vegetation, or to identify habitat edges or ecotones (i.e., transition zones between ecological communities) that could not otherwise be discerned effectively from maps created from field observations (Kerr and Ostrovsky 2003).

This point illustrates the unique importance of remote sensing as a data source for *GIS*, which are organized collections of computer hardware, software, geographic data, and personnel designed to efficiently capture, store, update, manipulate, and analyze all forms of geographically referenced information (Jensen 2005; ESRI 2001). In turn, *geographic information science* (*geomatics* or *geoinformatics*) is concerned with conceptual and scientific issues that arise from the use of GIS, or more broadly, with various forms of geographic information (Longley et al. 2001). Ultimately, the value of a GIS depends upon the quality of the data it contains (Jensen 2005; Longley et al. 2001). Because remote sensing serves as the primary source of GIS data, it is important for users to understand how these data are generated so they can evaluate subsequent geospatial analyses more critically.

Fig. 1.10 An illustration of the remote sensing concept: An instrument (i.e., sensor or scanner) mounted on an aircraft or satellite records information about objects and/or areas on the ground. Typically, these data are spectral in nature, meaning that they document the amount of electromagnetic energy associated with the targeted objects and/or areas. The extent, or footprint, of the geographic area captured in a single-sensor scene depends on the sensor's design and the altitude of the aircraft or spacecraft on which it is mounted

1.4 The Electromagnetic Spectrum

Remote sensing typically refers to technologies for recording electromagnetic energy that emanates from areas or objects. But what is electromagnetic energy? *Electromagnetic radiation (EMR)* is defined as all energy that moves with the velocity of light in a *harmonic wave pattern* (i.e., all waves are equally and repetitively spaced in time). Visible light is just one category of EMR; other types include radio waves, infrared, and gamma rays. Together, all of these types comprise the *electromagnetic spectrum* (Fig. 1.12). As illustrated in Fig. 1.12, the different forms of EMR vary across the spectrum in terms of both *wavelength* and *frequency*. Wavelength is the distance between one position in a wave cycle and the same position in the next wave, while frequency is the number of wave cycles passing through the same point in a given time period (1 cycle/s = 1 Hertz, or Hz).

Fig. 1.11 Visualization of the "swath" captured by a sensor aboard a hypothetical Earth-orbiting satellite. As the satellite orbits and the Earth rotates on its axis, the sensor images the planet's entire surface

The mathematical relationship between wavelength and frequency is expressed by the following equation:

$$c = \lambda \cdot \nu.$$

where λ is the wavelength, ν is the frequency, and c is the speed of light (which is constant at 300,000 km/s in a vacuum).

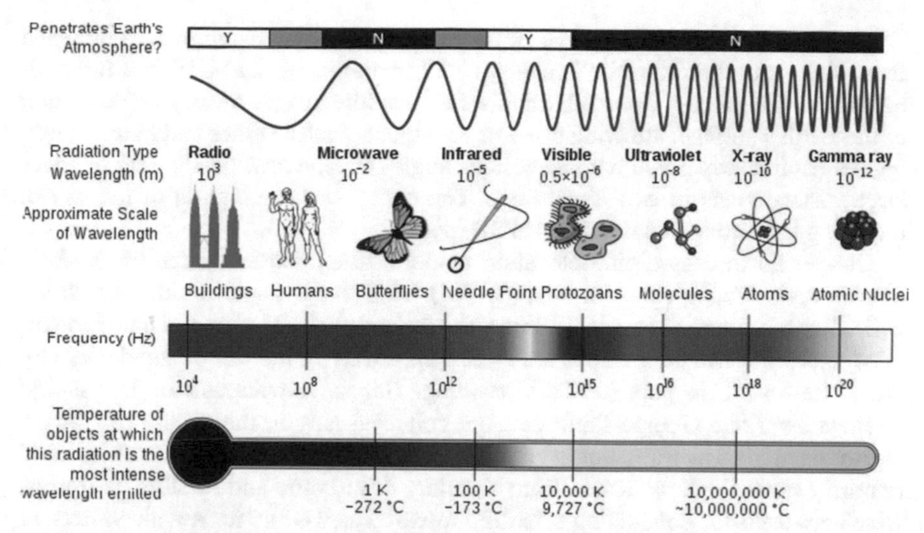

Fig. 1.12 The electromagnetic spectrum. (Illustration courtesy of NASA)

Visible light, representing only a small portion of the electromagnetic spectrum, ranges in wavelength from about 3.9×10^{-7} (violet) to 7.5×10^{-7} m (red), and has corresponding frequencies that range from 7.9×10^{14} to 4×10^{14} Hz (Fig. 1.12). Note that EMR wavelengths are commonly expressed in nanometers, where 1 nm $=10^{-9}$ m, or micrometers, where 1 μm $=10^{-6}$ m.

When EMR comes into contact with matter (i.e., any object or material, such as trees, water, or atmospheric gases), it interacts with it. The following interactions are possible: absorption, *reflection, scattering*, or *emission* of EMR by the matter, or *transmission* of EMR through the matter. Remote sensing is primarily based on detecting and recording reflected and emitted EMR. The ability to remotely sense features is possible only because every object or material has particular emission and/or reflectance properties, collectively known as its *spectral signature* or *profile*, which distinguishes it from other objects and materials. Remote sensors are designed to collect these "spectral" data.

Remote sensors record these data in either analog (e.g., *aerial photographs* collected with an aircraft-mounted film camera) or, now more commonly, digital format (e.g., a two-dimensional matrix, or *image*, composed of pixels that store EMR values recorded by a satellite-mounted array) (Jensen 2005). These sensors may be either *passive* or *active* in nature. Passive sensors—the predominant category of sensors currently operating around the world—record naturally occurring EMR that is either reflected or emitted from areas and objects of interest. In contrast, active sensors—such as microwave (i.e., *RA*dio *D*etection *A*nd *R*anging, or *RADAR*) systems—send human-made EMR towards the features of interest and then record how much of that EMR is reflected back to the system (Jensen 2005). Chapter 2 of this book provides details about many contemporary remote sensing systems, both active and passive. A commonly used example of a multispectral passive system is Google Earth.

Google Earth maps the surface of the Earth by superimposing images obtained from high-resolution satellite imagery, aerial photography, and GIS in a three-dimensional (3D) mode. Google Earth displays satellite images of varying resolution of the Earth's surface, allowing users to see objects such as cities and houses looking perpendicularly down or at an oblique angle (https://en.wikipedia.org) to cover large surface areas in two dimensions. The data cover some parts of the world, including the terrain and buildings in 3D mode.

Google Earth uses digital elevation model (DEM) data collected by NASA's Shuttle Radar Topography Mission (SRTM). This means one can view almost the entire Earth in three dimensions. Since November 2006, 3D views of many mountains, including Mount Everest, have been improved by the use of supplementary DEM data to fill the gaps in SRTM coverage (https://en.wikipedia.org). It should be mentioned that Google Earth data are collected only in the visible part of the electromagnetic spectrum, not to be confused with active and passive multispectral remotely sensed data collected from a variety of airborne and satellite platforms, which are used for conducting scientific research as well as for a wide variety of applications.

The most commonly used example of the multispectral systems for scientific research and applications include Landsat (http://landsat.gsfc.nasa.gov/) and SPOT (Satellites Pour l'Observation de la Terre or Earth-observing Satellites) satellites (http://eoedu.belspo.be/en/satellites/spot.htm). Landsat has provided data worldwide since July 1972 in various spectral and spatial resolutions which have been used for many applications including natural resources, environmental studies and concerns, episodal events, and disaster management. SPOT has also provided data worldwide since 1984 with applications similar to Landsat.

A common example of an active system is GPS, which is routinely and widely used for navigation purposes by the public, commercial, military, and scientific communities.

1.5 Photo Interpretation, Photogrammetry, and Image Processing

Prior to the widespread availability of satellite imagery, aerial photography served as the principal foundation for a wide variety of cartographic efforts and geographic analyses (Short 2010; also see the next section of this chapter). During its period of prominence, analytical techniques emerged that involved mostly nonspectral aspects of aerial photographs. *Air photo interpretation* uses characteristics such as tone, texture, pattern, shadow, shape, size, and site (location), to identify objects and areas in photographs. In contrast, *photogrammetry* uses aerial photographs to make reliable spatial measurements of objects. (The word photogrammetry is derived from three Greek roots meaning "light-writing-measurement.") The types of measurements that a photogrammetrist might collect include the distances between features in an area of interest or the heights of particular features. In recent decades, photogrammetric methods have been increasingly applied to digital rather than film-based aerial photographs, and in some cases, to imagery captured by satellite-based sensors (e.g., Konecny et al. 1987). However, this book focuses primarily on non-photo-interpretive and non-photogrammetric applications of remotely sensed images.

In particular, Chapter 3 discusses *digital image processing,* which is a concept that encompasses a wide variety of computer algorithms and approaches for visualization, enhancement, and interpretation of remotely sensed images. *Pattern recognition* is used in digital image processing algorithms with the goal of providing a reasonable answer for all possible inputs and to perform "most likely" matching of the inputs, taking into account the statistical properties of the image data. Key products of digital image processing include thematic maps and color-coded, classified images that depict the spatial pattern of certain characteristics of objects and features, such as the boundaries between the surface water and land. For instance, remote sensing is commonly applied to map the variety and extent of land cover; in turn, these maps serve as critical information for other applications such as ecological studies, urban planning, water resource monitoring, and environmental impact assessment, as well as for policy making (Dougherty et al. 2004; Hester et al. 2008; Khorram et al. 1996; Kerr and Ostrovsky 2003; Franklin and Wulder 2002).

1.6 The Importance of Accuracy Assessment

A key consideration regarding the use of remotely sensed data is that the resulting outputs must be evaluated appropriately. When the output is a classified map, it is critical to assess its accuracy. A map is an imperfect representation of the phenomena it is meant to portray. In other words, every map contains errors, and it is the responsibility of the remote sensing analyst to characterize these errors prior to a map's use in subsequent applications. The most widely accepted method for the accuracy assessment of remote-sensing-derived maps is by comparison to reference data (also known as "ground truth") collected by visiting an adequate number of sample sites in the field (Congalton and Green 1999; Goodchild et al. 1992; Khorram et al. 1999). The key instrument in this comparison is an error matrix, which quantifies the accuracy for each map class of interest as well as the overall map accuracy (i.e., combining all of the classes) and the concepts of Producer's Accuracy and User's Accuracy and the Kappa statistics. Accuracy assessment is discussed in greater detail in Chap. 3 of this book.

1.7 Cost Effectiveness and Reach Versus Richness of Remote Sensing Technology

Remotely sensed data can be collected over a very large area in a very short period. For example, Landsat collects a scene covering 185 km × 185 km in 26 s. With the advent of sophisticated computing hardware and software and image processing

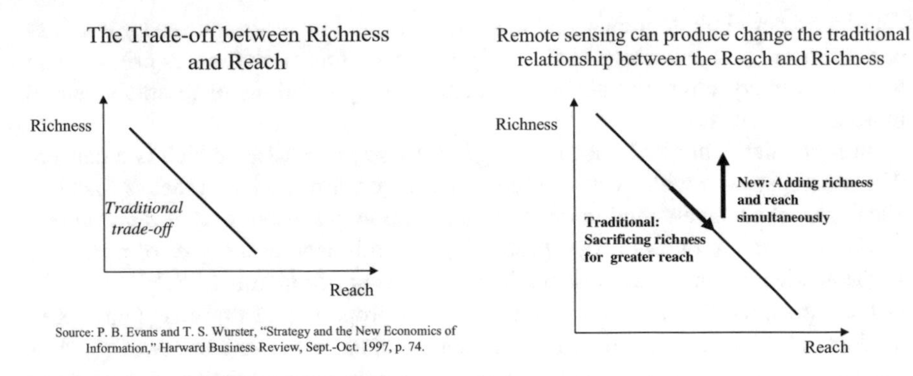

Fig. 1.13 The availability of rich remote sensing data can blow up the reach/richness trade-off

techniques, these data can be transformed into useful digital maps depicting a variety of applications in a geospatial context. However, one needs to be aware of not overselling this technology as a key to most solutions. Remember, all models are wrong, but some are useful. In order for a remote-sensing-derived product to be useful, one must verify the models being used, assess the accuracy of the results, and state the limitations associated with it. Questions such as who controls the data acquisition and distribution become concerns for the use of remotely sensed data.

It is possible to share extremely rich information with a very small number of people, and less rich information with a large number. In a traditional channel of distribution of information, there is a fundamental "trade-off" between "richness" and "reach" (Fig. 1.13). Reach simply means the number of people exchanging information. Richness can be defined as the amount of information that can be moved from a provider to a user in a given time.

The explosion of connectivity (thanks to open electronic networks), and the adoption of common information standards are now making possible for a very large number of people to exchange very rich information. This is fundamentally altering the traditional trade-off between richness and reach. Remote sensing provides data with high richness values (high spectral and spatial, spectral, and radiometric resolution) that reach a very broad user community (global). This is another example of disrupting the relationship between the reach and richness.

1.8 Organization of This Book

The primary goal of this book is to provide readers with a basic understanding of the principles and technologies behind remotely sensed data and their applications. While it includes some technical information, the book is not really a "how-to" guide for image processing and analysis. Instead, we hope that readers will walk away with the ability to confidently use remotely sensed data in their work or research and to think critically about key questions that users typically must answer:

- Have remotely sensed data been applied to my topic area in the past? In what ways? Could applications from other areas or disciplines inform my current project?
- What kind of remotely sensed data do I need? More specifically, what types of data are available from today's remote sensing instruments, and what are their strengths and limitations?
- How must my chosen data be prepared prior to analysis? What are the appropriate processing and/or analytical methods?
- What is the accuracy of the output products I have created? Is that accuracy sufficient for my ultimate objectives?
- What is the possibility that better remotely sensed data will become available in the near future? Realistically, what potential gains am I likely to see from these future data?

The remainder of the book is organized into the following chapters:

2. *Data Acquisition*. Topics: the concept of resolution for remotely sensed data, contemporary platforms and payloads
3. *Data Processing Tools*. Topics: data preprocessing, processing, and postprocessing
4. *4, 5, 6, and 7. Using Remote Sensing*. Topics: case studies and applications of remote sensing in terrestrial, atmospheric, marine/ocean, and planetary environments
8. *International Laws, Charters, and Policies*. Topics: international space policy, national policies, and implications for commercial remote sensing
9. *Future Trends in Remote Sensing*. Topics: forthcoming developments in information technology, payloads and platforms, and space policy

Each chapter ends with questions along with a list of additional literature and Web-based resources. While not essential, these materials are intended to enrich the reader's body of knowledge regarding the ever-growing field of remote sensing. In addition, we urge readers to supplement this book with their own research. A wealth of instructional and illustrative remote sensing examples can be found on the internet.

1.9 Review Questions

1. When and by whom was the first permanent photograph taken?
2. When and where and from what platform was the oldest existing aerial photograph taken?
3. What company and in what year produced the first camera type box for using rolled films?
4. Name the place and the year for the first successful flight by Wright Brothers.
5. In what year and by what agency and what satellite the first image of Earth was captured?
6. When and from what platform, the "Blue Marble" photograph of Earth was captured?

7. Define the electromagnetic spectrum and the relationship between wavelength and frequency.
8. What could result from the interaction between light and matter?
9. Name the best examples of the US and international satellite imagery used for scientific research and applications.
10. What role is played by remote sensing in disrupting the relationship between the reach and richness of data?
11. What are major breakthroughs in the development of platforms from which remotely sensed data can be acquired?
12. Name the three major advantages of using remote sensing data.

References

Baumann, P.R. 2009. Geo/SAT2. History of Remote Sensing, Department of Geography, State University of New York, College at Oneonta, Oneonta, NY

Cloud, J. 2001. Hidden in plain sight: the CORONA reconnaissance satellite programme and clandestine Cold War science. *Annals of Science* 58: 203–209.

Congalton, R.G., and K. Green. 1999. *Assessing the Accuracy of Remotely Sensed Data: Principles and Practices*. Boca Raton, FL: Lewis Publishers. 137 p.

Dougherty, M., R. Dymond, S. Goetz, C. Jantz, and N. Goulet. 2004. Evaluation of impervious surface estimates in a rapidly urbanizing watershed. *Photogrammetric Engineering and Remote Sensing* 70: 1275–1284.

Environmental Systems Research Institute (ESRI). 2001. *The ESRI Press Dictionary of GIS Terminology*. Redlands, CA: Environmental Systems Research Institute. 116 p.

European Space Agency (ESA). 2009. Space debris: evolution in pictures [web site]. European Space Agency, European Space Operations Centre. http://www.esa.int/esaMI/ESOC/SEMN2VM5NDF_mg_1.html

Franklin, S.E., and M.A. Wulder. 2002. Remote sensing methods in medium spatial resolution satellite data land cover classification of large areas. *Progress in Physical Geography* 26: 173–205.

Gates, D.M., H.J. Keegan, J.C. Schleter, and V.R. Weidner. 1965. Spectral properties of plants. *Applied Optics* 4: 11–20.

Goodchild, M.F., S. Guoqing, and Y. Shiren. 1992. Development and test of an error model for categorical data. *International Journal of Geographical Information Systems* 6: 87–104.

Hester, D.B., H.I. Cakir, S.A.C. Nelson, and S. Khorram. 2008. Per-pixel classification of high spatial resolution satellite imagery for urban land-cover mapping. *Photogrammetric Engineering and Remote Sensing* 74: 463–471.

Hirsch, R. 2008. *Seizing the Light: A Social History of Photography*, 2nd edition. New York: McGraw-Hill Higher Education. 480 p.

Jensen, J.R. 2005. *Introductory Digital Image Processing*, 3rd edition. Upper Saddle River, NJ: Prentice Hall. 316 p.

Kerr, J.T., and M. Ostrovsky. 2003. From space to species: ecological applications for remote sensing. *Trends in Ecology and Evolution* 18: 299–305.

Khorram, S., H. Cheshire, X. Dai, and J. Morisette. 1996. Land cover inventory and change detection of coastal North Carolina using Landsat Thematic Mapper data. In: Proceedings of the ASPRS/ACSM Annual Convention and Exposition, April 22–25, 1996, Baltimore, MD, Volume 1: Remote Sensing and Photogrammetry, pp. 245–250.

Khorram, S., G.S. Biging, N.R. Chrisman, D.R. Colby, R.G. Congalton, J.E. Dobson, R.L. Ferguson, M.F. Goodchild, J.R. Jensen, and T.H. Mace. 1999. *Accuracy Assessment of Remote Sensing-Derived Change Detection*. Bethesda, MD: American Society of Photogrammetry and Remote Sensing Monograph. 64 p.

Konecny, G., P. Lohmann, H. Engel, and E. Kruck. 1987. Evaluation of SPOT imagery on analytical photogrammetric instruments. *Photogrammetric Engineering and Remote Sensing* 53: 1223–1230.
Longley, P.A., M.F. Goodchild, D.J. Maguire, and D.W. Rhind. 2001. *Geographic Information Systems and Science*. Chichester, UK: John Wiley and Sons. 454 p.
Madry, S. 2013. Introduction to history of space remote sensing, In Handbook of Satellite Applications, 2nd ed. **Editors:** J.N. Pelton, S. Madry, L. Camacho, and S. Lara. Springer-Verlag, Chapt. 32, New York City, New York, 30 p, ISBN: 978-1-4419-7670-3
Mattison, D. 2008. Aerial photography. In: Hannavy, J., ed. *Encyclopedia of Nineteenth-century Photography*, Volume 1. New York: Routledge, pp. 12–15.
McCarthy, J.J. 2009. Reflections on: our planet and its life, origins, and futures. *Science* 326: 1646–1655.
Newhall, B. 1982. *The History of Photography: from 1839 to the Present*, 5th edition. New York: Museum of Modern Art. 319 p.
Professional Aerial Photographers Association (PAPA) International. 2011. History of aerial photography [web site]. PAPA International. http://www.papainternational.org/history.asp
Rees, G. 2001. *Physical Principles of Remote Sensing*. Cambridge, UK: Cambridge University Press. 343 p.
Short, N.M. 2010. The Remote Sensing Tutorial [web site]. National Aeronautics and Space Administration (NASA), Goddard Space Flight Center. http://rst.gsfc.nasa.gov/
Union of Concerned Scientists (UCS). 2011. UCS Satellite Database [web site]. Union of Concerned Scientists, Nuclear Weapons & Global Security. http://www.ucsusa.org/nuclear_weapons_and_global_security/space_weapons/technical_issues/ucs-satellite-database.html
Utterback, J.M. 1995. Developing technologies: the Eastman Kodak story. *McKinsey Quarterly* 1995(1): 130–144.
Witze, A. 2007. News: Briefing—A timeline of Earth observation [online document]. *Nature*. doi:10.1038/news.2007.320. http://www.nature.com/news/2007/071205/full/news.2007.320.html
Wood, R.D. 1997. A state pension for L.J.M. Daguerre for the secret of his Daguerreotype technique. *Annals of Science* 54: 489–506.

Suggested Reading

Morain, S.A. 1998. A brief history of remote sensing applications, with emphasis on Landsat. pp. 28–50 in Liverman, D., E.F. Moran, R.R. Rindfuss, and P.C. Stern, eds. *People and Pixels: Linking Remote Sensing and Social Science*. Washington, DC: National Academy Press. 256 p.
Paine, D.P. and J.D Kiser. 2003. *Aerial Photography and Image Interpretation*, 2nd edition. New York: John Wiley and Sons. 632 p. [This text also has a student companion web site, http://bcs.wiley.com/he-bcs/Books?action=index&itemId=0471204897&itemTypeId=BKS&bcsId=2050]
Manual of Remote Sensing. Second edition. Robert N. Colwell, editor. Falls Church, Virginia: American Society for Photogrammetry and Remote Sensing, 1983.
Remote Sensing Glossary, Earth Imaging Glossary

Relevant Websites

The First Photograph (documents the history behind Joseph Niépce's 1826 image, currently housed in the collection of the Harry Ransom Center at the University of Texas): http://www.hrc.utexas.edu/exhibitions/permanent/wfp/

Professional Aerial Photographers Association (PAPA) International—History of Aerial Photography: http://www.papainternational.org/history.asp
Aerial Photographs and Remote Sensing Images—Library of Congress Geography and Maps: An Illustrated Guide: http://www.loc.gov/rr/geogmap/guide/gmillapa.html
Project CORONA: Clandestine Roots of Modern Earth Science—University of California, Santa Barbara: http://www.geog.ucsb.edu/~kclarke/Corona/Corona.html
Landsat Program History: http://landsat.gsfc.nasa.gov/about/history.html
EOEdu, educational website for satellite Earth observation (in French, Dutch, and English): http://eoedu.belspo.be/index.html
http://store.geocomm.com/viewproduct.phtml?catid=121&productid=1332
http://store.geocomm.com/viewcat.phtml?catid=99
http://store.geocomm.com/viewproduct.phtml?catid=99&productid=2202

Chapter 2
Data Acquisition

As noted in Chapter 1, remotely sensed data are collected by a diverse array of passive and active sensors mounted on aircraft (including airplanes, helicopters, and uninhabited aerial systems) or spacecraft (usually satellites). During the latter half of the twentieth century, large-scale space programs such as National Aeronautics and Space Administration (NASA) and the European Space Agency (ESA) were primarily responsible for the development of new remote sensing technologies. Such programs remain key sources of technological innovations, but the expansion of commercial remote sensing has contributed greatly to the proliferation of sensors that provide high-quality data for a variety of applications. The main goal of this chapter is to provide an overview of the types of sensors currently available to analysts, including details about the data they acquire and some of their potential applications. Prior to this discussion, we first introduce the concept of *resolution* in remote sensing. One common definition of resolution is the ability to discern individual objects or features in a captured image or in the "real world." In this context, it is related to the concept of *scale* as used by cartographers: a large-scale map—whether constructed from remote sensing imagery or other geospatial data sources—shows more details than a small-scale map, thus making it possible to discern smaller objects or finer features. Nevertheless, the term resolution also relates to several other aspects of remotely sensed data. We illustrate these aspects with examples of commonly used satellite sensors and imagery.

2.1 Data Resolution

When we talk about "remotely sensed data," we are usually referring to digital images (i.e., two-dimensional arrays of pixels). These data are typically described by four types of resolution: *spatial, spectral, temporal,* and *radiometric*. *Spatial resolution,* which corresponds most closely to the definition of resolution provided in the preceding paragraph, is a measure of the clarity or fineness of detail of an image. For digital images, this translates to the ground area captured by a single

S. Khorram et al., *Principles of Applied Remote Sensing*,
DOI 10.1007/978-3-319-22560-9_2

image pixel; because pixels are typically square, resolution is generally expressed as the side length of a pixel. *Spectral resolution,* represented by the width of the wavelength interval and/or the number of spectral channels (or *bands*) captured by a sensor, defines the storage of recorded electromagnetic energy and the sensor's ability to detect wavelength differences between objects or areas of interest. The amount of time it takes for a sensor to revisit or reimage a particular geographic location is referred to as its *temporal resolution*. Finally, the sensitivity of a sensor to incoming electromagnetic energy (i.e., the smallest differences in intensity that the sensor can detect) is known as its *radiometric resolution*. This metric is usually expressed in terms of binary bit-depth (Jensen 2005), which refers to the number of tonal levels at which data for a given spectral band are recorded by a particular sensor. Digital images, like other forms of digital data, are stored as bits (i.e., 0 s and 1 s), so a sensor's bit-depth defines the number of unique values, on a binary scale, at which the incoming data can be stored for a pixel. This number is equal to 2^n, where n is the stated bit-depth. For example, a bit-depth of 3 means that incoming data can be stored as one of the eight unique values: 000, 001, 010, 011, 100, 101, 110, and 111. The binary bit-depth of most contemporary sensors is at least 8-bit, meaning that image pixels have at least 256 (0–255) possible values.

Each of these resolution types is discussed in subsequent paragraphs. In addition, the different resolution characteristics of some currently operational satellite sensors are presented in Table 2.1. The sensors described in this table are highly variable with respect to resolution. For example, the WorldView-3 satellite's onboard sensor has a spatial resolution of 31 cm for *panchromatic* (or black-and-white) imagery, while data collected by the "VEGETATION" sensor on the Système Probatoire d'Observation de la Terre (SPOT) 5 satellite are stored in 1150-m pixels. Temporally, the sensors listed in Table 2.1 have resolutions ranging from 15 min to 52 days; in other words, the satellite carrying a particular sensor—or the sensor itself, depending on certain aspects of its configuration—has the capacity to revisit a particular location on the Earth's surface every 15 min (i.e., Geostationary Operational Environmental Satellites, GOES) to every 52 days (i.e., the panchromatic and multispectral camera (PANMUX) sensor on the CBERS-4 satellite).

Spatial Resolution The ability to discern spatial structure is an important element of any remote sensing analysis. It is only in recent years that satellite imagery has supplanted aerial photography as the primary image data source for analyses such as the classification of urban land use (Barr and Barnsley 2000). In an urban landscape, surface features such as roads, office buildings, parks, and residential neighborhoods comprise a mosaic, where many small constituent units or pieces are interspersed by a few large ones. Roads and buildings are typically some of the smallest of these units. When viewed from above (e.g., from an airplane), the net visual effect of these units is an aggregate patchwork of various land uses and cover types, but the degree of patchwork detail portrayed by a remotely sensed image, and thus the level of specificity at which it can be classified, depends on its spatial resolution (Fig. 2.3). For example, a 30-m pixel (i.e., the spatial resolution provided by several sensors listed in Table 2.1) stores one digital number per spectral band of

Table 2.1 Characteristics of selected satellite sensors. (Adapted from Rogan and Chen 2004; updated for this publication)

Sensor (mission)	Organization[a]	Operation period	Swath width (km)	Spatial resolution (m)[b]	Temporal resolution	Radiometric resolution	Spectral resolution (μm)	Spectral bands
MSS (Landsat 1–5)	NASA, US	1972–1992	185	80 (MS), 240 (TIR)[c]	16–18 days	8-bit	0.5–1.1, 10.4–12.6[c]	4–5[c]
TM (Landsat 4, 5)	NASA, US	1982–2011	185	30 (MS), 120 (TIR)	16 days	8-bit	0.45–2.35, 10.4–12.5	7
ETM+ (Landsat 7)	NASA, US	1999	185	15 (PAN), 30 (MS), 60 (TIR)	16 days	8-bit	0.52–0.9 (PAN), 0.45–2.35, 10.4–12.5	7 + PAN
OLI (Landsat 8)	NASA, US	2013	185	15 (PAN), 30 (MS)	16 days	12-bit	0.5–0.68 (PAN), 0.433–0.453 (coastal/aerosol), 0.45–2.3, 1.36–1.39 (cirrus)	8 + PAN
TIRS (Landsat 8)	NASA, US	2013	185	100	16 days	12-bit	10.6–11.2, 11.5–12.5	2
MODIS (EOS Terra and Aqua)	NASA, US	1999	2300	250 (PAN), 500 (VNIR), 1000 (SWIR)	1–2 days	12-bit	0.620–2.155, 3.66–14.385	36
ASTER (EOS Terra)	NASA,US; METI, Japan	1999	60	15 (VNIR), 30 (SWIR), 90 (TIR)	4–16 days	8-bit (VNIR/SWIR), 12-bit (TIR)	0.52–0.86, 1.60–2.43, 8.125–11.65	14
Hyperion (EO-1)	NASA, US	2000	7.5	30	16 days	12-bit	0.353–2.577	220
ALI (EO-1)	NASA, US	2000	37	10 (PAN), 30 (MS)	16 days	12-bit	0.48–0.69 (PAN), 0.433–2.35	9 + PAN
OMI (EOS Aura)	NIVR, Netherlands; FMI, Finland; NASA, US	2004	2600	13,000 × 48,000, 13,000 × 24,000[d]	1 day	12-bit	0.27–0.5	740
CALIOP (CALIPSO)	NASA, US; CNES, France	2006	0.1	333	16 days	22-bit[e]	0.532, 1.064	2[f]

Table 2.1 (continued)

Sensor (mission)	Organization[a]	Operation period	Swath width (km)	Spatial resolution (m)[b]	Temporal resolution	Radiometric resolution	Spectral resolution (µm)	Spectral bands
VIIRS (Suomi NPP)	NASA, US; NOAA, US	2011	3000	375 or 750 depending on application	12 h	12-bit	0.41–12.5	21 + Day/Night PAN
Aquarius (SAC-D)	NASA, US; CONAE, Argentina	2011	390	150	7 days		3 L-band microwave radiometer beams + radar	N/A
AVHRR (NOAA 6–19; Metop-A, -B)	NOAA, US; EUMETSAT	1978[g]	2700	1100	12 h	10-bit	0.58–12.5	6[h]
I-M Imager (GOES 12–15)	NESDIS, US	1975[i]	8	1000 (VNIR), 4000 (SWIR), 8000 (moisture), 4000 (TIR)	0.25–3 h	10-bit	0.55–12.5	5
HICO (International Space Station)	Office of Naval Research (ONR), US; NASA, US	2009–2014	42 × 192[j]	90	~3 days	14-bit	0.353–1.08 (0.4–0.9)[k]	128 (87)[k]
SAR (RADARSAT-2)	CSA, Canada	2007	20–500[l]	3-100[l]	24 days[m]		C-band radar	N/A
VEGETATION (Proba-V)	ESA	2013	2250	100/350[n]	1–2 days[o]	12-bit	0.438–1.634	4
SAR (Sentinel-1 A)	ESA	2014	20–250[l]	5–40[l]	12 days		C-band radar	N/A
PMC (Gaofen-1)	CNSA, China	2013	69	2 (PAN), 8 (MS)	4 days	10-bit	0.45–0.9 (PAN), 0.45–0.89	4 + PAN
WFI (Gaofen-1)	CNSA, China	2013	830	16	4 days	10-bit	0.45–0.89	4
HROI (Gaofen-2)	CNSA, China	2014	48	0.8 (PAN), 3.2 (MS)	?[p]	?[p]	?[p]	?[p]
MUXCAM (CBERS-4)	CBERS, China/Brazil	2014	120	20	26 days	8-bit	0.45–0.89	4
PANMUX (CBERS-4)	CBERS, China/Brazil	2014	60	5 (PAN), 10 (MS)	52 days	8-bit	0.51–0.73 (PAN), 0.52–0.89	3 + PAN

Table 2.1 (continued)

Sensor (mission)	Organization[a]	Operation period	Swath width (km)	Spatial resolution (m)[b]	Temporal resolution	Radiometric resolution	Spectral resolution (μm)	Spectral bands
IRSCAM (CBERS-4)	CBERS, China/Brazil	2014	120	40 (NIR/SWIR), 80 (TIR)	26 days	8-bit	0.77–12.5	4
WFICAM (CBERS-4)	CBERS, China/Brazil	2014	866	64	5 days	10-bit	0.45–0.89	4
LISS-IV (RESOURCESAT-2)	ISRO, India	2011	70	5.8	24 days	10-bit	0.52–0.86	3
LISS-III (RESOURCESAT-2)	ISRO, India	2011	141	23.5	24 days	10-bit	0.52–1.70	4
AWiFS (RESOURCESAT-2)	ISRO, India	2011	740	56	24 days	12-bit	0.52–1.70	4
TANSO-FTS (GOSAT/Ibuki)	JAXA, Japan	2009	160[q]	10.5	3 days		0.758–14.3	4
Geoton-L1 (Resurs-P No. 1, No. 2)	Roscosmos, Russia	2013, 2014	32 (PAN), 38 (MS)	1 (PAN), 3–4 (MS)	3 days	10-bit	0.58–0.8 (PAN), 0.45–0.80	5 + PAN
KShMSA-VR/SR (Resurs-P No. 1, No. 2)	Roscosmos, Russia	2013, 2014	97 (VR mode), 440 (SR mode)	VR: 12 (PAN), 24 (MS); SR: 60 (PAN), 120 (MS)	3 days	12-bit	0.43–0.7 (PAN), 0.43–0.9	5 + PAN
GSA (Resurs-P No.1, No.2)	Roscosmos, Russia	2013, 2014	25	30	3 days	14-bit	0.4–1.1 nm band width	130
COSI (KOMPSAT-5)	KARI, South Korea	2013	5-100[l]	1–20[l]	28 days		X-band radar	N/A
AEISS-A (KOMPSAT-3 A)	KARI, South Korea	2015	12	0.55 (PAN), 2.2 (MS)	28 days	14-bit	0.45–0.9 (PAN), 0.45–0.9	4 + PAN
IIS (KOMPSAT-3 A)	KARI, South Korea	2015	12	5.5	28 days	14-bit	3.3–5.2	1
HRVIR (SPOT 4)	Airbus Group, France[r]	1998–2013	60	10 (PAN), 20 (MS)	26 days[s]	8-bit	0.61–0.68 (PAN), 0.50–1.75	4 + PAN

Table 2.1 (continued)

Sensor (mission)	Organization[a]	Operation period	Swath width (km)	Spatial resolution (m)[b]	Temporal resolution	Radiometric resolution	Spectral resolution (μm)	Spectral bands
VEGETATION (SPOT 4,5)	Airbus Group, France[r]	1998–2013, 2002–2014	2250	1150	26 days[s]	10-bit	0.43–1.75	4
HRG (SPOT 5)	Airbus Group, France[r]	2002	60	2.5–5 (PAN), 10 (VNIR), 20 (SWIR)	26 days[s]	8-bit	0.48–0.71 (PAN), 0.50–1.75	4+ PAN
SPOT 6, 7	Airbus Group, France/ Azercosmos, Azerbaijan[t]	2012, 2014	60	1.5 (PAN), 6 (MS)	26 days[u]	12-bit	0.45–0.745 (PAN), 0.45–0.89	4+ PAN
Pléiades 1 A, 1B	CNES/Airbus Group, France	2011, 2012	20	0.5 (PAN), 2 (MS)	26 days	12-bit	0.45–0.83 (PAN), 0.43–0.94	4+ PAN
IKONOS	DigitalGlobe, US[v]	1999	11.3	1 (PAN), 4 (MS)	3–5 days	11-bit	0.526–0.929 (PAN), 0.445–0.853	4+ PAN
QuickBird	DigitalGlobe, US	2001	18	0.65 (PAN), 2.62 (MS)	2.5–5.6 days	11-bit	0.405–1.053 (PAN), 0.43–0.918	4+ PAN
GeoEye-1	DigitalGlobe, US[v]	2008	15.2	0.41 (PAN), 1.65 (MS)	<3 days	11-bit	0.45–0.80 (PAN), 0.45–0.92	4+ PAN
WorldView-2	DigitalGlobe, US	2009	16.4	0.46 (PAN), 1.85 (MS)	1.1–3.7 days	11-bit	0.45–0.80 (PAN), 0.45–1.04	8+ PAN
WorldView-3	DigitalGlobe, US	2014	13.1	0.31 (PAN), 1.24 (MS), 3.7 (SWIR), 30 (CAVIS)[x]	<1 day	11-bit (PAN and MS), 14-bit (SWIR)	0.45–0.80 (PAN), 0.4–1.04, 1.195–2.365, 0.405–2.245 (CAVIS)	28+ PAN
RapidEye	BlackBridge, Germany	2009	77	5	1 day	12-bit	0.44–0.85	5
HiRAIS (Deimos-2)	Elecnor Deimos, Spain	2014	12	1 (PAN), 4 (MS)	4 days[y]	10-bit	0.45–0.9 (PAN), 0.42–0.89	4+ PAN

Table 2.1 (continued)

Sensor (mission)	Organization[a]	Operation period	Swath width (km)	Spatial resolution (m)[b]	Temporal resolution	Radiometric resolution	Spectral resolution (μm)	Spectral bands

[a] *NASA* National Aeronautics and Space Administration, *METI* Ministry of Economy, Trade, and Industry, *NIVR* The Netherlands Agency for Aerospace Programmes, *FMI* Finnish Meteorological Institute, *CNES* Centre National d'Études Spatiales, *NOAA* National Oceanic and Atmospheric Administration, *NESDIS* National Environmental Satellite, Data, and Information Service, *CSA* Canadian Space Agency, *ESA* European Space Agency, *CNSA* China National Space Administration, *CBERS* China-Brazil Earth Resources Satellite Program, *CONAE* Comisión Nacional de Actividades Espaciales, *EUMETSAT* European Organisation for the Exploitation of Meteorological Satellites, *JAXA* Japanese Aerospace Exploration Agency, *ISRO* Indian Space Research Organization, *KARI* Korean Aerospace Research Institute

[b] Acronyms used in describing sensor channels/configurations: *PAN* panchromatic, *MS* multispectral, *VNIR* visible and near-infrared, *SWIR* short-wave infrared *TIR* thermal infrared

[c] The MSS sensor on Landsat 3 had a fifth spectral band for thermal infrared. The MSS sensors on other Landsat missions had four-band configurations

[d] The OMI sensor aboard EOS Aura has one channel recording at 0.27–0.31 μm (in the ultraviolet spectral range) and another recording at 0.306-0.5 μm (ultraviolet and visible spectral range); the pixel size of the former channel is larger (13 × 48 km vs. 13 × 24 km)

[e] Each receiver channel on the CALIOP sensor, a spaceborne LiDAR system, has dual 14-bit digitizers that jointly provide a 22-bit dynamic range

[f] The CALIOP sensor produces simultaneous laser pulses at two wavelengths, 0.532 and 1.064 μm

[g] The NOAA satellite program began in 1978. Currently, NOAA-19 is designated as the program's operational satellite. The AVHRR sensor is also carried as a "heritage" instrument on the Metop-A and Metop-B satellites

[h] While the AVHRR sensor has six spectral channels, only five are transmitted to the ground at any time. The bands designated 3A and 3B are transmitted only during daytime and nighttime, respectively

[i] Although the GOES program has been active since 1975, only GOES-13 through GOES-15 are currently operational. GOES-14 was launched in 2009 but remains in on-orbit storage mode

Table 2.1 (continued)

Sensor (mission)	Organization[a]	Operation period	Swath width (km)	Spatial resolution (m)[b]	Temporal resolution	Radiometric resolution	Spectral resolution (μm)	Spectral bands

[j] Operated as a demonstration instrument, the HICO sensor captured only a single 42 × 92 km scene per orbit. Over its 5-year period of operation, HICO collected > 10,000 scenes worldwide

[k] Collected HICO data have 128 bands spanning the 0.353–1.08 μm range of wavelengths, but the data from outside the 0.4–0.9 μm range are less accurate and therefore only 87 bands are included in the output files made available to users

[l] The radar systems in this table (i.e., SAR/RADARSAT-2, SAR/Sentinel-1 A, and COSI/KOMPSAT-5) operate in a variety of scan modes with different swath widths and spatial resolutions

[m] RADARSAT-2 has a repeat cycle of 24 days, but its left- and right-looking modes may provide more rapid revisits

[n] The spatial resolution of the VEGETATION sensor on Proba-V is 100 m at nadir but 350 m over the full field of view (i.e., at the scene edges)

[o] The VEGETATION sensor provides daily coverage for the following latitudes: 35–75°N and 35–56°S. It offers full global coverage every 2 days

[p] Few details about Gaofen-2 are available, but it is likely similar to Gaofen-1 in terms of spectral resolution and revisit time

[q] The FTS sensor on the GOSAT satellite records 10.5 × 10.5 km images that are spaced 150 km apart in a grid (see Kuze et al. 2009)

[r] Spot Image, a company created by the French Space Agency (CNES), developed and operated the SPOT satellites. Spot Image became a subsidiary of Astrium, a European Aeronautic Defense and Space (EADS) company in 2008. EADS was reorganized as the Airbus Group in 2014

[s] The SPOT satellites have the capacity to record data off-nadir (i.e., to record data in areas that are not directly below the satellite). This may reduce the revisit time to 2–3 days (1 day for the wide-swath VEGETATION sensor), but the images will have different observation geometry, which may affect image processing

[t] Unlike the first five SPOT satellites, SPOT 6 and 7 were developed by the Airbus Group without financial support from the French government. In 2014, SPOT 7 was sold to Azerbaijan's space agency Azercosmos, who renamed it Azersky. Per agreement, Azercosmos has commercial rights to imagery from both satellites for the southern Asia region, while Airbus Group retains the rights elsewhere

[u] The SPOT 6 and 7 satellites are a constellation, operating on the same orbit but phased 180° apart. With the satellites' capacity to record data off-nadir, the constellation can provide 1-day revisit time

[v] GeoEye, the company that originally operated IKONOS and GeoEye-1, merged with DigitalGlobe in 2013

[x] *CAVIS* Clouds, Aerosols, Vapors, Ice, and Snow

[y] The 4-day revisit time of the HiRAIS sensor on Deimos-2 is enabled by a ±45° off-nadir viewing capability

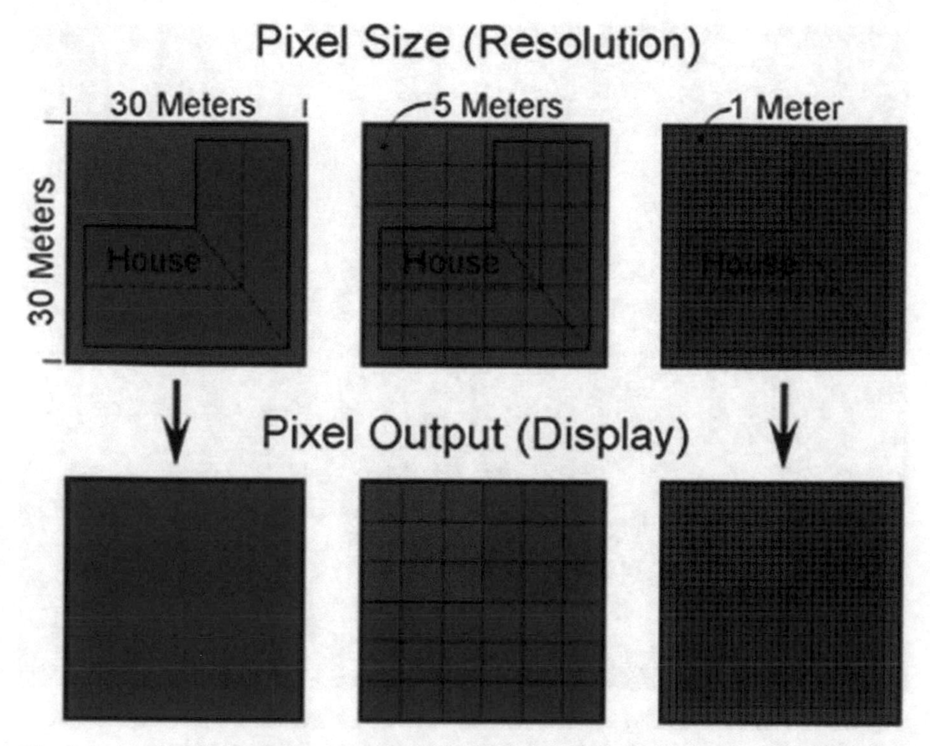

Fig. 2.1 An illustration of spatial resolution as it relates to pixel size. (Image courtesy of the EU Science Education through Earth Observation for High Schools (SEOS) Project and Satellite Imaging Corporation. (Source: www.seos-project.eu/modules))

information for any landscape feature smaller than 900 m^2. The pixel could, in fact, accommodate 6 one-story square houses, each having a 1500 ft^2 (139 m^2) footprint. In contrast, an Advanced Very High Resolution Radiometer (AVHRR) pixel, with its 1100-m resolution (see Table 2.1), could incorporate more than 8700 of these houses, if arranged regularly. Increases in spatial resolution have been a persistent trend in sensor innovation. Within a decade of the 1986 launch of the first SPOT satellite, whose 10-m panchromatic spatial resolution opened up new land cover mapping possibilities, space agencies in Japan, India, and Europe had produced satellites with finer spatial resolutions than any of the first five Landsat satellites. The latest generation of commercial satellite ventures, including Pléiades 1A and 1B, as well as WorldView-2 and WorldView-3, all provide spatial resolutions of 1 m or less.

Figure 2.1 is an illustration of the spatial resolution of a house as related to pixel size. The size of a pixel determines its spatial resolution (Khorram et al. 2012a, b), which in turn determines the degree of recognizable detail in an image. A practical example of four spatial resolutions of the same area is shown in Fig. 2.2, while Fig. 2.3 is a comparative example of Landsat ETM+ (30-m resolution) and QuickBird (fused at 0.8-m) images of Paris, France. Figure 2.3 demonstrates that the

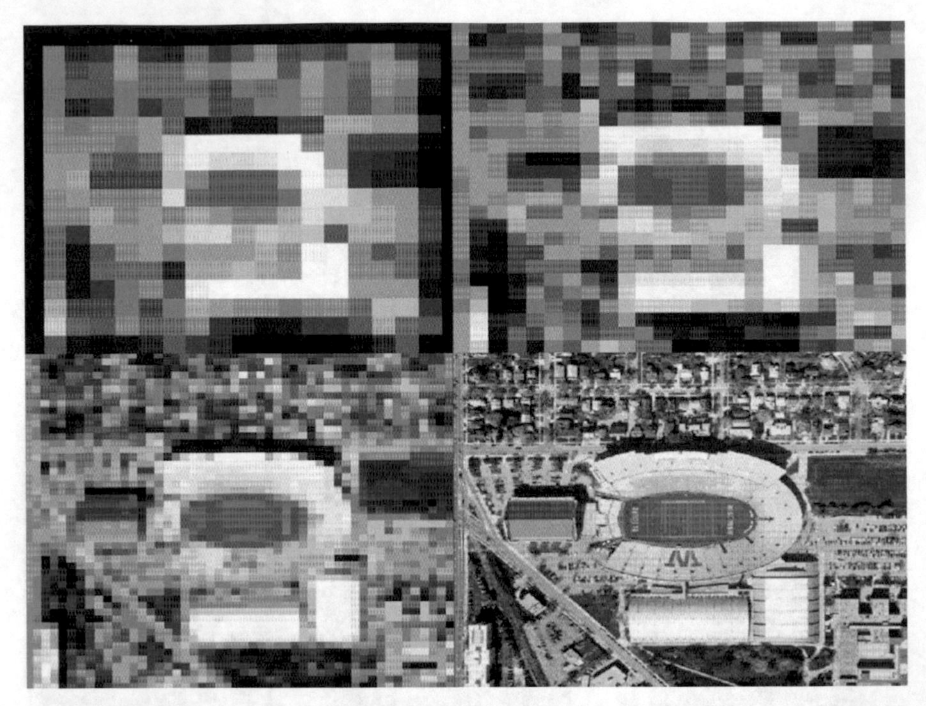

Fig. 2.2 Example images showing the University of Wisconsin stadium at different spatial resolutions: 30 m *(top left)*, 20 m *(top right)*, 10 m *(bottom left)*, and 1 m *(bottom right)*

Fig. 2.3 Comparison of Landsat ETM+ *(left)* and QuickBird *(right)* images of Paris, France. (Image courtesy of the EU Science Education through Earth Observation for High Schools (SEOS) Project, Landsat, and GeoEye (now DigitalGlobe))

Table 2.2 Comparison of the spectral resolutions of the Landsat Operational Land Imager (OLI) and WorldView-2 sensors

	Bandwidth (μm)	
Spectral band	*Landsat OLI*	*WorldView-2*
Panchromatic	0.50–0.68	0.45–0.8
Coastal (aerosol)	0.43–0.45	0.4–0.45
Blue	0.45–0.51	0.45–0.51
Green	0.53–0.59	0.51–0.58
Yellow		0.585–0.625
Red	0.64–0.67	0.63–0.69
Red edge		0.705–0.745
Near-infrared	0.85–0.88	0.77–0.895
Near-infrared		0.86–1.04
Short-wave infrared	1.57–1.65	
Short-wave infrared	2.11–2.29	
Cirrus	1.36–1.38	

medium-spatial-resolution Landsat ETM+ sensor has different optimal applications (e.g., continental- or regional-scale land cover mapping) than the applications that are possible with the high-resolution QuickBird sensor (e.g., extraction of buildings, trees, or other discrete features).

Spectral Resolution The spectral component of remote sensing data is also critical to appropriate data selection. The spectral resolution of a sensor is a description of the range and partitioning of the electromagnetic energy (or *reflectance*) recorded by the sensor. In this context, sensors can be divided into one of three loose classes of spectral sensitivity: panchromatic (one spectral band), *multispectral* (multiple spectral bands), or *hyperspectral* (many, possibly hundreds, of spectral bands) (Jensen 2005). Fundamentally, both multispectral and hyperspectral images can be thought of as "stacks" of identically geo-referenced (i.e., covering the exact same geographic area) panchromatic images, where each image in the stack corresponds to a particular portion of the electromagnetic spectrum. Most satellite-derived remote sensing data are captured in a panchromatic and/or multispectral format (although spaceborne hyperspectral sensors are becoming more commonplace). For example, the Operational Land Imager (OLI) onboard NASA's Landsat-8 satellite records data in nine bands, including one panchromatic band. Each of these bands is sensitive to different wavelengths of visible and infrared radiation. The sensor on the WorldView-2 satellite also records data in nine bands, including a panchromatic band. Table 2.2 compares the spectral bandwidths of the two sensors. Most multispectral sensors have several bands targeted at the *visible-to-near-infrared (VNIR)* range (0.4–1.4 μm) of the electromagnetic spectrum, as do OLI (six bands in this range) and WorldView-2 (all eight of its multispectral bands). This is because, in many geographic areas, ambient electromagnetic reflectance is strongly shaped by the distribution and condition of vegetation and water features, particularly in the VNIR range. WorldView-2 is noteworthy for two specialized bands within the

VNIR range that are not found on any other spaceborne multispectral sensor: a "yellow" band, aimed at detecting the yellowness of terrestrial and aquatic vegetation, which may help distinguish between vegetation types; and a "red-edge" band, focused on the narrow range of wavelengths where vegetation reflectance transitions from low to high, which may be useful for measuring plant health. While the OLI sensor does not have as many bands in the VNIR range as WorldView-2, it does have two bands in the *short-wave infrared (SWIR)* portion of the electromagnetic spectrum, as well as a band at the edge of the VNIR range intended specifically for cirrus cloud detection (see Table 2.2).

Unlike these multispectral sensors, NASA's Hyperion, the first hyperspectral sensor carried aboard a satellite, provides a virtually continuous recording of electromagnetic reflectance across 220 narrow spectral bands in the VNIR and SWIR wavelengths. Although the processing complexity of hyperspectral data has hindered its widespread applicability in many land cover studies, these data provide a great deal of information about the unique reflectance behavior of vegetation stress and diversity. The high spectral resolution also provides an increased ability to discriminate between features that have similar response profiles, or *signatures,* across the electromagnetic spectrum, such as forest stands composed of different deciduous tree species (Chen and Hepner 2001). It is important to note, however, that the first generation of satellite-based, "true" hyperspectral sensors (i.e., sensors with >100 bands), such as Hyperion or the now-defunct Hyperspectral Imager for the Coastal Ocean (HICO) that operated from the International Space Station, have typically been too coarse in terms of spatial resolution to be used for fine-scale analyses. Such applications remain the purview of airborne hyperspectral sensors (see Sect. 2.2.1), although this trend is likely to change with future advances in hyperspectral imaging technology.

Temporal Resolution Temporal resolution, or revisit time, is usually expressed in terms of days. For example, Landsat-8 has a 16-day orbit cycle, meaning that the satellite, and thus the OLI sensor and the thermal infrared sensor (TIRS), returns to a given location on the Earth's surface every 16 days. However, some satellites, such as the GOES, have a revisit cycle of less than an hour in certain locations. The GOES have a coarse, 8-km spatial resolution, which gives them limited utility for local land cover studies, but their temporal resolution makes them highly suitable for meteorological and climatological applications. By contrast, the balance of spatial and temporal resolutions achieved by NASA's Moderate Resolution Imaging Spectroradiometer (MODIS)—with a spatial resolution of 250–1000 m and a temporal resolution of 1–2 days—has made the sensor applicable to a variety of regional- to continental-scale research efforts. MODIS data make it possible to examine land cover changes and patterns affecting environmental quality that may not be obvious unless viewed at the daily or near-daily replication offered by this sensor (Ganguly et al. 2010). In addition, a number of satellites (e.g., SPOT 6 and 7) have the capacity to capture *off-nadir* imagery (i.e., to record data from areas that are not directly below the satellite sensor). This can reduce their typical revisit time, but it also affects the sensor geometry and subsequent image processing. Fur-

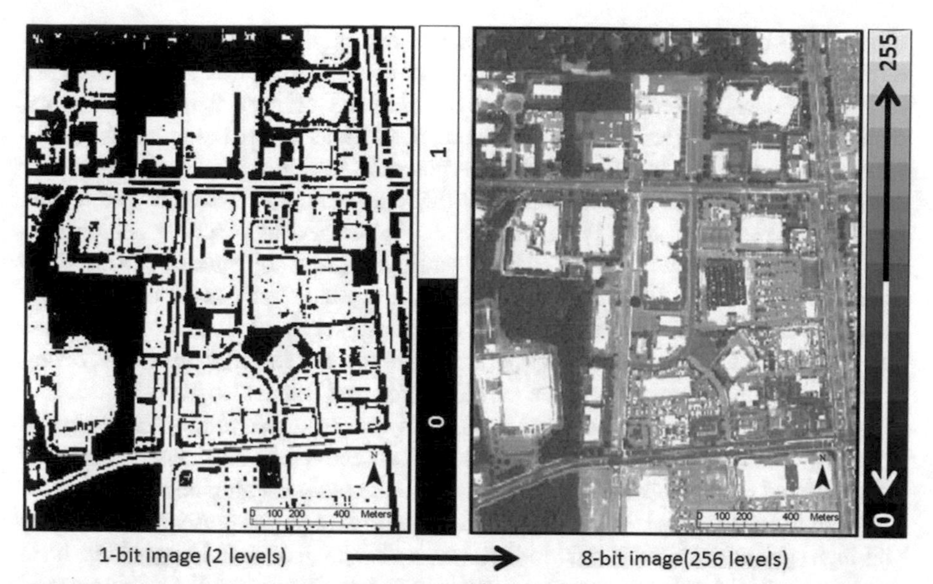

Fig. 2.4 The concept of radiometric resolution is illustrated by two images. The image on the *left* represents a 1-bit image, where two brightness values, or grayscale levels (i.e., *black and white*), are portrayed within the data. The image on the *right* represents an 8-bit image, where up to 256 grayscale levels are portrayed within the data. In contrast to these images, an 11-bit image holds up to 2048 grayscale levels

thermore, off-nadir image capture may also alter spatial resolution, particularly for high-resolution sensors; for instance, the spatial resolution of multispectral data from WorldView-3 is 1.38 m at 20° off-nadir versus 1.24 m at nadir.

Radiometric Resolution Many of the current generation of both government and commercial satellites have onboard sensors with high radiometric resolutions. Commonly, these sensors have 10-, 11-, or 12-bit collection depths (also known as their level of "data quantization"), which represents a substantial improvement over the 8-bit resolution typically exhibited by predecessors such as the older Landsat and SPOT sensors. Certain sensors, such as the Advanced Earth Imaging Sensor System-A (AEISS-A) aboard KOMPSAT-5, as well as the Cloud-Aerosol LiDAR with Orthogonal Polarization (CALIOP) instrument aboard NASA's CALIPSO satellite, have even higher radiometric resolutions (14- and 22-bit, respectively). Fundamentally, a sensor's radiometric resolution defines the *dynamic range* (i.e., the number of possible data values) recorded in each image band, so higher radiometric resolution means that a sensor is more sensitive to variations in electromagnetic reflectance (Fig. 2.4). This increased sensitivity has been shown to be very useful in the land cover classification of complex (e.g., urban and suburban) landscapes (Hester et al. 2008; Hester et al. 2010; Khorram et al. 2012b).

Ultimately, each of the four resolution types must be considered in light of the intended purpose of the data. For example, except near cities, most terrestrial landscapes are dominated by vegetation, water, or other natural surfaces, even amid production land uses such as agriculture or mining. Spatial resolution might be relatively less important for mapping or monitoring efforts in these settings because of high landscape homogeneity. In heavily forested or vegetated environments, data that are high in temporal and spectral resolutions rather than spatial resolution, such as AVHRR imagery, could be a powerful tool for evaluating seasonal or annual change in photosynthetic activity (Moulin et al. 1997; Stöckli and Vidale 2004). With respect to other types of analyses, hyperspectral data have demonstrated strong applicability for geological, mineralogical, and soil studies (e.g., Galvão et al. 2008; Weber et al. 2008). In addition, *radar* systems such as the C-band radar on the RADARSAT-2 satellite can be useful for observing sea surface conditions and other aspects of coastal areas (Moon et al. 2010), for detecting oil spills (Zhang et al. 2011), and for monitoring sea ice (Komarov and Barber 2014).

Regardless, in any remote sensing analysis, the appropriate input data should primarily be a function of the individual project objectives. For instance, the level of detail offered by high-spatial-resolution satellite imagery may not be well suited for mapping forested areas because of certain limitations (e.g., the presence of shadows and high spectral variability within forest type classes), but such images have been used successfully in large, homogeneous tracts of rain forest for the assessment of old-growth tree mortality (Clark et al. 2004; Sawaya et al. 2003). As this specialized application suggests, there is no one "best" sensor for mapping a particular type of landscape. Although the sensors listed in Table 2.1 are only a sample of the instruments currently producing remotely sensed data for use in scientific applications, they illustrate the wide diversity of options that an analyst should consider before choosing a particular sensor. While that ultimate selection will likely also be a function of other factors, such as budget and processing capabilities, it should be guided foremost by the analyst's evaluation about which choice will provide the most meaningful information.

2.2 Payloads and Platforms: An Overview

In remote sensing, the carrier of a sensor (e.g., an Earth-orbiting satellite) is known as its *platform,* while the sensor itself is the platform's *payload* (e.g., the Advanced Spaceborne Thermal Emission and Reflection Radiometer, or ASTER, a sensor mounted on the NASA EOS Terra satellite). Remote sensing platforms are roughly split into two categories: airborne and spaceborne. During the last two decades, there has been rapid proliferation of platforms in both categories. First, improvements in the accuracy, image quality, and available data formats of digital cameras have allowed them to supplant film cameras for most airborne applications. With respect to spaceborne platforms, a number of commercial imaging satellites have been launched recently, supplementing an existing assortment of platforms operated by NASA, the ESA, and other government agencies. Here, we provide a brief overview of major platforms and sensor formats in use historically and today.

Fig. 2.5 Comparison between vertical and oblique aerial photographs covering a portion of Orange County in southern California. (Image courtesy of Eagle Aerial Solutions)

2.2.1 Airborne Platforms

Although the primary focus of this book is satellite-based remote sensing and the processing of associated digital data, until the emergence of satellite imagery in the 1970s, aerial photography served as the main data source when mapping land cover and other phenomena on the Earth's surface (Short 2006). Through time, "true" color and color-infrared films became economically feasible as alternatives to panchromatic film and offered better capability for tasks such as vegetation classification or characterizing water quality (Caylor 2000). Until recently, digital cameras did not provide the same level of fine detail as film-based cameras, but many currently available digital cameras have comparable spatial resolutions to film cameras, with similar formatting and scale (Morgan et al. 2010).

The use of airborne video photography, or *videography,* for remote sensing largely began during the early 1980s. It has been applied in many contexts, including for precision agriculture (e.g., Metternicht 2003; Yang and Anderson 2000) and wildlife management (Coops and Catling 1997). Videography has several advantages over still photography, particularly the near-simultaneous capture of multiple views of a target area, although the image quality of a videotape is generally inferior to that of a film (Mausel et al. 1992); however, as with photographic cameras, analog video cameras have been largely replaced by digital versions in recent years (Wulder et al. 2007).

Aerial photography and videography may be either *vertical* or *oblique* (Fig. 2.5), depending on the orientation of the camera's optical axis relative to the Earth (Avery and Berlin 1992). Vertical photos and videos are captured with the camera's axis at an angle that is perpendicular or near-perpendicular ($90 \pm 3°$) to the ground. Oblique photos and videos are typically captured with the camera's axis tilted more than 20° from vertical.

Because airborne platforms are not suited to capturing large geographic areas (e.g., 10,000 km^2) at once, they have been replaced by spaceborne platforms for most broad-scale remote sensing projects (Short 2006). Nevertheless, aerial photography and videography remain viable in large part because they can be tailored to specific project needs (e.g., a particular spatial scale or resolution) in ways that satellite imagery cannot (Morgan et al. 2010). Aerial photography continues to be the foundation of many national-scale mapping efforts, such as the National Agricultural Imagery Program (NAIP) administered by the US Department of Agriculture (USDA) Farm Service Agency, which has the goal of providing regularly updated, geo-referenced imagery to the public (USDA FSA 2010). In addition, the development of image processing techniques for satellite imagery has in turn expanded the range of automated techniques that may be applied to digital aerial photos (Morgan et al. 2010), including both *pixel-based* and *object-based classification* (see Chap. 3 of this book). Furthermore, videography will continue to be relevant for studies related to animal movement and behavior (Kross and Nelson 2011).

Various non-photographic sensors have also been implemented via airborne platforms, including *active remote sensing systems* such as radar and *LiDAR*. Active systems differ from *passive remote sensing systems* (i.e., the vast majority of sensors) in that—rather than recording the electromagnetic energy that is naturally radiated by all objects—they generate and send their own "pulses" of electromagnetic energy toward features of interest, and then record the strength of the signals returned from the features. For instance, radar (short for RAdio Detection And Ranging) systems operate within the microwave portion of the electromagnetic spectrum (Lillesand et al. 2008). Essentially, a radar sensor works by emitting radio waves from an antenna that bounce off target features on the Earth, and then the sensor records the returned energy. The target features may be distinguished from one another by their differing effects on the returned signals. Radar sensors can capture images day or night and under all weather conditions (e.g., through cloud cover or smoke). For this reason, radar imaging is often used in disaster management, for instance in detecting ocean oil spills (Jha et al. 2008; Klemas 2010) or assessing earthquake damage (Wang et al. 2009). Radar *interferometry* involves the simultaneous use of two antennae to generate radio waves and collect returned signals; incorporating the distance between these antennae during data processing makes it possible to determine the three-dimensional positions of target features, and thus the topography of the imaged area (Massonnet and Fiegl 1998). For instance, Fig. 2.6 shows an application of airborne radar interferometry to develop a *digital elevation model,* or *DEM,* of the area surrounding the Ray Mine in central Arizona.

Although the term LiDAR (also written as LIDAR, lidar, or LADAR) is commonly believed to be an acronym of Light Detection And Ranging, it actually originated in the early 1960s as a blend of the words "light" and "radar" (see Ring 1963). Any LiDAR instrument principally consists of a laser, a scanner, and a specialized Global Positioning System (GPS) receiver; the instrument measures the distance between itself and a target feature by illuminating the feature with its laser and then analyzing the reflected light. A conceptual illustration of an airborne LiDAR system is shown in Fig. 2.7. Airborne LiDAR systems resemble airborne radar in that the output images are generated based on the amount of sensor-emitted energy that is returned from features on the ground. However, unlike radar, LiDAR sensors emit laser pulses (typically in the near-infrared portion of the electromagnetic system) at a very high rate (i.e., 10,000–100,000 pulses per second) (Reutebuch

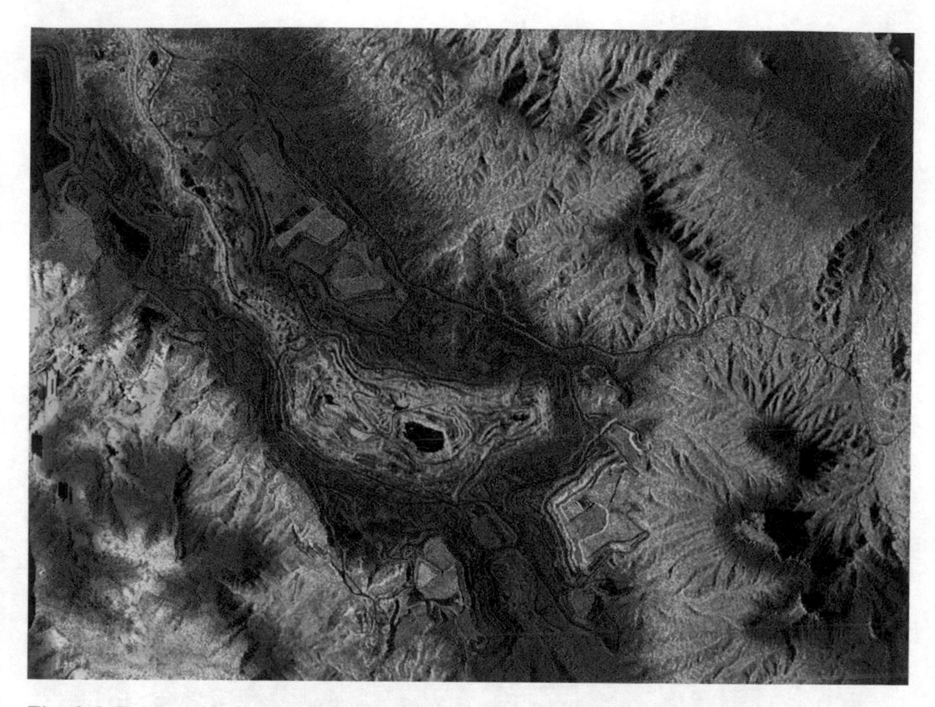

Fig. 2.6 Topographic image of the area surrounding the Ray Mine (image center) in Arizona. The image was captured by the National Aeronautics and Space Administration (NASA) Airborne Synthetic Aperture Radar (AIRSAR) sensor, operating in interferometric mode. Colors represent elevation contours, with the lowest elevations found near the mine and the highest elevations in the surrounding mountains. (Image courtesy of NASA Jet Propulsion Laboratory)

et al. 2005). In turn, the three-dimensional positions of targeted objects are determined based on the time it takes for pulses to return the sensor (which is a function of distance). The typical airborne LiDAR output is an elevation (i.e., topographic) data set with very accurate vertical measurements (±10–15 cm error when recorded under optimal conditions) (Charlton et al. 2003; Reutebuch et al. 2005). A noteworthy feature of most airborne LiDAR systems is that they can record multiple returns (Fig. 2.8a) from the same laser pulse in cases where an object does not completely block the pulse's path, allowing it to continue to another object closer to the ground (Reutebuch et al. 2005). This feature makes LiDAR systems especially useful for applications such as the characterization of forest canopy structure (Fig. 2.8b; Lim et al. 2003; Reutebuch et al. 2005). However, compared to radar systems, LiDAR systems are much more sensitive to weather and other conditions that interfere with the laser pulses.

Airborne LiDAR remote sensing systems first emerged in the late 1970s and early 1980s (Browell et al. 1998; Krabill et al. 1984), but broader adoption of the technology did not occur until the 1990s, when airborne LiDAR systems began to be applied in a wide variety of disciplines, including geography, geomorphology, agriculture, forestry, and civil engineering. In particular, the systems are well suited

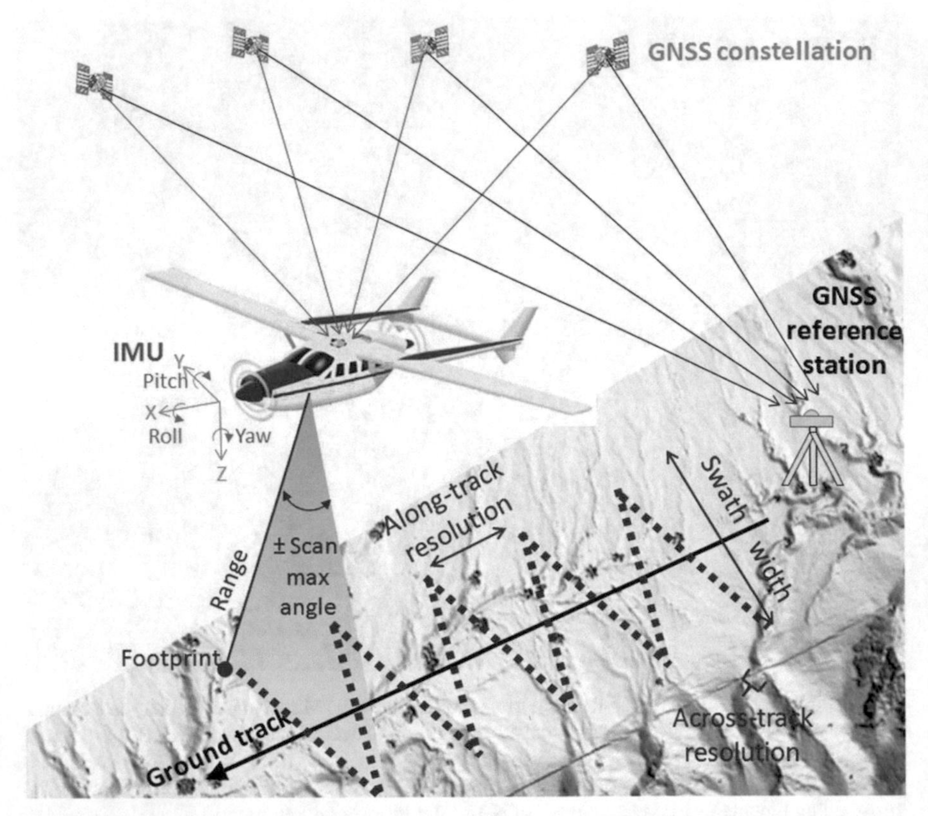

Fig. 2.7 Conceptual diagram of an airborne LiDAR system. *GPS* Global Positioning System. (Image source: Juan Carlos Fernandez-Diaz, University of Houston/NSF National Center for Airborne Laser Mapping)

to applications in dynamic coastal and environments, including surveys of coral reefs, benthic habitats, dune vegetation, and beach erosion and accretion (Brock et al. 2002; Wright and Brock 2002). For instance, the US Geological Survey's Experimental Advanced Airborne Research LiDAR (EAARL) system is able to measure submerged topography and adjacent coastal land elevations seamlessly in a single scan of transmitted laser pulses, operating in the blue-green portion of the electromagnetic spectrum, as illustrated in Fig. 2.9. Likewise, the precise three-dimensional information provided by these systems can be used to examine both natural and man-made features. Figure 2.10 shows such an application for the Bixby Bridge in the Big Sur area of California.

For the past few decades, multispectral scanners (MSSs) resembling the passive sensors found on many Earth-observing satellites have been mounted on airborne platforms, primarily in an effort to provide better spectral and/or spatial resolutions than their spaceborne counterparts or to record data at times outside of scheduled satellite overpasses (Jensen 2005; Jensen et al. 1984). Like their satellite-mounted counterparts, these airborne MSSs record ranges of spectral radiation in several

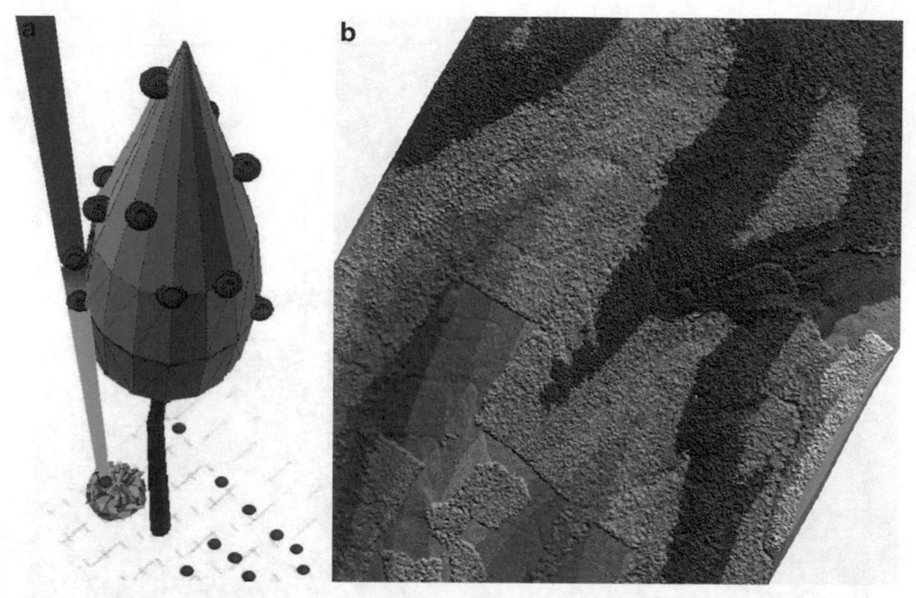

Fig. 2.8 **a** Conceptual rendering of multiple returns from a single laser pulse from an airborne LiDAR system. **b** Forest canopy surface recorded by LiDAR; colors indicate elevation contours. (Image credit: Mississippi State University, College of Forest Resources, Measurements & Spatial Technologies Laboratory)

bands (usually 10–15) across the electromagnetic spectrum, from the visible (or possibly ultraviolet) to the *thermal infrared (TIR)* wavelengths (Grebby et al. 2011; Jensen 2005; Jensen et al. 1984; Short 2006). Also like their spaceborne counterparts, the spectral resolutions of the visible and near-infrared bands on these airborne scanners are commonly on the order of 30–80 μm. This is in contrast to hyperspectral sensors, which—as noted earlier in this chapter—record data using up to hundreds of narrow bands, each representing a very small range of wavelengths (Fig. 2.11). For example, the Airborne Visible Infrared Imaging Spectrometer (AVIRIS) instrument, operated by NASA, collects data from 224 contiguous bands spanning the ultraviolet to near-infrared; the spectral resolution of each AVIRIS band is approximately 0.01 μm. This high degree of spectral resolution enables analysts to make subtle distinctions regarding objects or areas of interest that are not possible with multispectral data. For instance, hyperspectral imagery has been used to detect early stages of forest damage and decline caused by air pollution (Campbell et al. 2004), changes in the marine environment (Kratzer et al. 2008), as well as the activity of an invasive insect pest (Pontius et al. 2005). However, because of the very large data volume, or dimensionality, associated with the typical hyperspectral image, an analyst must often use statistical techniques (e.g., principal components analysis) to reduce the dimensionality prior to further image processing (Harsanyi and Chang 1994).

During the last decade, the greatest area of growth in airborne remote sensing has been the use of unmanned aerial systems (UASs), also known as unmanned

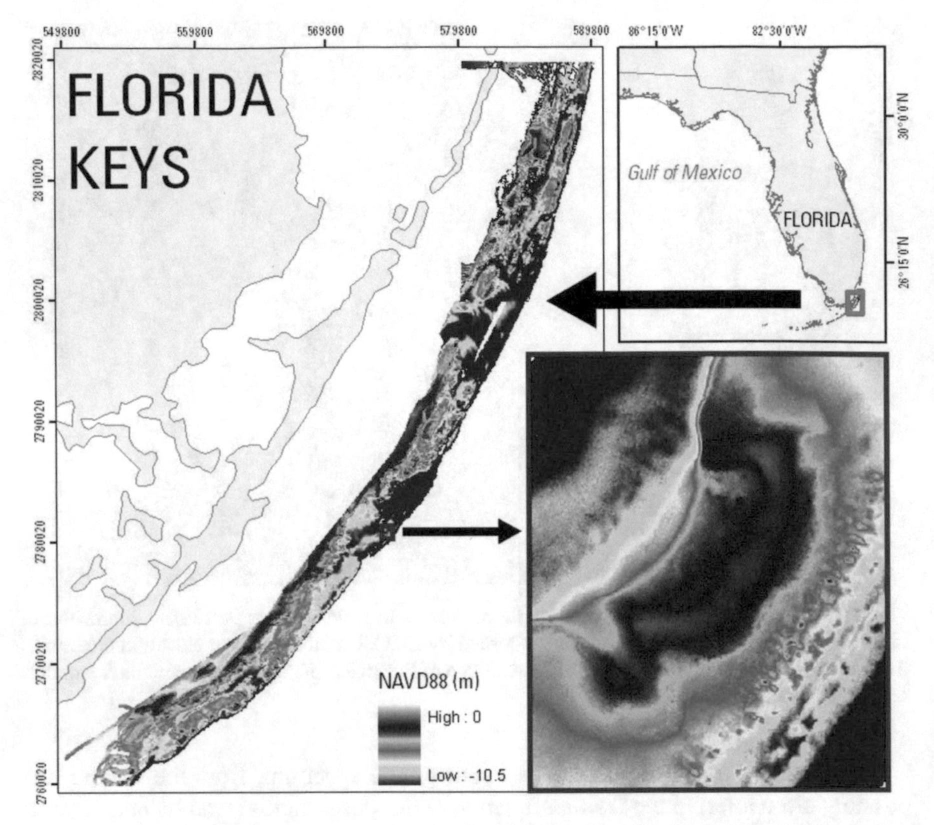

Fig. 2.9 A mosaic of 2001–2002 and 2006 Florida Keys submerged topography. (Image courtesy of the US Geological Survey, St. Petersburg Coastal and Marine Science Center)

Fig. 2.10 Airborne LiDAR data and a corresponding aerial photograph of the Bixby Bridge in Big Sur, California. Here, the LIDAR data provide vertical *(top left)* and profile *(bottom left)* views of the bridge. (Image courtesy of National Ocean Service, NOAA)

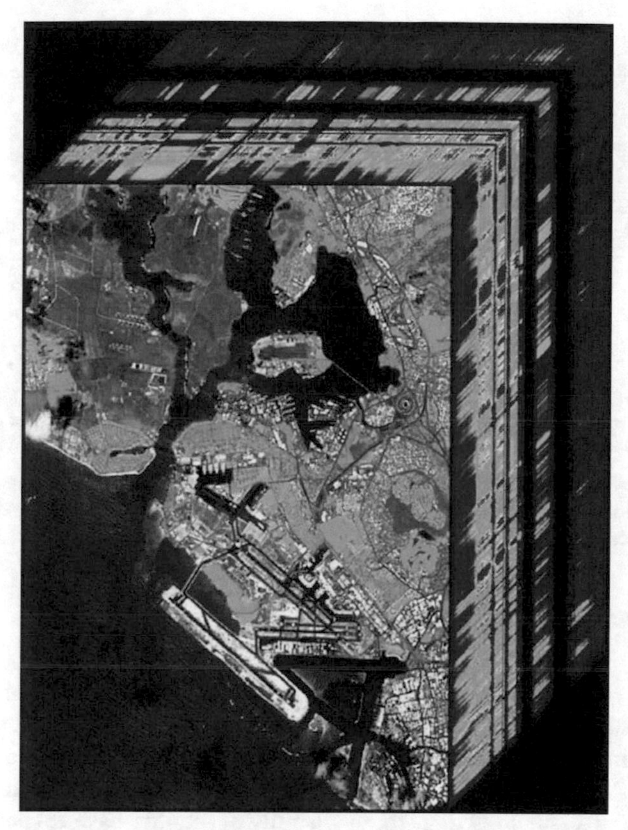

Fig. 2.11 A hyperspectral image data "cube" of Pearl Harbor, Hawaii, captured by the Airborne Visible Infrared Imaging Spectrometer (AVIRIS) sensor. The top of the image cube is a false-color composite. The sides of the cubes show the edges of each image corresponding to one of the sensor's 224 spectral bands. (Image courtesy of NASA Jet Propulsion Laboratory)

aerial vehicles (UAVs) or drones, as sensing platforms. Virtually every major category of sensor, including hyperspectral, radar, and LiDAR, has been implemented on a UAS, and UAS-based work has become the subject of an increasingly large proportion of the remote sensing literature (Colomina and Molina 2014). For example, NASA's unmanned Ikhana aircraft, mounted with a TIR sensor, was used to image wildfires in several locations in the western US during 2007 (see Fig. 2.12). Although manned aircraft can be comparatively less expensive and efficient for some applications—UASs are restricted in terms of sensor size, vehicle weight, and available power supply—this is likely to change as UAS technology evolves. Furthermore, UASs already have a number of specific advantages over manned aircraft, including their ability to take off and land in areas without adequate surface runways, as well as the fact that pilot and/or crew safety need not be concerns when flying in dangerous conditions (Watts et al. 2012).

Fig. 2.12 The Ikhana unmanned aerial system (UAS), flying over California during testing of the wildfire monitoring capabilities of an onboard thermal infrared (TIR) sensor. (Image courtesy of NASA, Jim Ross)

2.2.2 Spaceborne Platforms

An exhaustive list of all Earth-observing satellites that are currently in operation is beyond the scope of this book. Instead, we provide basic details about some prominent government and commercial satellites from the US and elsewhere (mostly listed in Table 2.1). In addition, we highlight promising developments in spaceborne remote sensing technology (e.g., the availability of satellite-based hyperspectral and LiDAR sensors).

2.2.2.1 NASA Satellites and Satellite Programs

Essentially, satellite-based environmental remote sensing began with NASA's Landsat Program. Landsat-1 (originally named ERTS-A and then ERTS-1; see Chap. 1), launched in July 1972, was the first satellite designed for multispectral imaging. Landsat-1 was the first in a series of satellites associated with the long-running program, the latest being Landsat-8, launched in February 2013 (Short 2006). Besides Landsat-8, only Landsat-7 (launched in April 1999) remains operational. Both Landsat-7 and Landsat-8 cover the same ground track (i.e., revisit a particular geographic area) every 16 days.

Landsat-5 had two onboard sensors, the MSS and the thematic mapper (TM), that were predecessors of the Enhanced Thematic Mapper (ETM+) sensor on Landsat-7. The ETM+ sensor incorporated two major improvements on the TM sensor: a higher-resolution TIR band (60 m vs. 120 m for the TM sensor) as well as the novel inclusion of a 15-m panchromatic band. The latter offered the possibility of *pan-sharpening,* a type of *data fusion* (see Chap. 3) intended to increase the spatial resolution of the multispectral bands (Liu 2000). This technique has become commonplace; indeed, although many current commercial satellites (see Table 2.1)

Table 2.3 Capabilities and applications of Operational Land Imager (OLI) and thermal infrared sensor (TIRS) spectral bands. (Material adapted from Jensen 2005 and the US Geological Survey's Landsat Missions Web site, http://landsat.usgs.gov/)

Sensor	Band	Capabilities/applications
OLI	Band 1 (coastal/aerosol)	Analysis of coastal waters and atmospheric aerosol concentrations
OLI	Band 2 (blue)	Penetrating water bodies; analysis of land use, soil, and vegetation
OLI	Band 3 (green)	Discriminating healthy green vegetation
OLI	Band 4 (red)	Vegetation discrimination; delineation of soil and geologic boundaries
OLI	Band 5 (near-infrared)	Crop identification; emphasizes soil–crop and land–water contrasts
OLI	Band 6 (mid-infrared)	Drought studies; discrimination between clouds, snow, and ice
OLI	Band 7 (mid-infrared)	Discrimination of geologic rock formations
OLI	Band 8 (panchromatic)	Pan-sharpening to improve spatial resolution of multi-spectral data
OLI	Band 9 (cirrus)	Determining cirrus cloud contamination of a scene
TIRS	Band 10 (thermal infrared)	Thermal mapping; soil moisture studies
TIRS	Band 11 (thermal infrared)	Improved thermal mapping; soil moisture studies

provide high-resolution (≤4 m) multispectral imagery, they still regularly include an even higher-resolution panchromatic band to facilitate pan-sharpening.

In May 2003, the ETM+ sensor aboard Landsat-7 experienced a failure of its scan-line corrector (SLC) mechanism. This uncorrectable failure resulted in data gaps on the left and right sides of each recorded image (Williams et al. 2006). Several methods have been developed to fill these data gaps, including merging two Landsat-7 scenes to create a mostly filled image. Nevertheless, the SLC problem added some urgency to the Landsat-8 mission, particularly when Landsat-5 stopped collecting TM data in 2011.

Landsat-8 has two onboard sensors, the OLI sensor and the TIRS, which together improve on the capability of the ETM+ sensor. In particular, both OLI and TIRS have better radiometric resolution than ETM+ (12-bit vs. 8-bit), and the OLI has a specialized band for analysis of coastal waters and atmospheric aerosols (also known as the "ultra-blue" band, as well as a previously noted band for cirrus cloud detection (see Sect. 2.1 and Table 2.2). With an approximate scene size of 170 × 183 km, the ETM+, OLI, and TIRS sensors require roughly 11,000 scenes to completely image the Earth, excluding the polar regions (Short 2006).

The importance of the Landsat Program in advancing the use of spaceborne remote sensing technology cannot be overstated. Together, the sensors aboard the Landsat satellites have provided the longest—by far—continuous and comprehensive record of Earth imagery. (Furthermore, NASA and the US Geological Survey have already started work on Landsat 9, with a planned launch in 2023.) Scientific journal articles involving Landsat data products number in the tens of thousands. Each spectral band of the OLI and TIRS sensors has utility for emphasizing and analyzing particular environmental phenomena (Table 2.3). For instance, Fig. 2.13 is an

Fig. 2.13 Landsat Operational Land Imager (OLI) image from Western Australia, captured in May 2013. The image has been enhanced; see the main text for details. (Image courtesy of the US Geological Survey)

OLI image from a tropical estuary area in Western Australia. This image was constructed by separating water and land areas, creating images for each using distinct band combinations, and then reassembling the two. The water areas of the image are displayed using the red, blue, and coastal/aerosol bands (bands 4, 2, and 1), while the land areas of the image are displayed using mid-infrared (MIR), near-infrared (NIR), and green bands (bands 6, 5, and 4). In addition, the entire Landsat image archive has been made available to the public via the Internet, facilitating long-term change analyses (see the list of relevant Web sites at the end of this chapter).

Landsat-7 and Landsat-8 are also considered part of NASA's Earth Observing System (EOS) mission. While there are several active satellites in EOS, perhaps the two most prominent are the paired satellites Terra (launched December 1999) and Aqua (launched May 2002). Together, the Aqua and Terra satellites cover the mid to higher latitudes of the Earth four times daily. Both satellites carry the MODIS sensor. The MODIS sensor records data in 36 spectral bands spanning the visible to TIR, with a unique configuration: bands 1 and 2 have a spatial resolution of 250 m, bands 3–7 have a spatial resolution of 500 m, and bands 8–36 have a spatial resolution of 1 km. Table 2.4 lists some of the major uses of these bands. As the list suggests, MODIS data have wide applicability, primarily for broad-scale (i.e.,

Table 2.4 MODIS spectral band characteristics and potential uses (from the MODIS mission Web site, http://modis.gsfc.nasa.gov/about/specifications.php)

Band	Wavelength (nm)	Primary Use
1	620–670	Land/Cloud/Aerosols
2	841–876	Boundaries
3	459–479	
4	545–565	Land/Cloud/Aerosols
5	1230–1250	Properties
6	1628–1652	
7	2105–2155	
8	405–420	
9	438–448	
10	483–493	
11	526–536	Ocean Color/
12	546–556	Phytoplankton/
13	662–672	Biogeochemistry
14	673–683	
15	743–753	
16	862–877	
17	890–920	Atmospheric
18	931–941	Water Vapor
19	915–965	
Band	**Wavelength (µm)**	**Primary Use**
20	3.660–3.840	Surface/Cloud
21	3.929–3.989	Temperature
22	3.929–3.989	
23	4.020–4.080	
24	4.433–4.498	Atmospheric
25	4.482–4.549	Temperature
26	1.360–1.390	Cirrus Clouds
27	6.535–6.895	Water Vapor
28	7.175–7.475	
29	8.400–8.700	Cloud Properties
30	9.580–9.880	Ozone
31	10.780–11.280	Surface/Cloud
32	11.770–12.270	Temperature
33	13.185–13.485	Cloud Top
34	13.485–13.785	Altitude
35	13.785–14.085	
36	14.085–14.385	

Fig. 2.14 Moderate Resolution Imaging Spectroradiometer (MODIS) Aqua image of a dust storm blowing over the Persian Gulf from Saudi Arabia in November 2014. Administrative boundaries *(black lines)* have been added to the image. (Image courtesy of Jeff Schmaltz, MODIS Rapid Response Team, NASA GSFC)

continental-scale) analyses. For instance, Fig. 2.14 shows a "true"-color MODIS Aqua image of a dust storm blowing over the Persian Gulf from the coastline of Saudi Arabia. Another example is an ongoing collaboration between the US Forest Service, NASA, and the University of Maryland to produce daily active fire occurrence data for the US using MODIS TIR bands (Giglio et al. 2003). Similarly, Mildrexler et al. (2007) outlined a methodology that uses MODIS land surface temperature and vegetation data to map continental-scale disturbances from wildfire, insect outbreaks, and other phenomena.

The ASTER sensor aboard EOS Terra is actually a combination of three radiometers that yield simultaneous and co-registered image data of differing spatial resolutions: three VNIR bands with a spatial resolution of 15 m; six SWIR bands with a spatial resolution of 30 m; and five TIR bands with a spatial resolution of 90 m. While ASTER imagery has regularly been used to study land surface temperature and thermal phenomena such as surface emissivity (e.g., Dash et al. 2002; Schmugge et al. 2002), it has also been applied for urban land cover classification (Chen et al. 2007). In addition, the sensor's SWIR bands are particularly useful for applications such as mapping the mineral composition of rocks and other geological features (Mars and Rowan 2010). Another illustration of the sensor's capabilities is an image, captured in March 2011, which reveals debris rafts that arose after a major earthquake and subsequent tsunami struck the northeastern coast of Japan (Fig. 2.15). One limitation for users is that ASTER is an on-demand instrument, meaning that imagery is only acquired over a location in response to a specific request for observation, which can be submitted via the Internet (see the list of Web sites at the end of this chapter).

NASA launched the Earth-Observing-1 (EO-1) satellite into the orbit in November 2000. The EO-1 satellite represents the first platform in the agency's New Millennium Program, which was initiated to test emerging technologies for future Earth-observing and deep space satellite missions. Of the various instruments on board EO-1, the Hyperion imaging spectrometer, with its 220 spectral bands, is of particular note. Hyperion data have been used for applications such as mineral mapping (Kruse et al. 2003), analysis of desertification (Asner and Heidebrecht 2003), soil salinization (Weng et al. 2008), and the detection of fungal disease in sugarcane (Apan et al. 2004). For example, a matching pair of natural- and false-color Hyperion images, which demonstrate how the sensor may be used to differentiate minerals cannot be distinguished using natural light, is shown in Fig. 2.16. In addition to their moderate spatial resolution (30 m), Hyperion data tend to have a poorer *signal-to-noise ratio* (SNR) relative to data from airborne hyperspectral sensors such as AVIRIS, which further limits their utility for fine-scale mapping and classification (Apan 2004; Kruse et al. 2003). Along with improved spatial resolution, future spaceborne hyperspectral sensors can be expected to include improvements to address this SNR limitation (also see the HICO description later in this chapter).

Another sensor on board the EO-1 satellite, the Advanced Land Imager (ALI), was developed as a prototype for sensors that could be carried on future Landsat missions (Chander et al. 2004; Ungar et al. 2003). It ultimately served as a template for the OLI sensor carried on Landsat-8. Although the sensor was primarily intended as a technology demonstration, ALI data have been applied, for example, in vegetation studies (e.g., Elmore and Mustard 2003; Liang et al. 2003).

Launched in 2004, Aura is the third of NASA's EOS satellites. Unlike Terra and Aqua, Aura specifically focuses on atmospheric science. The satellite carries a suite of conventional, downward-looking sensors, as well as horizon-viewing, or *limb,* instruments, which are used to study various aspects of atmospheric chemistry. One of its downward-looking instruments is the Ozone Monitoring Instrument (OMI). The OMI mission is managed by the Netherlands Agency for Aerospace Programmes (NIVR) in cooperation with the Finnish Meteorological Institute (FMI). This hyperspectral sensor captures the range of ultraviolet, visible, and near-infrared wavelengths with 740 spectral bands, which represents the highest spectral resolution of any current satellite-based sensor. One of the main objectives of the OMI has been to assess the status of the Earth's ozone (O_3) layer after the Montreal Protocol and related international agreements led to a global reduction in the release of chlorofluorocarbons (CFCs) during the late twentieth century (Levelt et al. 2006a). In addition to ozone, the OMI sensor measures a number of other trace gases, including nitrogen dioxide (NO_2) and sulfur dioxide (SO_2), which are both significant air pollutants (Levelt et al. 2006b). Figure 2.17 depicts tropospheric NO_2 levels in central eastern China, as recorded by the OMI sensor in March 2008. The

Fig. 2.15 Advanced Spaceborne Thermal Emission and Reflection Radiometer (ASTER) image of Sendai, Japan, after a major earthquake and tsunami struck the northeastern Japanese coast in March 2011. The image reveals large debris rafts that were created offshore as the tsunami withdrew from inland areas. (Image courtesy of NASA/GSFC/METI/ERSDAC/JAROS and the US/Japan ASTER Science Team)

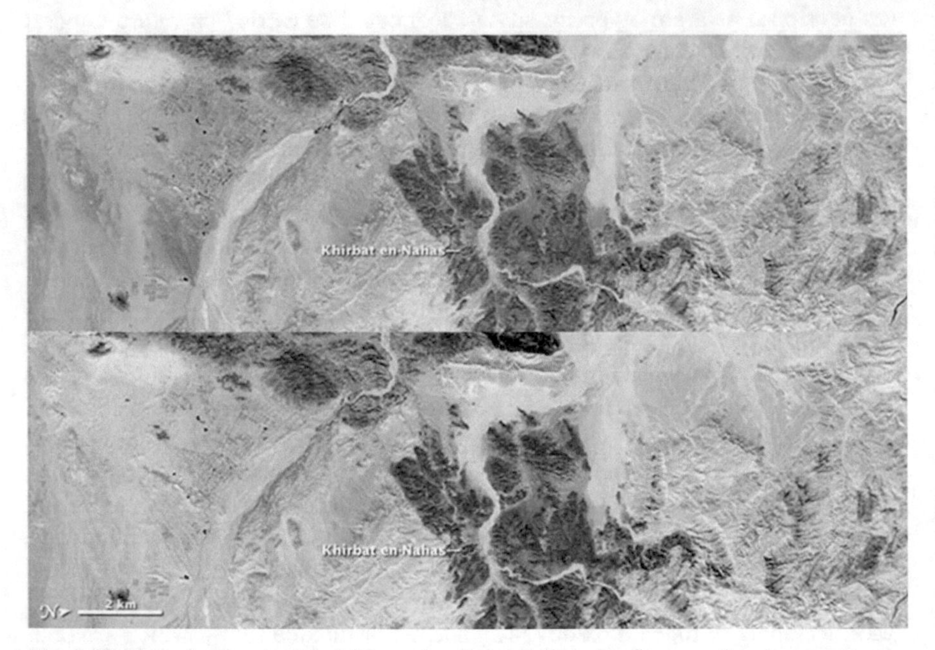

Fig. 2.16 Natural-color *(top)* and false-color *(bottom)* Hyperion images of rock outcrops near Kirbhat en-Nahas. These images highlight the capacity of the Hyperion sensor to differentiate minerals that appear similar under natural light. (Image courtesy of the NASA EO-1 Team, Robert Simmon)

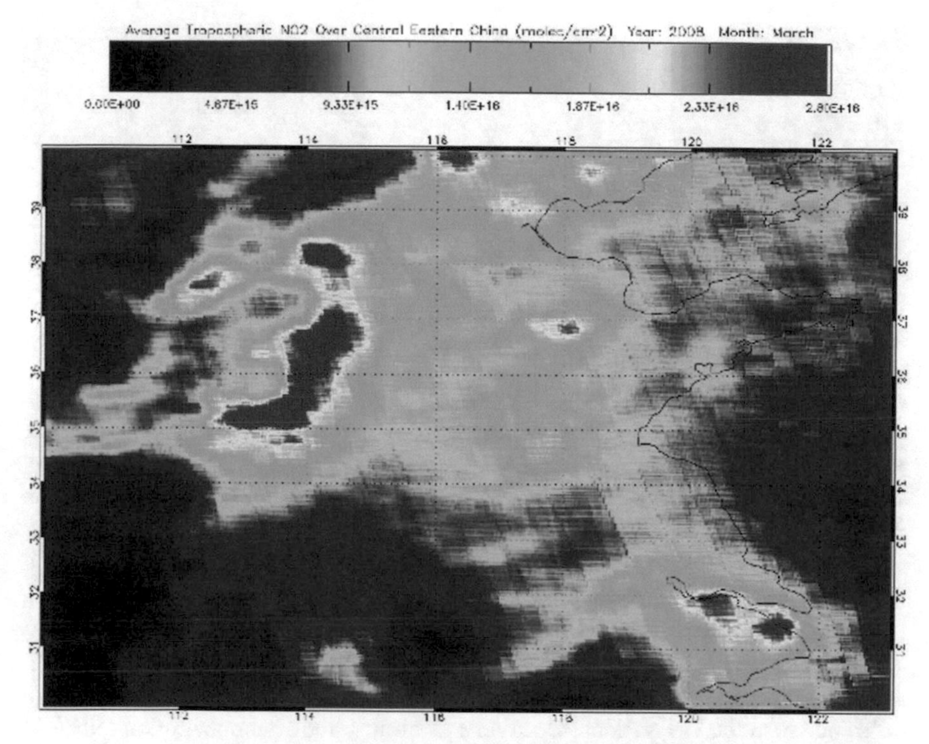

Fig. 2.17 Tropospheric nitrogen dioxide (NO_2) measurements in central eastern China, recorded March 2008 by the NASA/KNMI Ozone Monitoring Instrument (OMI) aboard the AQUA satellite. (Image courtesy of the Air Quality Group, University of Leicester, http://www.leos.le.ac.uk/aq)

image shows several areas with high NO_2 levels, particularly over large cities such as Shanghai (in the southeastern corner of image).

The CALIPSO satellite is jointly operated by NASA and Centre National d'Études Spatiales (CNES), the French government space agency (Winker et al. 2009). A key instrument on CALIPSO is the Cloud-Aerosol LiDAR with Orthogonal Polarization (CALIOP) sensor, which represents one of very few operational spaceborne LiDAR systems. As its name suggests, the CALIOP sensor is used for analysis of a variety of atmospheric phenomena, including dust storms (Liu et al. 2008). Figure 2.18 is a characteristic "vertical profile" image captured by CALIOP in May 2010 over the site of the Deepwater Horizon oil spill in the Gulf of Mexico. The spill, which was the largest accidental spill in history, was caused by the catastrophic failure of the Deepwater Horizon offshore platform the month before (Ottaviani et al. 2012). The CALIOP image shows the presence of volatile organic compounds over the spill (red areas in the center of the profile image). These compounds were released when oil reached the water's surface and made contact with air, thereby indicating the spill's extent.

Fig. 2.18 Cloud-Aerosol LiDAR with Orthogonal Polarization (CALIOP) image, recorded in May 2010, showing vertical profile aerosols above the site of the Deepwater Horizon oil spill. The low-lying *red bands* in the center of the profile indicate the presence of volatile organic compounds above the spill. (Image courtesy of NASA)

The Suomi National Polar-orbiting Partnership (NPP) is intended to serve as a bridge mission between the aging satellites that comprise the EOS and a future generation of satellites, known as the Joint Polar Satellite System (JPSS), which will eventually replace them. Like the JPSS, the Suomi NPP mission is a collaborative effort between the US National Oceanic and Atmospheric Administration (NOAA) and NASA. The instruments on the Suomi NPP satellite, which was launched in 2011, are essentially the same as those that will be carried on the JPSS spacecraft. In particular, the Visible Infrared Imaging Radiometer Suite (VIIRS) is a multipurpose sensor intended to collect terrestrial, oceanic, atmospheric, and *cryospheric* (permafrost, snow, and ice) measurements. An especially unique feature of the VIIRS sensor is a panchromatic Day/Night Band (DNB), which is able to record VNIR reflectance at night that is 10 million times fainter than reflected sunlight (Miller et al. 2013). This is an important advantage over most satellite-based sensors, which are only able to record TIR reflectance at night; for instance, while semitransparent clouds may appear opaque at TIR wavelengths, the DNB can penetrate these clouds and record the surface features beneath them (Miller et al. 2013). Figure 2.19 shows the Great Lakes and mid-Atlantic regions of the US as recorded in January 2015 by the VIIRS DNB. The mostly clear image shows snow on the ground over the northern portions of both regions. Notably, VIIRS (both on Suomi NPP and later on the JPSS satellites) is also expected to serve as a replacement for the widely utilized AVHRR sensor (described later in this chapter).

Aquarius, the primary instrument aboard the Satélite de Aplicaciones Cientificas (SAC)-D satellite, was developed by NASA to provide global observations of ocean surface salinity. The Aquarius/SAC-D mission represents a collaborative effort between NASA and Comisión Nacional de Actividades Espaciales (CONAE), the Argentinian space agency, which provided a number of complementary instruments in addition to the SAC-D spacecraft, which was launched successfully in 2011. The Aquarius sensor is an unusual, combined passive/active instrument that

Fig. 2.19 The Great Lakes and mid-Atlantic regions of the US as recorded with the Visible Infrared Imaging Radiometer Suite (VIIRS) Day/Night Band (DNB) in January 2015. (Image courtesy of NOAA)

includes three L-band radiometric beams (operating at 1.413 GHz) as well as radar scatterometers (1.26 GHz) for measuring surface backscatter (see Fig. 2.20). The radiometric beams measure oceanic microwave emissions, which are affected by the electrical conductivity of seawater and, when combined with knowledge of the sea surface temperature, can subsequently be translated into salinity estimates (Le Vine et al. 2010). The scatterometer accounts for the impact of ocean surface roughness on these measurements.

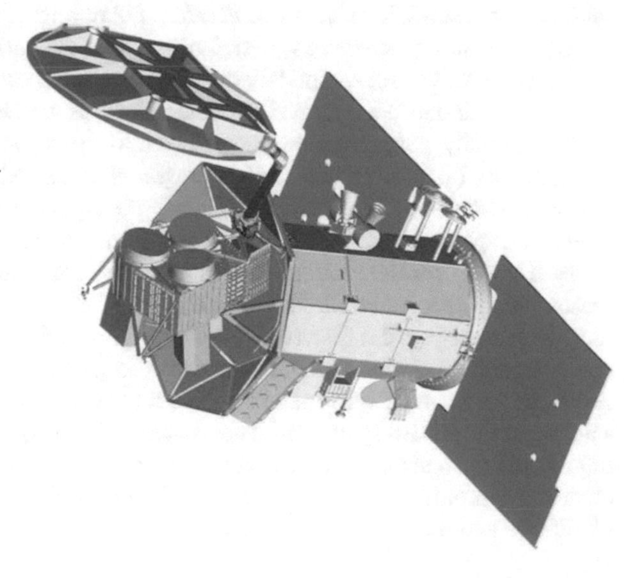

Fig. 2.20 Diagram of the Satélite de Aplicaciones Cientificas (SAC-D) spacecraft, showing the Aquarius instrument at the *left,* including its large reflector antenna. (Image courtesy of NASA Aquarius)

Table 2.5 Characteristics and typical uses of AVHRR spectral bands

Band	Bandwidth (μm)	Typical use
1	0.58–0.68	Daytime cloud and surface mapping
2	0.725–1	Land–water boundaries
3A	1.58–1.64	Snow and ice detection
3B	3.55–3.93	Night cloud mapping, sea surface temperature
4	10.30–11.30	Night cloud mapping, sea surface temperature
5	11.50–12.50	Sea surface temperature

2.2.2.2 Other Government Satellites and Satellite Programs

The AVHRR is a sensor mounted on a series of satellites operated by the US National Oceanic and Atmospheric Administration (NOAA). The latest in the series, NOAA-19, was launched in 2009. NOAA-19 carries the third version of the AVHRR sensor, with six spectral bands (Table 2.5). Only five bands are transmitted to the ground at any given time; the 3A and 3B bands are transmitted during daytime and nighttime, respectively. The AVHRR sensor was primarily developed for oceanographic and meteorological applications, but it has also been used for land cover mapping at broad scales (e.g., Loveland et al. 2000), especially because of its temporal coverage: AVHRR sensors image the entire Earth twice a day (Jensen 2005). However, the sensor's capacity in that regard has largely been superseded by the MODIS sensors on the EOS Aqua and Terra satellites, as well as the VIIRS sensor aboard Suomi NPP. Still, there was enough interest in maintaining the continuity of AVHRR imagery that the sensor was included on EUMETSAT's MetOp-A (launched in 2006) and MetOp-B (launched in 2012) satellites as a heritage instrument.

The NOAA and MetOp satellites have *polar orbits,* meaning that they pass over or near the North and South Poles during each pass around the Earth. In other words, their orbits are perpendicular to the equator. Many other satellites, such as Landsat-7 and Landsat-8, have *near-polar orbits*. Polar- and near-polar-orbiting satellites are typically *sun-synchronous,* which means that the satellite passes over a given location on Earth at approximately the same local time each day. This characteristic enables regular data collection at consistent times as well as long-term comparisons.

Alternatively, some satellites are placed into a *geostationary orbit*. A satellite in geostationary (also sometimes called *geosynchronous*) orbit remains fixed in a particular location above the Earth (commonly the equator) and orbits in the direction (and at the speed) of the planet's rotation. One advantage of geostationary satellites is that it is easy for the ground-based antennae to communicate with them. For this reason, this is the preferred orbit for many weather monitoring satellites, including the GOES system. Administered by the US National Environmental Satellite, Data, and Information Service (NESDIS), which is part of NOAA, the GOES system serves as the primary information source for the US National Weather Service. A satellite in the GOES system orbits at roughly 35,800 km above the equator, allowing its onboard sensor (i.e., the I-M Imager instrument) to record individual images covering approximately 25 % of the Earth's surface (Jensen 2005). Currently, three GOES are active: GOES-13, designated "GOES East," which is situated at 75° W;

GOES-14, which is in on-orbit storage status; and GOES-15, designated "GOES West," which is situated at 135° W (NOAA 2011).

Although most Earth-observing sensors are found on dedicated and unmanned satellites, one exception is the HICO, which operated from the International Space Station between 2009 and 2014. Designed and initially operated by the US Office of Naval Research, and eventually operated by NASA during the latter part of its lifetime, HICO captured only one image per orbital pass, but with a high spatial resolution (90 m) and a high SNR, addressing some of the limitations of the Hyperion sensor (see the description earlier in this chapter). In fact, its high SNR was necessary for successful imaging; because coastal ocean water is dark compared to the atmosphere, a substantial portion of the at-sensor signal is sacrificed for atmospheric correction, leaving little signal for detecting ocean features (Corson et al. 2008). Unfortunately, radiation from a solar storm in September 2010 rendered HICO inoperable. Nevertheless, the sensor far exceeded its initial mission goals, collecting roughly 10,000 scenes over 5 years.

The Canadian Space Agency (CSA) launched the RADARSAT-2 satellite in December 2007, as a follow-up to RADARSAT-1, which operated between 1995 and 2013. The primary payload on RADARSAT-2 is a *synthetic aperture radar (SAR)* imaging system. An SAR system utilizes the motion of the platform that carries it (e.g., a satellite) to simulate a large antenna, or aperture. In short, as the platform moves, the system transmits multiple, successive radio pulses toward targeted objects or areas. The returned signals are then synthesized into a single image with higher spatial resolution than could be captured with a smaller, physical antenna. The SAR system on RADARSAT-2 operates in the C-band range of the electromagnetic spectrum, at a wavelength of 5.6 cm (5.405 GHz frequency). By comparison, the X-band system on KOMPSAT-5 (see below) operates at a wavelength of 3.2 cm (9.66 GHz frequency).

As with airborne radar systems, RADARSAT-2 may record data day or night, and under all weather conditions, but has the added advantage of frequent revisits (every 24 days) of targeted areas. (RADARSAT-2 has left- and right-looking modes that can further reduce the revisit time.) The SAR system on RADARSAT-2 also has multiple beam modes (driven by an active phased array antenna) that essentially trade swath size for improved spatial resolution. A particular strength of satellite-based SAR imagery is the ability to measure ocean waves and currents over large areas (Goldstein et al. 1989); for instance, Fig. 2.21 is a RADARSAT image showing wave patterns in the Pacific Ocean around Point Reyes, California.

The ESA launched its first Earth-observing satellite, ERS-1, in 1991. Since that initial launch, ESA has developed a broad-scale Earth observation program, Copernicus, exemplified by the Sentinel family of satellites. Each of the six planned Sentinel missions involves a two-satellite constellation. Sentinel-1 A, which was launched in April 2014, is similar to RADARSAT-2, carrying a C-band SAR system that can be operated in multiple modes. Figure 2.22 shows the Villarica volcano in southern Chile, which erupted on March 3, 2015. The image was constructed from two Sentinel-1 A scans, one on February 20 and another on March 4; the colored areas highlight surface changes between the two dates, while unchanged areas appear grey. Sentinel-1 A has a revisit time of 12 days, but once its sister satellite, Sentinel-

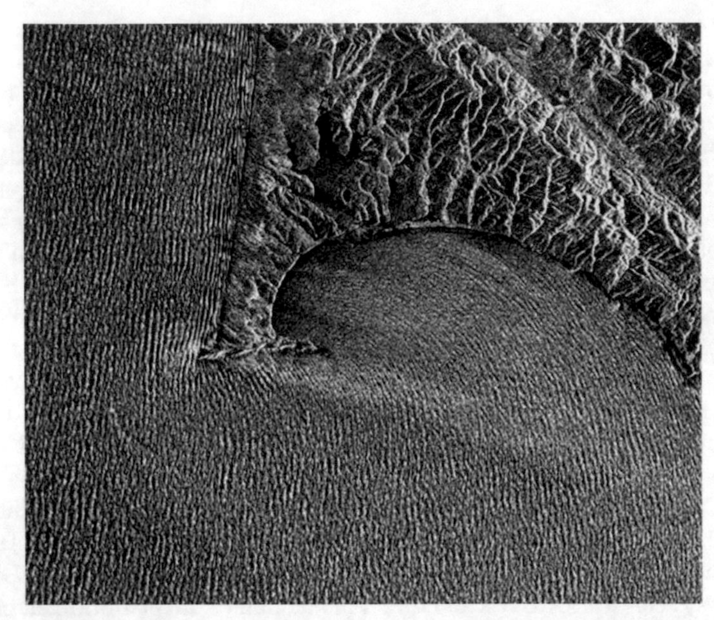

Fig. 2.21 RADARSAT image recorded near Point Reyes, California. (Image courtesy of NASA)

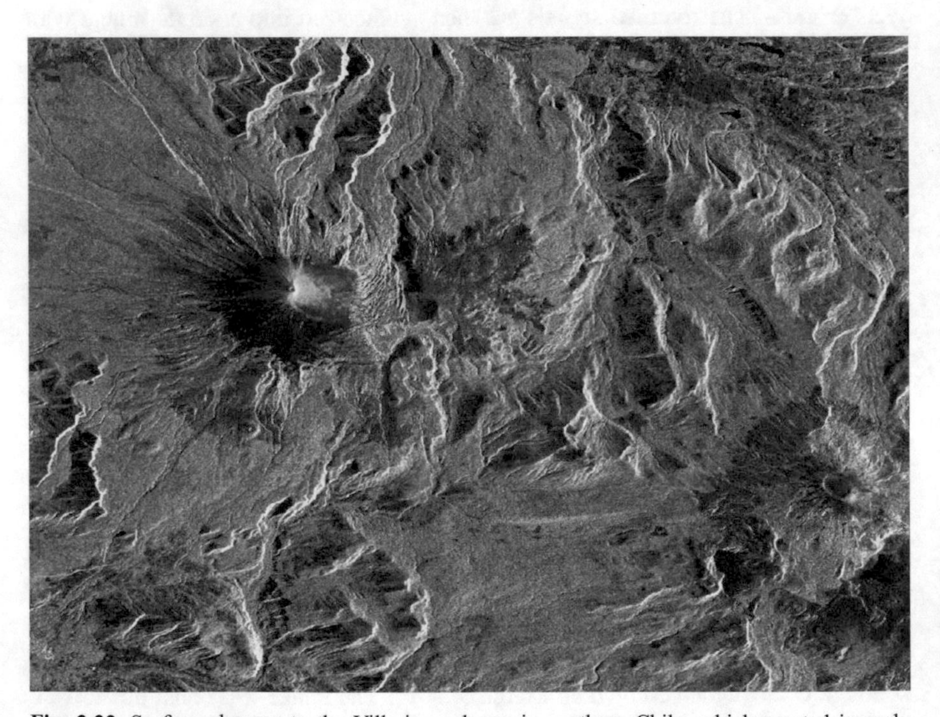

Fig. 2.22 Surface changes to the Villarica volcano in southern Chile, which erupted in early March 2015. The image was developed from scans by the Sentinel-1A synthetic aperture radar (SAR) instrument, taken before and after the eruption. (Image courtesy of the European Space Agency and the German Remote Sensing Data Center)

1B, is launched (planned for 2016) and operational, they will together provide a 6-day revisit time (Torres et al. 2012).

The ESA has also launched a series of small, low-cost micro-satellites as part of its Project for Onboard Autonomy (Proba). The most recent of these is Proba-V, which was launched in 2013. The initial purpose of Proba-V was technology demonstration, but it has since taken on a more significant role because of its primary instrument, the VEGETATION sensor, which is an improved version of an instrument by the same name carried on the SPOT 4 and SPOT 5 satellites (discussed later in this chapter). The VEGETATION sensor was developed to provide daily global coverage of vegetation dynamics at a 1-km spatial resolution, and the resulting data products have a large user community (Maisongrande et al. 2004). Proba-V ensures the continuity of the VEGETATION data stream, having taken over the responsibility for data collection from SPOT 5 in 2014 (after SPOT 4 had ceased to operate in 2013).

Several nations outside North America and Europe have also established successful satellite programs. For example, the China National Space Administration (CNSA) has launched two satellites, Gaofen-1 and Gaofen-2 ("gao fen" = high resolution) as part of its civilian High-Definition Earth Observation Satellite (HDEOS) program (Zhang et al. 2014). Gaofen-1 (launched in 2013) carries two sensors, the high-resolution (2-m/8-m) panchromatic and multispectral camera (PMC) and the medium-resolution (16-m) wide field imager (WFI), while Gaofen-2 (launched in late 2014) carries a high-resolution optical imager (HROI) that improves substantially (0.8-m panchromatic, 3.2-m multispectral) on Gaofen-1's PMC sensor. The ultimate goal of the HDEOS is to provide a near-real-time image support for a variety of applications, including disaster relief and damage assessment; in this context, Gaofen-1 data have been used to analyze urban infrastructural conditions after the 2014 Yutian earthquake in Xinjiang, China (Zhang et al. 2014). In addition, WFI data from Gaofen-1 have also been utilized to study suspended particulates in Poyang Lake, the largest freshwater lake in China (Li et al. 2015). Figure 2.23 is an image of China's Yellow River delta, captured by the WFI sensor aboard Gaofen-1 in May 2013.

Fig. 2.23 Image of the Yellow River delta in China, captured by Gaofen-1's wide field imager (WFI) in May 2013. (Image courtesy of CNSA)

The China–Brazil Earth Resources Satellite (CBERS) program was established in 1988 by a cooperative agreement between the two countries. The program has developed five satellites since its inception: CBERS-1 (1999–2003), CBERS-2 (2003–2007), CBERS-2B (2007–2010), CBERS-3 (lost during launch in 2013), and CBERS-4 (launched in late 2014). The CBERS-4 satellite carries four sensors (see Table 2.1) that build upon the instrument suites of its predecessors: the medium-resolution (20-m) multispectral camera (MUXCAM), the high-resolution (5-m/10-m) panchromatic and multispectral camera (PANMUX), the infrared medium resolution scanner (IRSCAM), and the wide-field imaging camera (WFICAM). These sensors provide users with a variety of choices in terms of spectral coverage as well as spatial and temporal resolutions. Starting in 2004, Brazil's National Institute for Space Research (INPE) has made CBERS data freely available, which has caused the number of registered data users to swell to more than 50,000 (Fonseca et al. 2014). Perhaps most notably, CBERS images have been utilized by the Amazon Deforestation Monitoring Program. Although this monitoring program largely relies on Landsat data, CBERS data have been useful for filling in gaps in the Landsat coverage due to cloud cover, which is a common problem for remote sensing in tropical forests (Fonseca et al. 2014).

The Indian Space Research Organization (ISRO) has constructed and launched 18 Earth-observing satellites, starting with Indian Resources Satellite (IRS) 1 A in 1988. Its latest Earth-observing satellite, Resourcesat-2, was successfully launched in 2011. Resourcesat-2 is intended to continue the mission of its predecessor, Resourcesat-1, which has been in orbit since 2003 (Seshadri et al. 2005). Like its predecessor, Resourcesat-2 carries three multispectral sensors with differing spatial resolutions (see Table 2.1): the 5.8-m, three-spectral-band linear imaging self-scanner (LISS)-IV; the 23.4-m, four-band linear imaging self-scanner (LISS)-III; and the 56-m, four-band advanced wide field sensor (AWiFS). Together, these three sensors offer a high degree of analytical flexibility (Seshadri et al. 2005), making the satellite's data potentially suitable for a variety of tasks, as demonstrated by the use of Resourcesat-1 imagery for studying coastal dynamics and water quality (Rajawat et al. 2007), monitoring snow cover (Kulkarni et al. 2006), and crop discrimination (Sesha Sai and Narasimha Rao 2008). Figure 2.24 is a close-up from one of the first images captured by a Resourcesat-2's LISS-IV sensor, which shows the city of New Delhi.

The Japanese Aerospace Exploration Agency (JAXA) was formed in 2003 from a merger of three organizations: the National Space Development Agency of Japan (NASDA), the National Aerospace Laboratory (NAL), and the Institute of Space and Aeronautical Science (ISAS). Prior to the merger, NASDA developed and launched a number of satellites, including Japanese Earth Resources Satellite (JERS)-1, which operated from 1992 to 1998. JAXA's Advanced Land Observing Satellite (ALOS), or "DAICHI," operated from 2006 until April 2011 (JAXA 2011); similar to JERS-1, ALOS carried three optical instruments including an L-band SAR. One of JAXA's ongoing missions is the Greenhouse Gases Observing Satellite (GOSAT), also known by its Japanese name, "Ibuki." Launched in 2009, GOSAT is the first satellite designed to remotely measure atmospheric concentra-

Fig. 2.24 Close-up image of New Delhi, India, captured in May 2011 by the linear imaging self-scanner (LISS)-IV sensor aboard Resourcesat-2. Connaught Place can be seen at the top center of the image, while the Rajpath, the ceremonial boulevard between the presidential palace and the India Gate, stretches across the center of the image. (Image courtesy of the Indian Space Research Organization)

tions of carbon dioxide (CO_2) and methane (CH_4; Kuze et al. 2009; Morino et al. 2010). The satellite carries two instruments, the Thermal and Near-Infrared Sensor for Carbon Observation Cloud and Aerosol Imager (TANSO-CAI) and Fourier Transform Spectrometer (TANSO-FTS). The TANSO-FTS sensor's four bands operate in the SWIR and TIR portions of the electromagnetic spectrum, collecting data on surface reflectance as well as surface and atmospheric thermal radiance (Kuze et al. 2009). Unlike many other remote sensors, the TANSO-FTS instrument does not provide continuous spatial coverage; instead, it captures 10.5 × 10.5-km images that are spaced roughly 150 km apart in a grid. Over GOSAT's 3-day orbital period, the TANSO-FTS sensor records approximately 56,000 points globally (Kuze et al. 2009). By comparison, less than 200 ground monitoring stations worldwide are used for systematic greenhouse gas analysis (Morino et al. 2010).

A recently launched pair of satellites, Resurs-P ("Resource-Prospecting") No. 1 and No. 2, are the latest in a series of Earth-observing satellites developed by Russia or the Soviet Union; the first in the series, Resurs 01–1, operated for a little over a year (October 1985–November 1986), although the subsequent Resurs 02 satellite lasted 11 years (1988–1999) (Belward and Skøien 2015). The Resurs-P satellites are operated by the Russian Federal Space Agency (Roscosmos). Each satellite carries a payload of three instruments: the high-resolution Geoton-L1 sensor, the dual-mode KShMSA sensor, and perhaps most notably, the 130-band GSA hyperspectral sensor (see Table 2.1 for specifications).

The Korean Aerospace Research Institute (KARI) has launched a number of Earth-observing satellites since 1999 as part of South Korea's "Mid- to Long-Term Plan for National Space Development" (Kim et al. 2015). The KOMPSAT-5 satellite (launched in 2013) carries the Corea SAR Instrument (COSI), an X-band system with multiple imaging modes. Figure 2.25 shows the city of Paris, as captured by COSI operating in high-resolution mode. The recently (March 2015) launched

Fig. 2.25 The city of Paris, France, as captured by the Corea synthetic aperture radar (SAR) Instrument (COSI) aboard KOMPSAT-5, operating in high-resolution mode. The Eiffel Tower is visible in the image's center-right. (Image courtesy of SI Imaging Services)

KOMPSAT-3 A satellite carries the 0.55-m panchromatic/2.2-m multispectral Advanced Earth Imaging Sensor System (AEISS)-A, which is comparable to the sensors aboard many of today's high-resolution commercial satellites (see Table 2.1). KOMPSAT-3 A also carries the 5.5-m Infrared Imaging System (IIS), which operates in the lower (i.e., mid-infrared) portion of the TIR range (Kim et al. 2015). A South Korean company, SI Imaging Services (SIIS), has worldwide rights to market and distribute KOMPSAT imagery as well as imagery from a few commercial satellites such as Deimos-2 (see Table 2.1 and the list of links at the end of this chapter).

2.2.2.3 Commercial Satellites

Other than the aforementioned Deimos-2 (with its High-Resolution Advanced Imaging System, or HiRAIS) and the five-satellite RapidEye constellation operated by BlackBridge (see Table 2.1), two companies have become the primary worldwide providers of commercial satellite imagery: the Airbus Group, based in France, and DigitalGlobe, based in the US. Each of these companies has achieved its current position in the industry through a series of mergers, acquisitions, and/or name changes. For instance, the Airbus Group can be traced back to Spot Image, a public company started by the French Space Agency (CNES) in 1982 to manage the SPOT satellites. In 2008, Spot Image became a subsidiary of Astrium, a European Aeronautic Defense and Space (EADS) company. In 2014, EADS renamed themselves as the Airbus Group, which includes Airbus Defence and Space, the division that manages its current satellite programs.

Historically, the SPOT satellites that were operated by Spot Image, including SPOT-4 (1998–2013) and SPOT-5 (2002–current), often served as useful alternatives to the Landsat satellites. For example, the still-active SPOT-5 carries two High-Resolution Geometrical (HRG) instruments that generate 10-m resolution VNIR imagery; in some vegetation mapping studies, this higher spatial resolution relative to the Landsat TM and ETM+ sensors was found to be a sufficient trade-off for the HRG sensors' comparatively lower spectral resolution (White et al. 2006). In addition, 5-m panchromatic images captured with the HRG sensors have proven sufficient for fine-scale analyses such as the automatic detection of ships for monitoring of fisheries (Corbane et al. 2008). Furthermore, the two HRG sensors can be operated together in "super mode" to generate 2.5-m panchromatic imagery (Pasqualini et al. 2005). Finally, the utility of the coarse-resolution VEGETATION sensor aboard SPOT-4 and SPOT-5 for regional mapping studies (Stibig et al. 2007) was mentioned earlier in the chapter, but it is worth noting further that, unlike various Landsat sensors with their 16-day revisit times, the SPOT sensors, including the sensors on SPOT-6 and SPOT-7, are configured to permit off-nadir viewing (Jensen 2005), which can reduce the revisit time from 26 days (i.e., the satellites' orbit cycle) to 2–3 days (and one day for the VEGETATION sensors on SPOT-4 and SPOT-5).

In the 1990s and early 2000s, one of the primary issues with SPOT data was that they cost substantially more than similarly performing Landsat data. For instance, in an analysis of the utility of satellite imagery for mapping mountain pine beetle *(Dendroctronus ponderosae)* damage in British Columbia, Canada, White et al. (2006) determined that 10-m HRG imagery from SPOT-5 yielded similar accuracies to Landsat TM/ETM+ imagery, but cost 200 times (CAN$1200 vs. CAN$6) as much for their study area. However, it should be noted that images from the SPOT-4/5 sensors were substantially less expensive than 10 m HRG imagery; see the Web site listed at the end of this chapter). The relatively high cost of commercial satellite imagery remains an issue, but regardless, the success of Spot Image laid the groundwork for other commercially operated satellites that, starting with IKONOS in 1999, have provided imagery with very high spatial resolutions. Indeed, the latest SPOT satellites, SPOT-6 and SPOT-7, offer 1.5-m panchromatic and 6-m multispectral imagery, while the similarly configured Pléiades satellites (developed by Airbus Defence and Space in cooperation with CNES) capture 0.5-m panchromatic and 2-m multispectral images. Figure 2.26 is a close-up view, extracted from

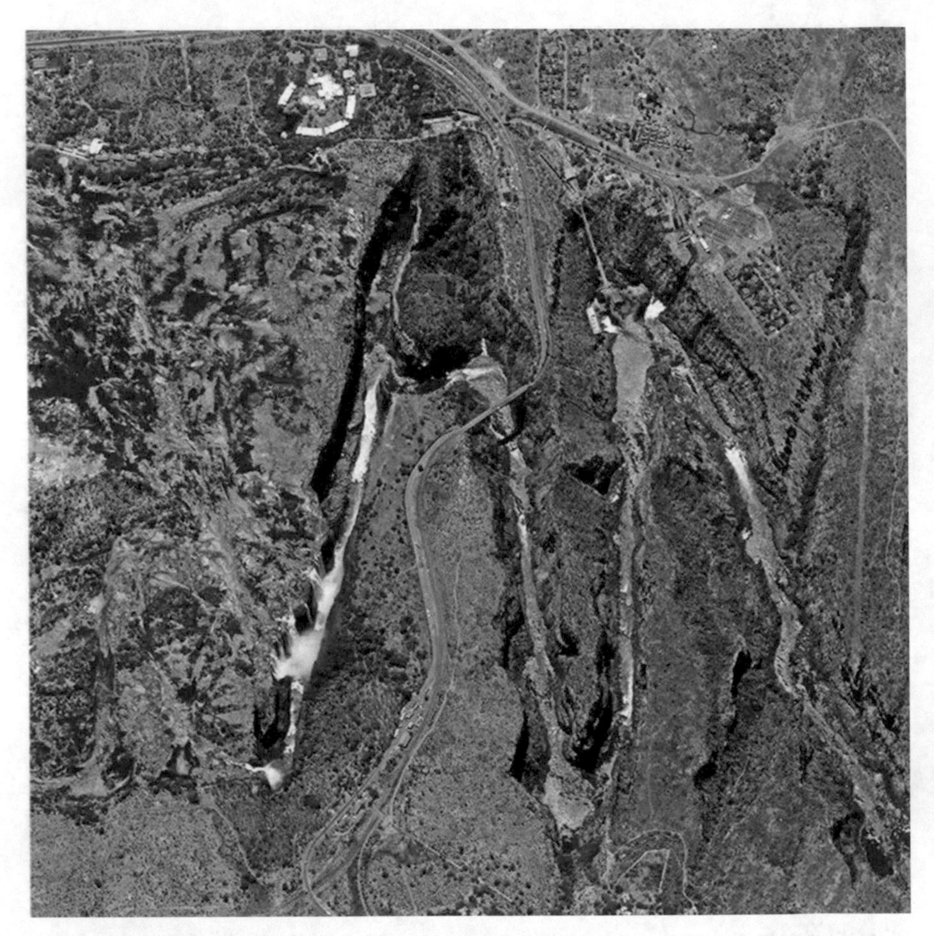

Fig. 2.26 Close-up from a Pléiades-1B image showing the area around Victoria Falls in southern Africa. The falls sit on the border between Zambia and Zimbabwe. (Image courtesy of CNES and Airbus Defence and Space)

a Pléiades-1B image, showing the area around Victoria Falls in southern Africa. Nevertheless, the commercial remote sensing company that has arguably made the greatest advances is DigitalGlobe. It currently operates five satellites that produce high-spatial-resolution imagery: IKONOS and GeoEye-1, acquired when the company that originally operated them (GeoEye) merged with DigitalGlobe, as well as QuickBird, WorldView-2, and WorldView-3 (see Table 2.1 for specifications). WorldView-3, which was launched in 2014, is particularly noteworthy due to its extremely high spatial resolution (0.31-m panchromatic, 1.24-m VNIR, 3.7-m SWIR), as well as an array of spectral bands previously only seen in sensors developed by large government space agencies such as NASA and ESA: eight VNIR bands, eight SWIR bands, and 12 moderate-resolution (30-m) CAVIS bands (CAVIS, clouds, aerosols, vapors, ice, and snow).

As noted earlier in this chapter, there are clearly many potential applications for high-spatial-resolution imagery such as that provided by the latest commercial satellites. Still, it is important to acknowledge that these images are not well suited to standard classification techniques (e.g., maximum likelihood classification, see Chap. 3), primarily because objects and areas of interest are often represented by multiple pixels with high variability in their spectral values (e.g., due to shadows) (Blaschke 2010; Yu et al. 2006). Instead, object-based classification techniques (see Chap. 3) are increasingly popular for application to high-spatial-resolution images. Another issue for commercial satellite imagery is the cost to the user, especially given the free availability of other satellite imagery (e.g., Landsat) that, while of coarser spatial resolution, may be sufficient for analysis. In any case, users of remotely sensed data should carefully consider the objectives of their particular projects when selecting the imagery most likely to meet those objectives.

The various space agencies, as well as their commercial counterparts, are in a constant state of research and development. A list of planned satellite missions is not feasible, since economic, political, and technological realities often lead to modifications, launch delays, and in some cases, complete cancellation of missions. Nevertheless, it is possible to identify a few trends regarding the future of spaceborne remote sensing. First, the ongoing commercialization of satellite operations is likely to continue, although NASA and other space agencies will maintain important roles in instrument development. Second, all four types of resolution (i.e., spatial, spectral, radiometric, and temporal) should continue to see steady improvement with future sensors, although "moderate-resolution" instruments will still be in demand, particularly because they tend to offset their resolution limitations with a wide geographic scope. Finally, there appears to be at least a partial shift away from large and expensive, multi-instrument platforms toward smaller, dedicated satellites (or "constellations" of satellites) that permit developers to keep pace with emerging user needs, provide better temporal coverage, and reduce technology development time. The net result of these trends should be a proliferation of data sources and analytical opportunities for future users of satellite-based data.

2.3 Review Questions

1. What is the benefit of higher radiometric resolution (e.g., 12-bit versus 8-bit) for a remotely sensed data user?
2. Describe one of the challenges of using high-spatial-resolution satellite image data.
3. What was the first environmental remote sensing satellite, and when was it launched?
4. What spectral wavelengths (or image bands) are important for remote sensing of coastal areas?
5. Despite tremendous advances in satellite-based sensors, airborne remote sensing is still regularly used for a variety of applications. List some of the reasons why a user might opt for the latter.

6. Describe the two primary instruments aboard Landsat-8, and contrast them with the ETM+ sensor on Landsat-7.
7. Besides NASA and the European Space Agency, which governmental space organization has been the most active in environmental remote sensing?
8. Distinguish between a sun-synchronous and a geostationary orbit.
9. The WorldView-2 and WorldView-3 satellites both have what is known as a "red edge" spectral band. What is meant by the red edge, and how might a red edge band be used?
10. Compare and contrast radar and LiDAR.

References

Apan, A., A. Held, S. Phinn, and J. Markley. 2004. Detecting sugarcane 'orange rust' disease using EO-1 Hyperion hyperspectral imagery. *International Journal of Remote Sensing* 25: 489–498.

Asner, G.P. and K.B. Heidebrecht. 2003. Imaging spectroscopy for desertification studies: comparing AVIRIS and EO-1 Hyperion in Argentina drylands. *IEEE Transactions on Geoscience and Remote Sensing* 41: 1283–1296.

Avery, T.E. and G.L.L. Berlin. 1992. *Fundamentals of Remote Sensing and Airphoto Interpretation*, 5th Edition. New York: Prentice Hall. 472 p.

Barr, S. and M. Barnsley. 2000. Reducing structural clutter in land cover classifications of high spatial resolution remotely-sensed images for urban land use mapping. *Computers and Geosciences* 26: 433–449.

Belward, A.S. and J.O. Skøien. 2015. Who launched what, when and why; trends in global land-cover observation capacity from civilian earth observation satellites. *ISPRS Journal of Photogrammetry and Remote Sensing* 103: 115–128.

Blaschke, T. 2010. Object based image analysis for remote sensing. *ISPRS Journal of Photogrammetry and Remote Sensing* 65: 2–16.

Brock, J.C., C.W. Wright, A.H. Sallenger, W.B. Krabill, and R.N. Swift. 2002. Basis and methods of NASA Airborne Topographic Mapper lidar surveys for coastal studies. *Journal of Coastal Research* 18(1): 1–13.

Browell, E.V., S. Ismail, and W.B. Grant. 1998. Differential absorption lidar (DIAL) measurements from air and space. *Applied Physics B* 67: 399–410.

Campbell, P.K.E., B.N. Rock, M.E. Martin, C.D. Neefus, J.R. Irons, E.M. Middleton, and J. Albrechtova. 2004. Detection of initial damage in Norway spruce canopies using hyperspectral airborne data. *International Journal of Remote Sensing* 25: 5557–5583.

Caylor, J. 2000. Aerial photography in the next decade. *Journal of Forestry* 98(6): 17–19.

Chander, G., D.J. Meyer, and D.L. Helder. 2004. Cross calibration of the Landsat-7 ETM+ and EO-1 ALI sensor. *IEEE Transactions on Geoscience and Remote Sensing* 42: 2821–2831.

Charlton, M.E., A.R.G. Large, and I.C. Fuller. 2003. Application of airborne LiDAR in river environments: the river Coquet, Northumberland, UK. *Earth Surface Processes and Landforms* 28: 299–306.

Chen, J. and G.F. Hepner. 2001. Investigation of imaging spectroscopy for discriminating urban land covers and surface materials. Proceedings of the 2001 AVIRIS Earth Science and Applications Workshop, Palo Alto, CA. NASA JPL Publication 02-1. http://popo.jpl.nasa.gov/docs/work-shops/01_docs/2001Chen_web.pdf

Chen, Y., P. Shi, T. Fung, J. Wang, and X. Li. 2007. Object-oriented classification for urban land cover mapping with ASTER imagery. *International Journal of Remote Sensing* 28: 4645–4651.

Clark, D.B., C.S. Castro, L.D.A. Alvarado, and J.M. Read. 2004. Quantifying mortality of tropical rain forest trees using high-spatial-resolution satellite data. *Ecology Letters* 7: 52–59.

Colomina, I. and P, Molina. 2014. Unmanned aerial systems for photogrammetry and remote sensing: a review. *ISPRS Journal of Photogrammetry and Remote Sensing* 92: 79–97.

Coops, N.C. and P.C. Catling. 1997. Predicting the complexity of habitat in forests from airborne videography for wildlife management. *International Journal of Remote Sensing* 18: 2677–2682.

Corbane, C., F. Marre, and M. Petit. 2008. Using SPOT-5 HRG data in panchromatic mode for operational detection of small ships in tropical area. *Sensors* 8: 2959–2973.

Corson, M.R., D.R. Korwan, R.L. Lucke, and W.A. Snyder. 2008. The Hyperspectral Imager for the Coastal Ocean (HICO) on the International Space Station. Proceedings of the IEEE International Geoscience and Remote Sensing Symposium (IGARSS), 7–11 July 2008, Vol. 4: IV-101-IV-104

Dash, P., F.M. Gottsche, F.S. Olesen, and H. Fischer. 2002. Land surface temperature and emissivity estimation from passive sensor data: theory and practice–current trends. *International Journal of Remote Sensing* 23: 2563–2594.

Elmore, A.J., and J.F. Mustard. 2003. Precision and accuracy of EO-1 Advanced Land Imager (ALI) data for semiarid vegetation studies. *IEEE Transactions on Geoscience and Remote Sensing* 41: 1311–1320.

Fonseca, L.M.G., J.C.N. Epiphanio, D.M. Valeriano, J.V. Soares, J.C.L. Dalge, and M.A. Santos. 2014. Earth observation applications in Brazil with focus on the CBERS Program. *IEEE Geoscience and Remote Sensing Magazine* 2: 53–55.

Galvão, L.S., A.R. Formaggio, E.G. Couto, and D.A. Roberts. 2008. Relationships between the mineralogical and chemical composition of tropical soils and topography from hyperspectral remote sensing data. *ISPRS Journal of Photogrammetry & Remote Sensing* 63: 259–271.

Ganguly, S., M.A. Friedl, B. Tan, X. Zhang, and M. Verma. 2010. Land surface phenology from MODIS: Characterization of the Collection 5 global land cover dynamics product. *Remote Sensing of the Environment* 114: 1805–1816.

Giglio, L., J. Descloitres, C.O. Justice, and Y. Kaufman. 2003. An enhanced contextual fire detection algorithm for MODIS. *Remote Sensing of Environment* 87: 273–282.

Goldstein, R.M., T.P. Barnett, and H.A. Zebker. 1989. Remote sensing of ocean currents. *Science* 246: 1282–1285.

Grebby, S., J. Naden, D. Cunningham, and K. Tansey. 2011. Integrating airborne multispectral imagery and airborne LiDAR data for enhanced lithological mapping in vegetated terrain. *Remote Sensing of Environment* 115: 214–226.

Harsanyi, J.C., and C.-I. Chang. 1994. Hyperspectral image classification and dimensionality reduction: an orthogonal subspace projection approach. *IEEE Transactions on Geoscience and Remote Sensing* 32: 779–785.

Hester, D.B., S.A.C. Nelson, H.I. Cakir, S. Khorram, and H. Cheshire, 2010. High-resolution land cover change detection based on fuzzy uncertainty analysis and change reasoning. *International Journal of Remote Sensing* 31: 455–475.

Hester, D.B., H.I. Cakir, S.A.C. Nelson, and S. Khorram. 2008. Per-pixel classification of high spatial resolution satellite imagery for urban land-cover mapping. *Photogrammetric Engineering and Remote Sensing* 74: 463–471.

Japanese Aerospace Exploration Agency (JAXA). 2011. Japanese Earth Resources Satellite-1 "FUYO-1" (JERS-1) [web site]. http://global.jaxa.jp/projects/sat/jers1/index.html

Jensen, J. R. 2005. *Introductory Digital Image Processing: A Remote Sensing Perspective*, 3rd Edition. Upper Saddle River, NJ: Prentice Hall. 526 p.

Jensen, J.R., E.J. Christensen, and R. Sharitz. 1984. Nontidal wetland mapping in South Carolina using airborne multispectral scanner data. *Remote Sensing of Environment* 16: 1–12.

Jha, M.N., J. Levy, and Y. Gao. 2008. Advances in remote sensing for oil spill disaster management: state-of-the-art sensors technology for oil spill surveillance. *Sensors* 8: 236–255.

Khorram, S., F.H. Koch, C.F. van der Wiele, and S.A.C. Nelson. 2012a. *Remote Sensing*. New York: Springer. 134 p.

Khorram, S., S.A.C. Nelson, H.I. Cakir, and C.F. van der Wiele. 2012b. Digital image acquisition: preprocessing and data reduction. In: Pelton, J.N., S. Madry, and S. Camacho-Lara, eds. Handbook of Satellite Applications, 2nd Edition. New York: Springer, pp. 809–837.

Kim, H.-O., H.-S. Kim, H.-S. Kim, and H.-J. Choi. 2015. Space-based Earth observation activities in South Korea. *IEEE Geoscience and Remote Sensing Magazine* 3: 34–39.

Klemas, V. 2010. Tracking oil slicks and predicting their trajectories using remote sensors and models: case studies of the Sea Princess and Deepwater Horizon oil spills. *Journal of Coastal Research* 26: 789–797.

Komarov, A.S. and D.G. Barber. 2014. Sea ice motion tracking from sequential dual-polarization RADARSAT-2 images. *IEEE Transactions on Geoscience and Remote Sensing* 52: 121–136.

Krabill, W.B., J.G. Collins, L.E. Link, R.L. Swift, and M.L. Butler. 1984. Airborne laser topographic mapping results. *Photogrammetric Engineering and Remote Sensing* 50: 685–694.

Kross, S.M. and X.J. Nelson. 2011. A portable low-cost remote videography system for monitoring wildlife. *Methods in Ecology and Evolution* 2: 191–196.

Kruse, F.A., J.W. Boardman, and J.F. Huntington. 2003. Comparison of airborne hyperspectral data and EO-1 Hyperion for mineral mapping. *IEEE Transactions on Geoscience and Remote Sensing* 41: 1388–1400.

Kulkarni, A.V., S.K. Singh, P. Mathur, and V.D. Mishra. 2006. Algorithm to monitor snow cover using AWiFS data of Resourcesat-1 for the Himalayan region. *International Journal of Remote Sensing* 27: 2449–2457.

Kuze, A., H. Suto, M. Nakajima, and T. Hamazaki. 2009. Thermal and near infrared sensor for carbon observation Fourier-transform spectrometer on the Greenhouse Gases Observing Satellite for greenhouse gases monitoring. *Applied Optics* 48: 6716–6733.

Le Vine, D.M., G.S.E. Lagerloef, and S.E. Torrusio. 2010. Aquarius and remote sensing of sea surface salinity from space. *Proceedings of the IEEE* 98: 688–703.

Levelt, P.F., E. Hilsenrath, G.W. Leppelmeier, G.H.J. van den Oord, P.K. Bhartia, J. Tamminen, J. F. de Haan, and J.P. Veefkind. 2006a. Science objectives of the Ozone Monitoring Instrument. *IEEE Transactions on Geoscience and Remote Sensing* 44: 1199–1208.

Levelt, P.F., G.H.J. van den Oord, M.R. Dobber, A. Mälkki, H. Visser, J. de Vries, P. Stammes, J.O.V. Lundell, and H. Saari. 2006b. The Ozone Monitoring Instrument. *IEEE Transactions on Geoscience and Remote Sensing* 44: 1093–1101.

Li, J., X. Chen, L. Tian, J. Huang, and L. Feng. 2015. Improved capabilities of the Chinese high-resolution remote sensing satellite GF-1 for monitoring suspended particulate matter (SPM) in inland waters: radiometric and spatial considerations. *ISPRS Journal of Photogrammetry and Remote Sensing* 106: 145–156.

Liang, S., H. Fang, M. Kaul, T.G. Van Niel, T.R. McVicar, J.S. Pearlman, C.L. Walthall, C.S.T. Daughtry, and K.F. Huemmrich. 2003. Estimation and validation of land surface broadband albedos and leaf area index from EO-1 ALI data. *IEEE Transactions on Geoscience and Remote Sensing* 41: 1260–1267.

Lillesand, T., R. Kiefer, and J. Chipman. 2008. *Remote Sensing and Image Interpretation*, 6th Edition. New York: John Wiley and Sons. 763 p.

Lim, K., P. Treitz, M. Wulder, B. St-Onge, and M. Flood. 2003. LiDAR remote sensing of forest structure. *Progress in Physical Geography* 27: 88–106.

Liu, J.G. 2000. Evaluation of Landsat-7 ETM + panchromatic band for image fusion with multi-spectral bands. *Natural Resources Research* 9: 269–276.

Liu, Z., A. Omar, M. Vaughan, J. Hair, C. Kittaka, Y. Hu, K. Powell, C. Trepte, D. Winker, C. Hostetler, R. Ferrare, and R. Pierce. 2008. CALIPSO lidar observations of the optical properties of Saharan Dust: a case study of long-range transport. *Journal of Geophysical Research* 113: D07207.

Loveland, T.R., B.C. Reed, J.F. Brown, D.O. Ohlen, Z. Zhu, L. Yang, and J.W. Merchant. 2000. Development of a global land cover characteristics database and IGBP DISCover from 1 km AVHRR data. *International Journal of Remote Sensing* 21: 1303–1330.

Maisongrande, P., B. Duchemin, and G. Dedieu. 2004. VEGETATION/SPOT: an operational mission for the Earth monitoring; presentation of new standard products. *International Journal of Remote Sensing* 25: 9–14.

Mars, J.C. and L.C. Rowan. 2010. Spectral assessment of new ASTER SWIR surface reflectance data products for spectroscopic mapping of rocks and minerals. *Remote Sensing of Environment* 114: 2011–2025.

Massonnet, D. and K.L. Feigl. 1998. Radar interferometry and its application to changes in the Earth's surface. *Reviews of Geophysics* 36: 441–500.

Mausel, P.W., J.H. Everitt, D.E. Escobar, and D.J. King. 1992. Airborne videography: current status and future perspectives. *Photogrammetric Engineering and Remote Sensing* 58: 1189–1195.

Metternicht, G. 2003. Vegetation indices derived from high-resolution airborne videography for precision crop management. *International Journal of Remote Sensing* 24: 2855–2877.

Mildrexler, D.J., M. Zhao, F.A. Heinsch, and S.W. Running. 2007. A new satellite-based methodology for continental-scale disturbance detection. *Ecological Applications* 17: 235–250.

Miller, S.D., W. Straka, S.P. Mills, C.D. Elvidge, T.F. Lee, J. Solbrig, A. Walther, A.K. Heidinger, and S.C. Weiss. 2013. Illuminating the capabilities of the Suomi National Polar-Orbiting Partnership (NPP) Visible Infrared Imaging Radiometer Suite (VIIRS) Day/Night Band. *Remote Sensing* 5: 6717–6766.

Moon, W.M., G. Staples, D.-J. Kim, S.-E. Park, and K.-A. Park. 2010. RADARSAT-2 and coastal applications: surface wind, waterline, and intertidal flat roughness. *Proceedings of the IEEE* 98: 800–815.

Morgan, J.L., S.E. Gergel, and N.C. Coops. 2010. Aerial photography: a rapidly evolving tool for ecological management. *BioScience* 60: 47–59.

Morino, I., O. Uchino, M. Inoue, Y. Yoshida, T. Yokota, P.O. Wennberg, G.C. Toon, D. Wunch, C.M. Roehl, J. Notholt, T. Warneke, J. Messerschmidt, D.W.T. Griffith, N.M. Deutscher, V. Sherlock, B. Connor, J. Robinson, R. Sussman, and M. Rettinger. 2010. Preliminary validation of column-averaged volume mixing ratios of carbon dioxide and methane retrieved from GOSAT short-wavelength infrared spectra. *Atmospheric Measurement Techniques Discussions* 3: 5613–5643.

Moulin, S., L. Kergoat, N. Viovy, and G. Dedieu. 1997. Global-scale assessment of vegetation phenology using NOAA/AVHRR satellite measurements. *Journal of Climate* 10: 1154–1170.

National Oceanic and Atmospheric Administration (NOAA). 2014. GOES Status [web page]. NOAA National Environmental Satellite, Data, and Information Service (NESDIS), Office of Satellite and Product Operations. http://www.ospo.noaa.gov/Operations/GOES/status.html

Ottaviani, M., B. Cairns, J. Chowdhary, B. Van Diedenhoven, K. Knobelpiesse, C. Hostetler, R. Ferrare, S. Burton, J. Hair, M.D. Obland, and R. Rogers. 2012. Polarimetric retrievals of surface and cirrus clouds properties in the region affected by the Deepwater Horizon oil spill. *Remote Sensing of Environment* 121: 389–403.

Pasqualini, V., C. Pergent-Martini, G. Pergent, M. Agreil, G. Skoufas, L. Sourbes, and A. Tsirika. 2005. Use of SPOT 5 for mapping seagrasses: An application to *Posidonia oceanica*. *Remote Sensing of Environment* 94: 39–45.

Pontius, J., R. Hallett, and M. Martin. 2005. Using AVIRIS to assess hemlock abundance and early decline in the Catskills, New York. *Remote Sensing of Environment* 97: 163–173.

Rajawat, A.S., M. Gupta, B.C. Acharya, and S. Nayak. 2007. Impact of new mouth opening on morphology and water quality of the Chilika Lagoon—a study based on Resourcesat-1 LISS-III and AWiFS and IRS-1D LISS-III data. *International Journal of Remote Sensing* 28: 905–923.

Reutebuch, S.E., H.-E. Andersen, and R.J. McGaughey. 2005. Light detection and ranging (LIDAR): an emerging tool for multiple resource inventory. *Journal of Forestry* 103: 286–292.

Ring, J. 1963. The laser in astronomy. *New Scientist* 344: 672–673.

Rogan, J. and D.M. Chen. 2004. Remote sensing technology for mapping and monitoring land-cover and land-use change. *Progress in Planning* 61: 301–325.

Sawaya, K.E., L.G. Olmanson, and N.J. Heinert. 2003. Extending satellite remote sensing to local scales: land and water resource monitoring using high-resolution imagery. *Remote Sensing of Environment* 88: 144–156.

Schmugge, T., A. French, J.C. Ritchie, A. Rango, and H. Pelgrum. 2002. Temperature and emissivity separation from multispectral thermal infrared observations. *Remote Sensing of Environment* 79: 189–198.

Sesha Sai, M.V.R. and P.V. Narasimha Rao. 2008. Utilization of Resourcesat-1 data for improved crop discrimination. *International Journal of Applied Earth Observation and Geoinformation* 10: 206–210.

Seshadri, K.S.V., M. Rao, V. Jayaraman, K. Thyagarajan, and K.R. Sridhara Murthi. 2005. *Acta Astronautica* 57: 534–539.

Short, N.M. 2006. The Remote Sensing Tutorial [web site]. National Aeronautics and Space Administration (NASA), Goddard Space Flight Center. http://fas.org/irp/imint/docs/rst/

Stibig, H.-J., A.S. Belward, P.S. Roy, U. Rosalina-Wasrin, S. Agrawal, P.K. Joshi, Hildanus, R. Beuchle, S. Fritz, S. Mubareka, and C. Giri. 2007. A land-cover map for South and Southeast Asia derived from SPOT-VEGETATION data. *Journal of Biogeography* 34: 625–637.

Stöckli, R. and P.L. Vidale. 2004. European plant phenology and climate as seen in a 20-year AVHRR land-surface parameter dataset. *International Journal of Remote Sensing* 25: 3303–3330.

Torres, R., P. Snoeij, D. Geudtner, D. Bibby, M. Davidson, E. Attema, P. Potin, B. Rommen, N. Floury, M. Brown, I.N. Traver, P. Deghaye, B. Duesmann, B. Rosich, N. Miranda, C. Bruno, M. L'Abbate, R. Croci, A. Pietropaolo, M. Huchler, and F. Rostan. 2012. GMES Sentinel-1 mission. *Remote Sensing of Environment* 120: 9–24

Ungar, S.G., J.S. Pearlman, J.A. Mendenhall, and D. Reuter. 2003. Overview of the Earth Observing One (EO-1) mission. *IEEE Transactions on Geoscience and Remote Sensing* 41: 1149–1159.

USDA Farm Service Agency (FSA). 2010. Imagery Programs—NAIP imagery [web site]. U.S. Department of Agriculture, Farm Service Agency, Aerial Photography Field Office. http://www.fsa.usda.gov/FSA/apfoapp?area=home&subject=prog&topic=nai

Wang, C., H. Zhang, B. Zhang, Y. Tang, H. Wu, X. Wen, and D. Yan. 2009. Disaster phenomena of Wenchuan earthquake in high resolution airborne synthetic aperture radar images. *Journal of Applied Remote Sensing* 3: 031690.

Watts, A.C., V.G. Ambrosia, and E.A. Hinkley. 2012. Unmanned aircraft systems in remote sensing and scientific research: classification and considerations of use. *Remote Sensing* 4: 1671–1692.

Weber, B., C. Olehowski, T. Knerr, J. Hill, K. Deutschewitz, D.C.J. Wessels, B. Eitel, and B. Büdel. 2008. A new approach for mapping of Biological Soil Crusts in semidesert areas with hyperspectral imagery. *Remote Sensing of Environment* 112: 2187–2201.

Weng, Y., P. Gong, and Z. Zhu. 2008. Soil salt content estimation in the Yellow River delta with satellite hyperspectral data. *Canadian Journal of Remote Sensing* 34: 259–270.

White, J.C., M.A. Wulder, and D. Grills. 2006. Detecting and mapping mountain pine beetle red-attack damage with SPOT-5 10-m multispectral imagery. *BC Journal of Ecosystems and Management* 7: 105–118.

Williams, D.L., S. Goward, and T. Arvidson. 2006. Landsat: yesterday, today, and tomorrow. *Photogrammetric Engineering and Remote Sensing* 72: 1171–1178.

Winker, D.M., M.A. Vaughan, A. Omar, Y. Hu, K.A. Powell, Z. Liu, W.H. Hunt, and S.A. Young. 2009. Overview of the CALIPSO Mission and CALIOP data processing algorithms. *Journal of Atmospheric and Oceanic Technology* 26: 2310–2323.

Wright, C.W. and J.C. Brock. 2002. EAARL: A LiDAR for mapping shallow coral reefs and other coastal environments. In: Proceedings of the 7th International Conference on Remote Sensing for Marine and Coastal Environments, Miami, FL, 20–22 May 2002 (CD-ROM).

Wulder, M.A. J.C. White, S. Magnussen, and S. McDonald. 2007. Validation of a large area land cover product using purpose-acquired airborne video. *Remote Sensing of Environment* 106: 480–491.

Yang, C. and G.L. Anderson. 2000. Mapping grain sorghum yield variability using airborne digital videography. *Precision Agriculture* 2: 7–23.

Yu, Q., P. Gong, N. Clinton, G. Biging, M. Kelly, and D. Schirokauer. 2006. Object-based detailed vegetation classification with airborne high spatial resolution remote sensing imagery. *Photogrammetric Engineering and Remote Sensing* 72: 799–811.

Zhang, B., W. Perrie, X. Li, and W.G. Pichel. 2011. Mapping sea surface oil slicks using RADARSAT-2 quad-polarization SAR image. *Geophysical Research Letters* 38: L10602.

Zhang, Y., H. Huang, Z. Jiang, Y. Fang, and X. Cheng. 2014. Thermal anomaly before earthquake and damage assessment using remote sensing data for 2014 Yutian earthquake. Proceedings of *SPIE* 9260: 92604D.

Suggested Reading

Belward, A.S., and J.O. Skøien. 2015. Who launched what, when and why; trends in global land-cover observation capacity from civilian earth observation satellites. *ISPRS Journal of Photogrammetry and Remote Sensing* 103: 115–128.

National Research Council (U.S.). 2013. Landsat and beyond: sustaining and enhancing the nation's land imaging program. Washington, DC: The National Academies Press.

Relevant Websites

NASA's Visible Earth, highlighting numerous application examples of satellite-based sensors: http://visibleearth.nasa.gov

Science Education through Earth Observation for High Schools (SEOS) Project, a joint project of the European Commission and ESA: http://www.seos-project.eu/modules/remotesensing/remotesensing-c03-p01.html

U.S. National Oceanic and Atmospheric Administration (NOAA), National Ocean Service: What is LIDAR? http://oceanservice.noaa.gov/facts/lidar.html

Landsat image archive: http://landsat.usgs.gov/

ASTER Data Acquisition Request Tool: http://asterweb.jpl.nasa.gov/NewReq.asp

Canadian Space Agency (CSA): http://www.asc-csa.gc.ca

European Space Agency (ESA): http://www.esa.int/ESA

China National Space Administration (CNSA): http://www.cnsa.gov.cn/n615709/cindex.html

China-Brazil Earth Resources Satellite (CBERS) Program: http://www.cbers.inpe.br/ingles/

Indian Space Research Organization (ISRO): http://www.isro.org/

Japan Aerospace Exploration Agency (JAXA): http://global.jaxa.jp/

Russian Federal Space Agency (Roscosmos), Research Center for Earth Operative Monitoring: http://eng.ntsomz.ru/

SI Imaging Services (SIIS), commercial distributor of KOMPSAT, Deimos-2, and other satellite imagery: http://www.si-imaging.com

SPOT/Proba-V VEGETATION Product Distribution Site (VITO): http://www.vito-eodata.be/PDF/portal/Application.html#Home

Airbus Defence and Space: http://airbusdefenceandspace.com

DigitalGlobe: http://www.digitalglobe.com

RapidEye (BlackBridge): http://www.blackbridge.com/rapideye/

Chapter 3
Data Processing Tools

Techniques that are conventionally used to process remotely sensed data into geospatial input or output products (e.g., classified maps) for further analysis and application are discussed in this chapter. This chapter focuses on the processing of moderate-spatial-resolution, multispectral digital image data, such as the imagery captured by the Landsat Enhanced Thematic Mapper (ETM)+ or Landsat Operational Land Imager (OLI) sensors. This type of data is still utilized in a large percentage of applied remote sensing efforts, especially in terrestrial environments. In the latter portion of the chapter, we briefly cover the processing of other categories of remotely sensed data such as hyperspectral and Light Detection and Ranging (LiDAR) data, which require specialized processing approaches. Readers interested in learning more about the processing of these other data types should consult the suggested reading list at the end of the chapter.

The processing of multispectral image data is typically divided into three stages: preprocessing, processing, and post-processing. Each of these stages has its own distinct set of common tools and techniques. However, before discussing any of the stages of data processing, it is first important to understand the more fundamental task of multispectral image display.

3.1 Display of Multispectral Image Data

Raw image data are commonly displayed (i.e., on a computer monitor using geographic information system (GIS) or remote sensing software) in the form of individual bands or as "true" color composites (TCC) or "false" color composites (FCC). Any individual band of a multispectral digital image can be displayed as grayscale (panchromatic) image, where the lowest-value pixels are displayed as black, the highest-value pixels are displayed as white, and pixels with intermediate values are displayed in corresponding shades of gray. Alternatively, TCC and FCC work under the premise that a computer monitor will display no more than

S. Khorram et al., *Principles of Applied Remote Sensing*,
DOI 10.1007/978-3-319-22560-9_3

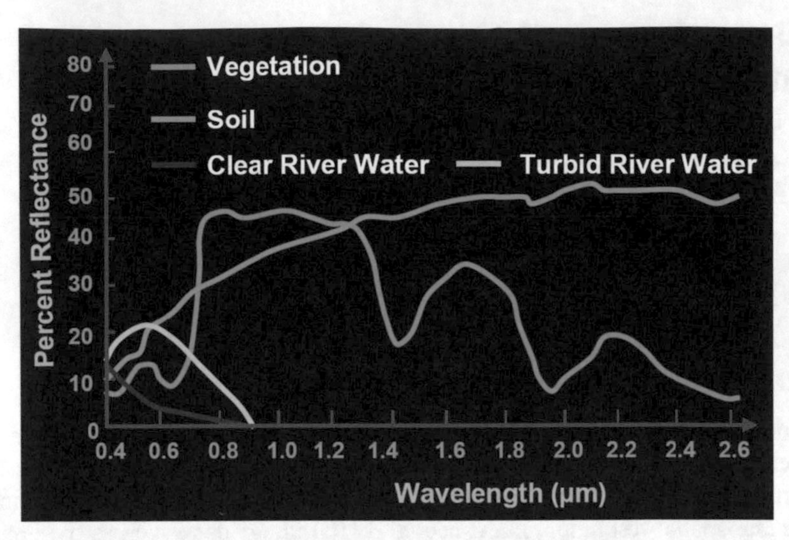

Fig. 3.1 Spectral signatures (i.e., spectral properties) of *vegetation*, *soils*, *clear water*, and *turbid water* ranging from visible to near-infrared portions of electromagnetic spectrum that can be used in classification of images for these features

three image bands at a time, each matched to one of three primary color ramps, or "guns": blue, green, and red. Thus, a TCC image displays the blue band from a raw multispectral image using the blue color ramp, the green band from the image using the green color ramp, and the red band using the red color ramp. Essentially, a TCC depicts its features in natural color. In contrast, an FCC displays the combination of any three bands from a multispectral image other than the true-color, blue/green/red combination. In the standard FCC, the green band of the input image is displayed using the blue color ramp, the red image band is displayed using the green color ramp, and a near-infrared band from the input image (e.g., Band 4 from a Landsat Thematic Mapper (TM) image) is displayed using the red color ramp. The reason that this particular FCC is so popular is because it can be used to highlight the presence of healthy green vegetation. Figure 3.1 displays line graphs of the reflectance properties (also called spectral response curves) of vegetation, soil-clear river water, and turbid river water over a range of wavelengths from the visible to infrared portions of the electromagnetic spectrum. Note that the two forest vegetation classes have much higher reflectance in near-infrared wavelengths than in visible ones. Thus, when an image's near-infrared band is displayed using a red color ramp in an FCC, live vegetation will appear red in the composite image. Figure 3.2 provides a side-by-side comparison of a TCC and FCC, constructed from Landsat 8 data, for a study area in California. Another example of TCC and FCC in the South San Francisco Bay Area is shown in Fig. 3.3. Various levels of salinity in salt flats and turbidity and suspended solids in the South Bay are visually apparent in this image.

Fig. 3.2 An example of true color composite (TCC; *left*) and standard false color composite (FCC; *right*) of a study area in Oxnard, coastal California based on Landsat 8 data. Note that vegetation appears in various shades of *green* in the TCC and in various shades of *red* in the standard FCC. (Image courtesy of Verkere and Khorram 2015)

3.2 Preprocessing Image Data

Typical image preprocessing operations include geometric correction, atmospheric correction, radiometric correction, and the calculation of various spectral band combinations and ratios, and data fusion. A remote sensing analyst employs these operations in order to prepare the best possible input data for the actual image processing stage. The operations essentially serve one of two purposes: (1) to minimize distortions and/or errors in an image that could hinder successful classification; or (2) to extract or enhance an image's most critical information, thus making classification more straightforward.

3.2.1 Geometric Correction

Remotely sensed images may contain two types of geometric distortions: systematic and nonsystematic. Systematic distortions are due to platform motion caused by forward movement of the aircraft or spacecraft, variations in mirror scanning rate, panoramic distortions, variations in platform velocity, and distortions due to the curvature of the Earth. Nonsystematic distortions are typically due to sensor malfunctions or variations in sensor altitude and attitude.

These distortions also affect the way the images from the sensors are captured. For example, an image suffering from systematic distortion may have coordinates that are consistently off by a certain amount across the whole image, in that each row might be shifted a few meters east or west relative to its true position. However

Fig. 3.3 An example of false color composite *(FCC)* and true color composite *(TCC)* from Landsat 8 in South San Francisco Bay. The color variations in various shades of *blue* reflect the levels of turbidity and suspended solids in water. The salt flats in various shades of *pink* in TCC corresponding to *green* in FCC reflect the levels of salinity. (Image courtesy of Khorram 2015)

not all systematic distortion errors are caused by factors that vary in a consistent manner. Nonsystematic distortion errors occur when random factors cause local variations in image scale and coordinate location. These errors are particularly common in airborne imagery, where turbulence and variations in ground topography can alter the image scale over short distances. These errors also occur in satellite imagery, in this case caused by the rate at which the Earth rotates beneath a satellite, or by the rate at which off-nadir scale changes across an image. Nonsystematic distortion errors may be evident within the imagery due to the direction in which the sensor is pointed and the amount it varies (e.g., aircraft pitch, yaw, and roll), which in turns causes the coordinates of the pixels' locations to change in a nonsystematic manner.

Many systematic distortion errors can be minimized using equations to adjust the pixel locations systematically across the entire image. In fact, systematic errors are often already removed in commercially available data. For example, it is common

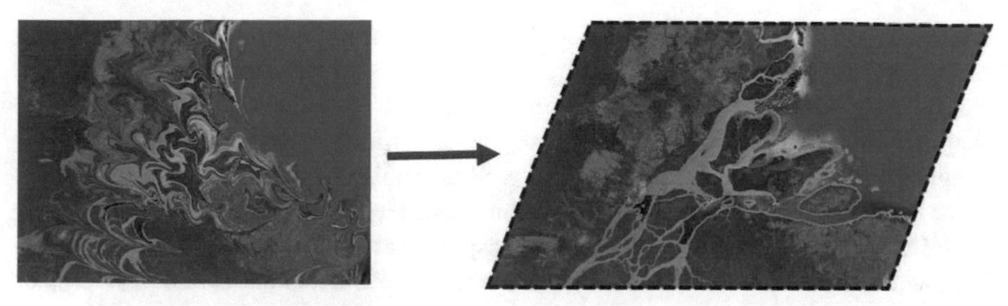

Fig. 3.4 An example of systematic distortion and correction from a composite Landsat 5 Thematic Mapper *(TM)* image of the Amazon River Delta, Brazil. The image on the *left* represents an image exhibiting systematic distortion due to platform motion caused by forward movement of the aircraft or spacecraft. The image on the *right* represents corrections applied to the image

to purchase Landsat imagery that has already been corrected to account for the rotation of the Earth beneath the satellite, which is why the image has the outline of a parallelogram, with pairs of obtuse and acute angles, rather than a rectangle (see Fig. 3.4). The most common techniques for removing the remaining systematic and nonsystematic distortions are image-to-map rectification and image-to-image registration (i.e., geographic matching to existing spatial data), which involve the selection of a large number of well-defined reference points (or *control points*) shared by the target image and the reference image or map. A variety of transformation models may be used to align the target image with the reference based on these control points, including linear transformations such as the affine transformation and nonrigid transformations such as splines (Fonseca and Manjunath 1996; Zitová and Flusser 2003).

3.2.2 Atmospheric Correction

Suspended particles or other materials in the atmosphere at the time of data acquisition may change the data (i.e., pixel values) recorded by the sensors aboard satellites and (less frequently) aircraft. In most terrestrial applications of remotely sensed data, analysts use virtually cloud-free days and/or multiple scenes to avoid this issue as well as the effects of haze, cloud cover, and other atmospheric effects. However, in coastal and near-shore ocean areas, analysts must be concerned with these atmospheric effects. In general, methods to remove or minimize atmospheric effects on an image can be categorized as either absolute or relative in nature (Khorram et al. 2013a).

Absolute radiometric correction of atmospheric attenuation takes into account the solar zenith angle at the time of satellite overpass, the total optical thickness caused by molecular scattering, the atmospheric transmittance for a given angle of incidence, the spectral irradiance at the top of the atmosphere, and the Rayleigh and Mie scattering laws (Forster 1984; Khorram et al. 2005; Turner and Spencer 1972). Many of the algorithms developed for absolute correction operate under the principle of *radiative transfer,* which assumes that the amount of electromagnetic radiation that travels through the Earth's atmosphere and ultimately reaches a location (i.e., an image pixel) can be computed based on atmospheric conditions at the time of image acquisition, as well as a season- and/or region-specific atmospheric model, which together address the factors outlined above. Some radiative-transfer-based algorithms have been adapted for use with hyperspectral imagery (e.g., Gao et al. 2009).

Relative radiometric correction of atmospheric attenuation normalizes the intensities among different bands within a scene to remove detector-related problems or corrects the intensities through a comparison with a standard reference surface on the same date and same scene (Jensen et al. 2005).

For example, *single-image normalization* uses a histogram adjustment for any shift in the histogram that may have been caused by atmospheric effects. This method is based on the fact that infrared data are largely free of atmospheric scattering effects as compared with the visible region. Thus, histogram shifts due to haze can be used to adjust for the atmospheric effects. This method involves establishing a subtractive bias for each band and is very simple to use.

Alternatively, *multiple image normalization* uses regression analysis for a number of dates. This method is primarily used to ensure that the spectral values from one date are comparable to other dates, thereby implicitly accounting for atmospheric differences between dates. The normalization process involves the identification of *pseudo-invariant features* (e.g., large building rooftops or asphalt parking lots) that undergo minimal change through time, which can thus serve as radiometric ground control points (Jensen et al. 2005; Schott et al. 1988). The spectral values of pseudo-invariant features in the target image are related to the corresponding values of these features in the reference images via linear regression. This method is primarily used for change detection purposes and is simple to use.

3.2.3 Radiometric Correction

Radiometric distortions vary among different sensors. Typically, solar elevation corrections and Earth–Sun distance corrections are applied to satellite data to remove the effects of the seasonal position of the Sun relative to the Earth and to normalize for seasonal variations in the distance between the Earth and the Sun. Furthermore, noise removal algorithms can be applied to remove any malfunctions in the sensors or detectors. The causes can include signal quantization, data line drops, and recording. Several de-striping algorithms are available to remove striping and banding effects in satellite data. Line drops can be corrected by replacing the spectral values in the missing band with the average of the line(s) above and below them. Nonsystematic variations in gray levels from pixel to pixel (i.e., bit errors) can be corrected by replacing these values with neighboring values that exceed threshold values established by the analyst, as shown in Fig. 3.5. In this example, dropped data lines that are recorded as zero values in the digital image (Fig. 3.5a) are replaced by the average of the values on the scanned lines above and below the bad data lines (Fig. 3.5b). However, when the dropped lines are more than a few

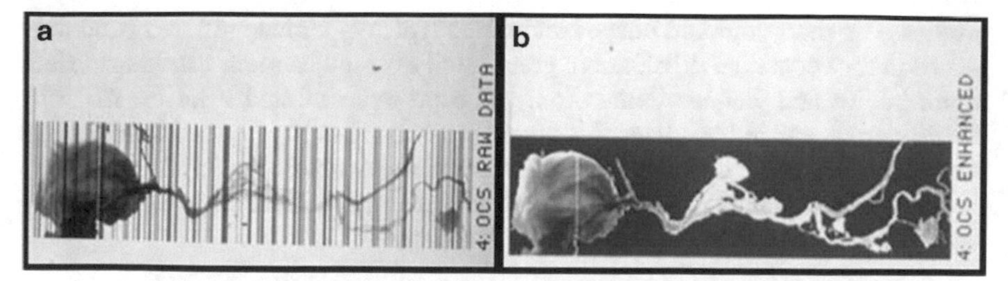

Fig. 3.5 An example of radiometric correction of dropped data lines in an Ocean Color Scanner *(OCS)* image of San Pablo Bay, CA, and its associated delta. (From Khorram 1985)

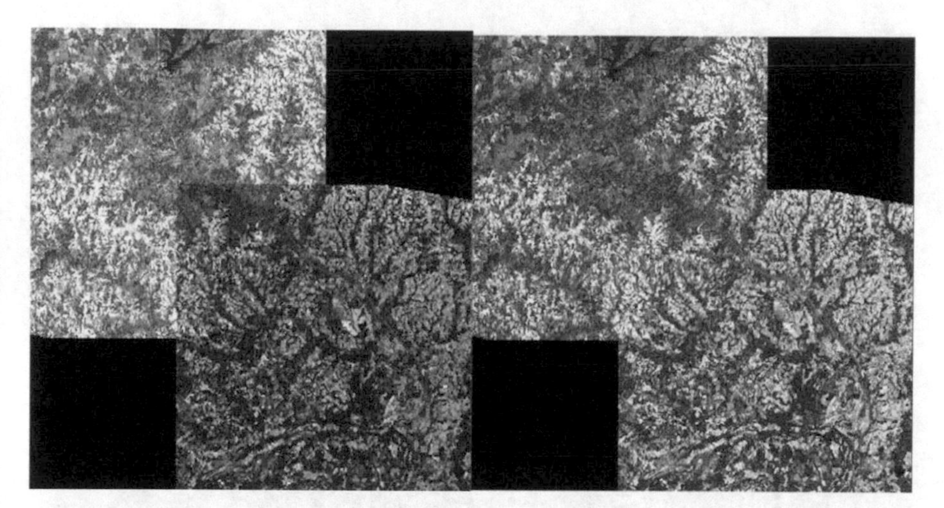

Fig. 3.6 Landsat Thematic Mapper *(TM)* scenes before *(left)* and after *(right)* normalization

pixels in width (usually 3–5 depending on the local scene dynamic), then linear averaging is not usually appropriate.

Typically, Sun elevation corrections and Earth–Sun distance corrections are applied to satellite data to remove the effects of the seasonal position of the Sun relative to the Earth, and to normalize for the seasonal variations in the distance between the Earth and the Sun. Figure 3.6 depicts the removal of radiometric differences due to atmospheric conditions at the time of data acquisition (two different dates) before and after normalization.

3.2.4 Band Combinations, Ratios, and Indices

For some applications it is often necessary to use combined bands during the preprocessing phase. Algebraic combinations of certain bands via division, addition, subtraction, or multiplication can lead to enhanced information. Differences and ratios in various bands are primarily used for change detection and spectral enhancement studies. The most common image band ratios typically include infrared band over red band for vegetation distribution; green band over red band for mapping surface water bodies and wetland delineation; red band over infrared band for mapping turbid waters; and red band over blue band or red band over green band for mineral mapping, such as the detection of iron-rich or iron-poor rocks.

The most frequently used index for vegetation mapping is the *Normalized Difference Vegetation Index (NDVI)*, which is defined as

$$\text{NDVI} = \frac{B2 - B1}{B2 + B1},$$

where $B2$ represents the brightness values (i.e., the digital numbers or pixel values) from a near-infrared band of an image, and $B1$ represents the corresponding values in the image's red band. NDVI images can be displayed in black and white, as shown in Fig. 3.7, or in color, as shown in Fig. 3.8, which illustrates the global distribution of vegetation in various shades of green. NDVI images are typically used for covering large geographic areas, thus reducing the cost of data processing for applications such as vegetation mapping.

Fig. 3.7 A *black*-and-*white* Normalized Difference Vegetation Index *(NDVI)* image of a study area in Oxnard, coastal California, created from Landsat data. (Image courtesy of Verkere and Khorram 2015)

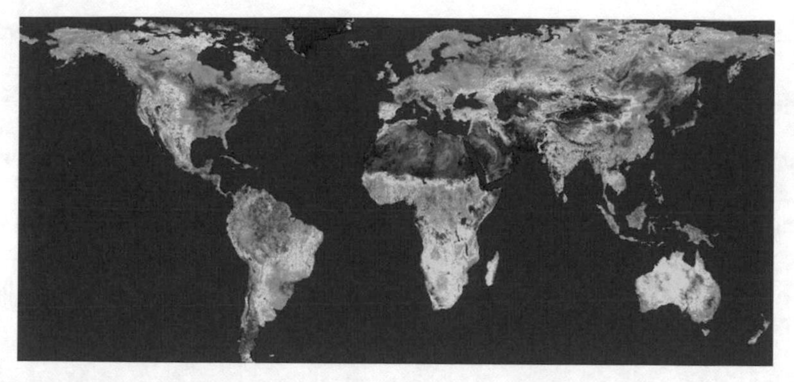

Fig. 3.8 Normalized Difference Vegetation Index *(NDVI)* image depicting global vegetation cover in various shades of *green*

An alternative to the NDVI, developed specifically for use with Moderate Resolution Imaging Spectroradiometer (MODIS) data, is the Enhanced Vegetation Index (EVI). The EVI is defined as follows:

$$\mathrm{EVI} = G \times \frac{B3 - B2}{B3 + C1 \times B2 - C2 \times B1 + L},$$

where $B3$ refers to atmospherically corrected reflectance values from the near-infrared band (centered at 0.858 μm) of a MODIS image, $B2$ refers to the image's corresponding red band, and $B1$ refers to its blue band. The two coefficients $C1$ and $C2$ correct for aerosol influences in the red band using the blue band, while L is a canopy background adjustment factor (i.e., a soil adjustment factor). The following coefficient values were established via empirical testing: $C1=6$, $C2=7.5$, $L=1$, and G (i.e., a gain factor)$=2.5$ (Huete et al. 2002). The EVI is perceived as a good option for vegetation mapping because it adjusts for an open canopy structure, which is far more common globally than the closed canopy structure exhibited by dense forest cover (Huete et al. 2002). It also avoids the signal saturation that regularly occurs with the NDVI in high-biomass conditions. In truth, many analysts consider the NDVI and EVI to be complementary, and they are regularly used together in studies (e.g., Schnur et al. 2010; Wardlow et al. 2007; Xiao et al. 2005).

Because the standard EVI formulation uses the blue band, it cannot be applied to sensors that lack this band, such as the Advanced Very High Resolution Radiometer (AVHRR). To permit application of the index to AVHRR and thus extend the length of the EVI historical record to before the MODIS period, a two-band formulation was developed that omits the blue band (Jiang et al. 2008):

$$\mathrm{EVI2} = G \times \frac{B2 - B1}{B2 + 2.4 \times B1 + 1}.$$

In this case, $B2$ refers to an image's near-infrared band, $B1$ refers to the red band, and G (the gain factor)$=2.5$. Differences *between* the EVI and EVI2 are generally minor for land cover types that are free of snow and ice, and the two-band formulation retains its advantage over NDVI in terms of accounting for canopy background reflectance (Jiang et al. 2008; Rocha and Shaver 2009).

Various bands can be combined to produce images highlighting different features. An example is an image displaying the short wavelength infrared bands 4, 6, and 8 as red-green-blue (RGB) of Advanced Spaceborne Thermal Emission and Reflection Radiometer (ASTER) data (representing all mid-infrared bands in red, green, and blue). Regions rich in clay, carbonate, and sulfate minerals have diagnostic absorption features, resulting in distinct colors on these image. For example, limestones are yellow-green, and purple areas are kaolinite-rich. (Fig. 3.9; http://asterweb.jpl.nasa.gov/).

3.2.5 Data Fusion

In a remote sensing context, *data fusion* typically refers to the process of integrating image data collected at different spatial, spectral, and/or temporal resolutions

Fig. 3.9 Bands 4, 6, and 8 of ASTER data displayed in *red, green,* and *blue* colors, respectively. (Image courtesy of JPL, NASA)

(Cakir and Khorram 2008; Pohl and Van Genderen 1998). A number of similar terms have been used to describe the process, including *image fusion, merging,* and *synergy* (Wald 1999). Commonly, fusion is done at the pixel level, although it may also be performed at the feature level (i.e., recognizable objects are extracted from multiple images and then fused) or the decision level (i.e., images are processed individually and subsequently combined via decision rules; Pohl and Van Genderen 1998). A primary objective of data fusion is to facilitate the interpretation or classification of a target geographic area in a manner, and with a degree of accuracy, that would not be possible given image data from a single source (Cakir et al. 1999; Solberg et al. 1994). This may also be seen as reducing the uncertainty associated with the initial data source.

With respect to image data, the fusion process may permit the use of photo interpretive techniques not typically feasible with moderate-resolution multispectral imagery, or it may improve feature extraction and image segmentation techniques associated with object-oriented classification (Alparone et al. 2008).

During the past two decades, fusion approaches have advanced to combine imagery from sensors with disparate operational characteristics and output data formats. For instance, several studies (Cakir et al. 1999; Simone et al. 2002; Solberg et al. 1994) have outlined techniques for fusing multispectral imagery (e.g., Landsat TM, Satellite Pour l'Observation de la Terre (SPOT) XS, and synthetic aperture radar, SAR) imagery for improved land cover classification. In addition, Lee et al. (2008) provided an example of the fusion of LiDAR and aerial imagery for automated identification and delineation of buildings, while Treuhaft et al. (2004) argued that the fusion of interferometric synthetic aperture radar (InSAR), LiDAR, and hyperspectral imagery might be ideal for determining aspects of forest structure. Nonetheless, one of the most common applications of data fusion in remote sensing is for *pan-sharpening* (Alparone 2008), which is the integration of a multispectral image and a corresponding, higher-resolution panchromatic image (Fig. 3.10).

Recently launched commercial Earth observing satellites such as GeoEye-1, WorldView-2, and WorldView-3 provide multispectral image data with high (~2 m) spatial resolutions, which may suggest little need for pan-sharpened imagery. However, two important factors serve as practical limits to the spatial resolution of a given multispectral sensor (Cakir and Khorram 2008). First, the sensor's ability to record incoming radiation energy is constrained by the size of its optics. More importantly, the volume of data that must be collected and stored grows exponentially as the spatial resolution increases. For this reason, many sensors, including those named above, bundle several multispectral bands with a higher-resolution panchromatic band. This bundling process allows a sensor operator to accommodate user demands for the best possible spectral and spatial resolutions while minimizing onboard data handling limitations (Cakir and Khorram 2008).

Visual comparison of results based on various data fusion techniques as applied to the same area based on multispectral 4-m-resolution multispectral IKONAS satellite data as fused with 1-m panchromatic IKONAS data are shown in Figs. 3.10 and 3.11. Both in the TCC and in the FCC comparisons, one must carefully examine the color comparison of various features with the original 4-m multispectral data.

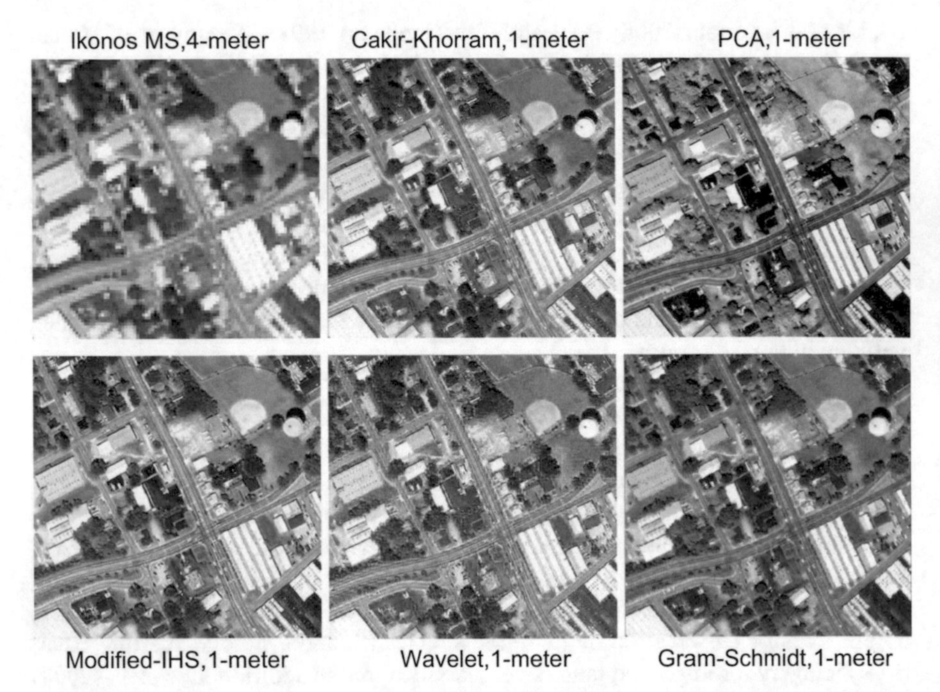

Fig. 3.10 The true color composite display of various fusion techniques as applied to the same area. *MS* multispectral; *PCA* principal components analysis; *IHS* intensity, hue, saturation

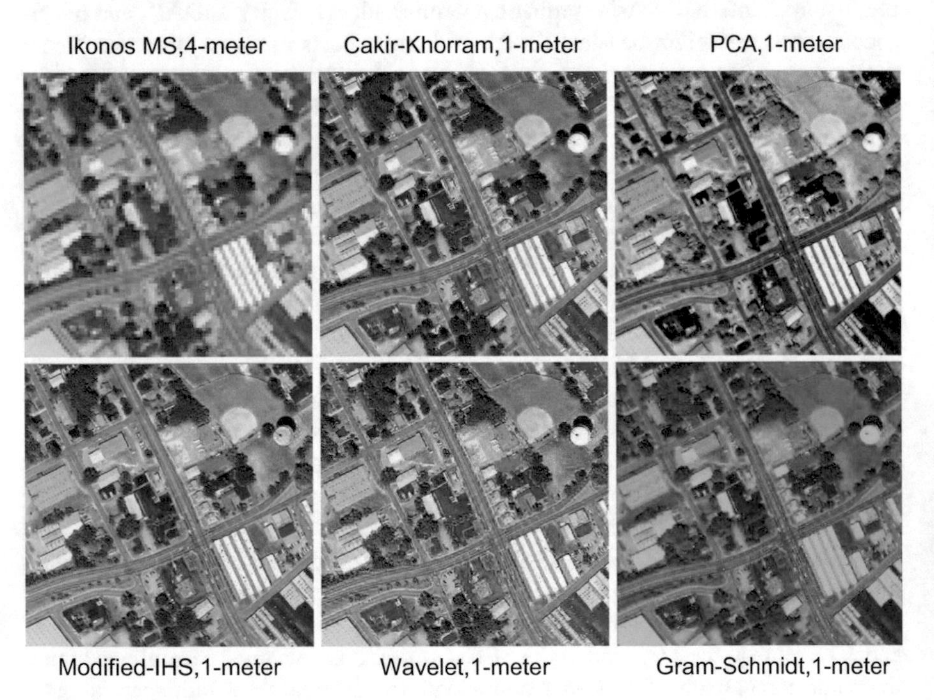

Fig. 3.11 The false color composite display of various fusion techniques as applied to the same area. *MS* multispectral; *PCA* principal components analysis; *IHS* intensity, hue, saturation

Fig. 3.12 Pan-sharpening through fusion of multispectral and panchromatic QuickBird images. A true color composite *(TCC)* of the original multispectral image is shown on the *left,* while a TCC of the pan-sharpened images is shown on the *right*

Further analyses can be performed by the comparison of histograms of original data versus fused data (Cakir and Khorram 2008) in various bands. The closest colors indicate a better technique since colors represent the spectral properties of these features before and after fusion. The sharpness of all features after fusion is apparent in all fusion techniques.

Figure 3.12 illustrates an image of an area adjacent to the pyramids in Egypt before and after applying Cakir–Khorram fusion technique.

There are various methodological approaches to data fusion. A commonly used approach in remote sensing is based on *principal components analysis (PCA)*. PCA is a multivariate statistical technique for reducing the dimensionality of a set of data. A comprehensive discussion of the mathematics of PCA is beyond the scope of this text. Briefly, PCA transforms an original data set (e.g., a multispectral image with four bands in the visible and near-infrared ranges) into a set of orthogonal (uncorrelated) axes or component variables. The number of these "principal" components is less than or equal to the number of variables (e.g., the number of image bands) in the original data. A critical aspect of the PCA transformation is that the first principal component explains as much of the variation in the original data as possible, while each subsequent component explains as much variation as possible while also remaining orthogonal to the previous component(s). For PCA-based data fusion (see Chavez et al. 1991 for examples), the first step is to apply PCA to a multispectral image. Then, the higher-spatial-resolution image (i.e., a corresponding panchromatic image) is modified to match the first principal component from the multispectral image, typically by "stretching" it to have the same mean and variance. Next, the first principal component of the PCA-transformed multispectral image is replaced by the modified high-resolution image, after which the PCA-transformed image is transformed back to the original image data space (Cakir and Khorram 2008).

A key assumption behind the PCA approach is that the first principal component of the multispectral image carries information common to all bands, while the other principal components contain information unique to each band (Chavez and Kwarteng 1989). Therefore, replacing the first spectral component should not substantially impact the spectral accuracy of the fused image (Cakir and Khorram 2008).

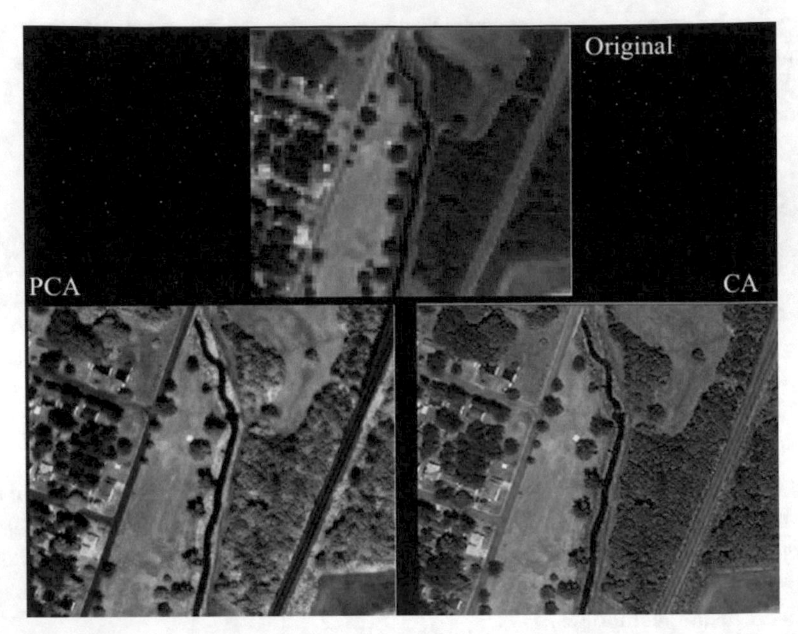

Fig. 3.13 Comparison of principal components analysis *(PCA)*-based and correspondence analysis *(CA)*-based data fusion from IKONOS satellite data

This is likely true if the high-resolution image and first principal component of the transformed multispectral image are highly correlated, but this may not always be the case. As an alternative, Cakir and Khorram (2008) developed the use of *correspondence analysis (CA)* for data fusion. While the approaches are similar, a key difference is that, in a CA-based data fusion approach, the last component, rather than the first component, is replaced with the modified high-resolution image. Because this last component represents extremely little of the variation in the original data, replacing it should theoretically have very little impact on the spectral content of the fused image. Figure 3.13 shows a comparison of the results of PCA-based and CA-based data fusion utilizing IKONOS satellite data for an area in central North Carolina (Khorram et al. 2005).

An example of the utility of data fusion for change detection in vegetation is shown in Fig. 3.14 (Cakir et al. 2006).

Another data fusion approach is based on *wavelet* theory. A *wavelet transform* decomposes a subject image into a sequence of new images, each with a distinct resolution; in other words, this sequence represents subsequent versions of the original image at increasing scales (Núñez et al. 1999). After applying the wavelet transform to a multispectral image and a corresponding high-resolution image, fusion can then be performed by replacing some of the wavelet images from the multispectral image with corresponding wavelets from the higher-resolution image or by adding the high-resolution wavelet images directly to the multispectral data (Núñez et al. 1999).

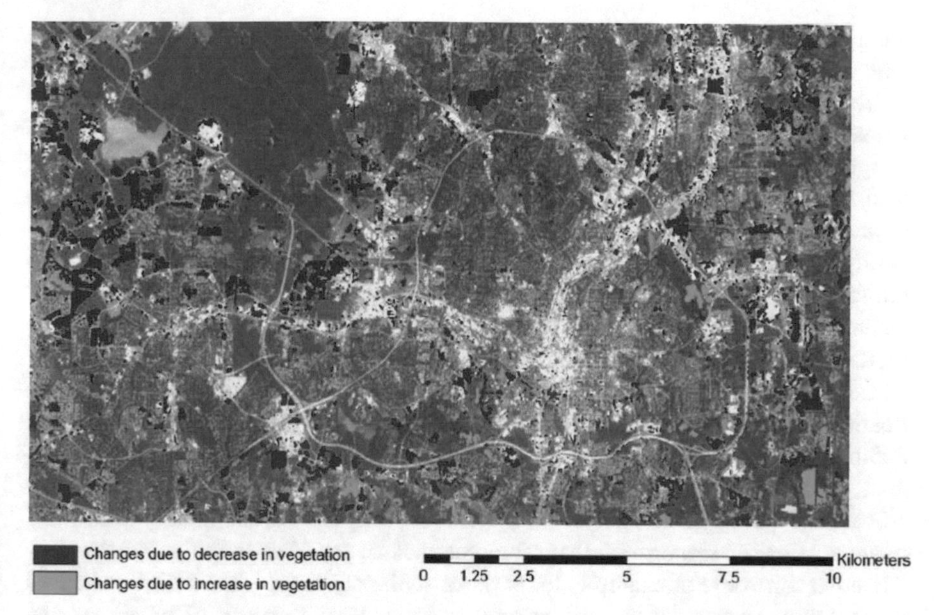

Fig. 3.14 An example of changes in vegetation in Raleigh, NC, using the Cakir–Khorram data fusion technique

Although fusion is potentially quite useful, it does represent an alteration of the original data. Moreover, successful fusion of images at the pixel level depends on the quality of the geocoding process (Pohl and Van Genderen 1998). Therefore, remote sensing analysts should carefully consider their objectives, data resolution needs, the data available, and the computing resources available for their projects before undertaking a fusion effort.

3.3 Image Processing

Multispectral satellite images that have been through preprocessing are then ready for processing, which essentially means they are ready for *image classification*. During classification, each pixel in an image is assigned to a particular category in a set of categories of interest, for example a set of land cover types. The classification process is generally based on the pixels' values (i.e., their spectral values) in each band of the image. A major aim of the classification process is to convert the original spectral data, which are variable and may exhibit complex relationships across several image bands, into a straightforward thematic map (typically a land cover map) that is easily understood by end users. Thus, image classification is a critical step in a remote sensing project because it pulls the most essential and meaningful information out of a multidimensional data set that would otherwise be difficult to interpret. In addition, the thematic map created from the image may then be easily integrated with other spatial data sets in a GIS for further analysis and integration with other data types.

The most prominent (i.e., readily available in commercial software) image classification techniques utilize hard logic, meaning that image pixels are definitively assigned to a single class in the selected classification scheme. This is in contrast to fuzzy logic, which acknowledges the variability of natural phenomena and the imprecision inherent to class definitions, and so aims to characterize a pixel's probability of membership in one of possibly several classes in the scheme (Fisher and Pathirana 1990; Hagen 2003). Fuzzy logic is an example of newer classification techniques that will be discussed briefly near the end of this section, but our primary focus is on traditionally used "hard-logic" classifiers.

Classification systems based on hard logic involve one of two types of classification: *supervised* or *unsupervised classification* (Fig. 3.15; Hester et al. 2008). Although supervised and unsupervised classification techniques have the same end goal, they follow fundamentally different conceptual strategies. Supervised classification depends on an analyst (i.e., a human interpreter of the image) to identify *training sites* (or *signatures*) that represent each of the classes in the chosen classification scheme. For example, for each particular cover type in a set of land cover types, the analyst must identify several areas, meaning groups of pixels, on the image being classified that represent that land cover type as distinctly as possible. Collectively, these signatures define a particular statistical profile for each class.

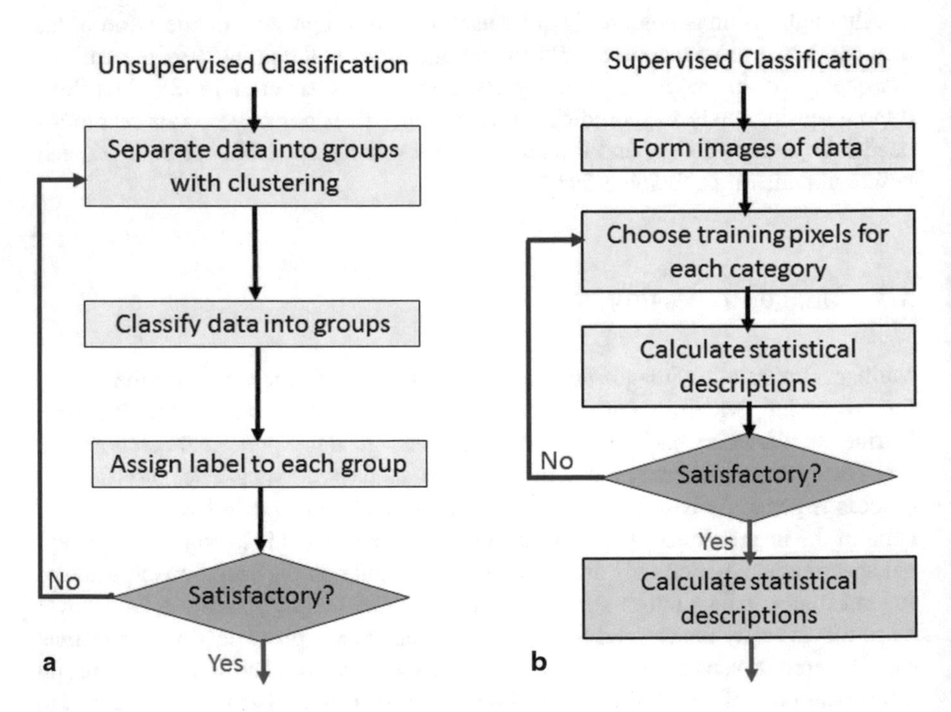

Fig. 3.15 General flowcharts of **a** unsupervised and **b** supervised classification methods. (Adapted from NASA *Remote Sensing Tutorial*)

Then analytical procedures are used to compare each pixel in the image with these statistical profiles and assign the pixel to the best matching class. Unsupervised classification involves little input from an analyst prior to the classification process. Instead, most of the work is done by an iterative (i.e., repetitive) computerized process that assigns pixels in an image to particular *clusters,* which represent natural groupings of pixels that are spectrally similar across the image bands. After this clustering process, the analyst then decides to which class in the selected classification scheme each cluster belongs.

3.3.1 Selection of a Classification Scheme

The first step in any image classification process is the selection of an appropriate *classification scheme* to denote the classes of interest to the analyst. The classes in the scheme must be defined by certain rules or criteria that distinguish them from one another and, just as importantly, must represent categories that are relevant to an end user. In this regard, certain applications may require highly specialized classification schemes; for instance, a scheme describing surface geology might include classes such as "basalt," "hematite," and "kaolinite." However, for most remote sensing projects in terrestrial environments, this typically involves a more generalized type of land cover (or rather land use and land cover, LULC) classification scheme. Three widely used classification schemes are the US Geological Survey (USGS) classification system (Anderson et al. 1976; USGS 2004), the US Fish and Wildlife Service (USFWS) Wetland Classification System (Cowardin et al. 1979), and the National Oceanic and Atmospheric Administration (NOAA) Coastal Change Analysis Program (C-CAP) system (NOAA 2004). Notably, each of these schemes was designed with specific end-user objectives in mind. For example, the USFWS scheme is designed to differentiate wetlands not only by the frequency of flooding, soils, and dominant vegetation but also by the type of aquatic system (e.g., riverine, estuarine) to which the wetland belongs (Cowardin et al. 1979; Jensen et al. 2005). The USGS classification scheme is widely used for general land cover mapping, in part because it was designed for application to remotely sensed data, and with four levels of progressively detailed classes, it offers flexibility given the input data and the end-user objectives. For classification of moderate-resolution multispectral imagery, the most frequently applied categories are from Level I and Level II (Table 3.1).

It is important to ensure that the selected scheme is not excessively detailed given the spatial and spectral resolution of the image being classified. For example, moderate-resolution imagery (e.g., Landsat ETM+, Landsat OLI) is unlikely to support accurate discrimination between forest types involving tree species with similar spectral signatures across image bands, such as loblolly pine *(Pinus taeda)* or shortleaf pine (*Pinus echinata;* van Aardt and Wynne 2001). In such cases, the use of hyperspectral imagery or ancillary data sets may represent reasonable alternatives to hard-logic classification of multispectral images.

Table 3.1 Levels I and II of the US Geological Survey (USGS) Classification Scheme. (Modified from Anderson et al. 1976)

Level I	Level II	
Urban or built-up land	11	Residential
	12	Commercial and services
	13	Industrial
	14	Transportation, communications, and utilities
	15	Industrial and commercial complexes
	16	Mixed urban or built-up land
	17	Other urban or built-up land
Agricultural land	21	Cropland and pasture
	22	Orchards, groves, vineyards, nurseries, and ornamental
	23	Horticultural areas
	24	Confined feeding operations
		Other agricultural land
Rangeland	31	Herbaceous rangeland
	32	Shrub and brush rangeland
	33	Mixed rangeland
Forest land	41	Deciduous forest land
	42	Evergreen forest land
	43	Mixed forest land
Water	51	Streams and canals
	52	Lakes
	53	Reservoirs
	54	Bays and estuaries
Wetland	61	Forested wetland
	62	Nonforested wetland
Barren land	71	Dry salt flats
	72	Beaches
	73	Sandy areas other than beaches
	74	Bare exposed rock
	75	Strip mines, quarries, and gravel pits
	76	Transitional areas
	77	Mixed barren land
Tundra	81	Shrub and brush tundra
	82	Herbaceous tundra
	83	Bare ground tundra
	84	Wet tundra
	85	Mixed tundra
Perennial snow or ice	91	Perennial snowfields
	92	Glaciers

3.3.2 Optimum Band Selection Prior to Classification

Multispectral images have *high dimensionality* in the sense that spectral information is recorded in several bands representing different ranges of wavelengths.

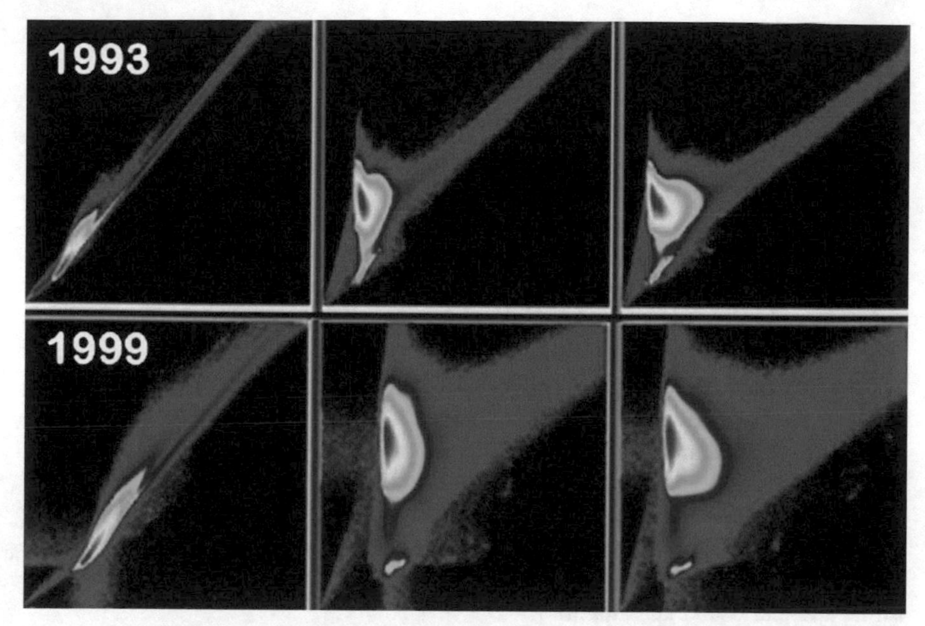

Fig. 3.16 Feature space plots from two different remotely sensed images, one captured in 1993 *(top)* and one captured in 1999 *(bottom)*. A greater level of dispersion can be seen in the feature space plots for the 1999 image

However, there are some redundancies of the information in these bands. Adjacent bands, in particular, are likely to be highly correlated; many recorded phenomena generate similar spectral response patterns in multiple bands. Prior to classification, especially supervised classification, it is advisable to identify a subset of bands that limits these redundancies while retaining the most important spectral information. Statistical separability measures such as divergence (Mausel and Lee 1990) predict which combinations of bands are most likely to discriminate accurately between different land cover categories. Some graphical analyses, such as feature space plots, similarly facilitate band selection, while also providing analysts with a greater understanding of the basic nature of the image data (Jensen 2005) .

Feature space plots (see example in Fig. 3.16) are two-dimensional scatterplots of the pixels in an image, where each dimension represents a single image band. Each pixel is plotted as a point according to its values from the two bands; essentially, the value in one band serves as a plot *x*-coordinate, while the value in the other band provides a *y*-coordinate. In the plot, the brightness of a given point indicates how many image pixels had the particular band value combination associated with that point. The primary aspect of interest in a feature space plot is the degree of dispersion among plot points. If the points are largely clustered along the plot diagonal, it indicates poor separability between the bands; conversely, if the points are widely dispersed in the plot space, it indicates a high degree of independence in the information recorded in the bands. Ideally, an analyst should examine feature space plots for all possible band combinations when selecting the optimal subset of bands. In addition, training sites (i.e., signatures) delineated for use in supervised classification may be plotted on top of the feature space plot, thus indicating the degree of spectral separability between the training sites.

3.3.3 Unsupervised Classification

Unsupervised classification methods involve limited initial input from an analyst, as a computer algorithm determines the most appropriate clusters from the image data. Nevertheless, the analyst will likely have to make certain decisions to guide the clustering algorithm, such as the number of desired output clusters, the number of iterations, and the statistical threshold values that are used to separate clusters. It is important to understand that, while most of the output clusters can be easily assigned to a particular class in the selected classification scheme, it is also common that some of the clusters will be ambiguous in that they apparently represent a mix of the classes in the scheme. One possible solution is a process known as *cluster busting* (Jensen 2005), which involves subsequent unsupervised classification—with a new set of clusters—of the pixels assigned to ambiguous clusters during an initial unsupervised classification.

The key message here is that analysts should be aware that unsupervised classification, like supervised classification, has limitations that affect the accuracy and interpretability of the output-classified image. It is also important to understand that unsupervised classification techniques are not necessarily computationally efficient. In some cases they require a large number of calculations while passing through the image data, but given current computing power, this is not usually a limitation.

Although there are many clustering methods, the two most widely used techniques for unsupervised classification of multispectral images are the *clustering algorithms* technique (also known as the chain method) and the *Iterative Self-Organizing Data Analysis Technique (ISODATA)*. With the clustering algorithms approach, the clustering process is executed in two steps. In the first step, a number of clusters (with the maximum number set by the analyst) are built through sequential evaluation of the pixels in the image. The clusters are composed of pixels that exhibit similar spectral values and therefore occupy a common spectral space (Celik 2009; Jain 1989). One characteristic of this first step is that the built clusters are evaluated in terms of the distance between them in spectral space, and any clusters that are separated by a distance less than an analyst-specified threshold value are merged. The cluster-building process is repeated until all clusters are separated by a distance greater than this specified threshold, which means that each cluster has a well-defined mean vector (i.e., a set of mean spectral values, one for each image band, of the pixels comprising the cluster). In the second step of this technique, a minimum-distance-to-means-classification algorithm is applied to image, whereby every pixel is assigned to one of the built clusters based on comparison to these mean vectors.

ISODATA (Sabins 1987; Tou and Gonzalez 1977) has seen especially widespread application, not least because it has been implemented in popular remote sensing and GIS software packages. It is considered to be an iterative technique because it may make many passes through the data, rather than just two, in order to develop an adequate final set of clusters. As with the clustering algorithms approach, the analyst may specify the maximum desired number of output clusters. In addition, the analyst may set the following other parameters: the maximum number of iterations; the minimum percentage of pixels assigned to each cluster; the maximum standard deviation; the minimum distance between cluster means, and most critically, the maximum percentage of pixels whose cluster labels can remain unchanged between iterations (Hester 2008; Jensen 2005). This latter parameter is important because when this percentage is reached, the algorithm stops processing.

The ISODATA algorithm begins with a creation of arbitrary clusters, and in the first pass through the image, each pixel is compared with the mean vectors of these clusters and assigned to the cluster to which it is closest in spectral space. In subsequent iterations, new mean vectors are calculated for the clusters resulting from the previous iterations and the image pixels are compared with these new means. This process is repeated until the specified criteria are satisfied, that is, until there is very little change in cluster labels between iterations or the maximum numbers of iterations has been reached (Jensen 2005). Note that any clusters that exceed the analyst-specified maximum standard deviation are merged prior to the next iteration. An example of an application of unsupervised classification using the ISODATA method for the city of San Francisco is shown in Fig. 3.17.

3.3.4 Supervised Classification

The training sites that an analyst selects during the initial stage of a supervised classification project serve to establish the relationships between the classes of interest and the image spectral data. To identify and delineate candidate sites, an analyst will typically rely on ancillary data sources, such as aerial photography, existing GIS coverages, or field visits. In addition, there are several rules of thumb that help to ensure the training sites provide an adequate foundation for successful classification (Jensen 2005). First, the number of training sites should be at least three times the number of categories of interest. The training sites should be selected in order to represent the spectral distribution of each class of interest as thoroughly as possible, and should be distributed randomly or systematically throughout the image area. Finally, each training site should be as homogeneous as possible, meaning that the group of pixels in a training site should have similar spectral values; a simple indication of training site homogeneity is that the associated pixels exhibit similar colors and/or tones in a displayed FCC (see Fig. 3.18).

Once the training sites are finalized, a number of different algorithms may be used for the classification itself. Three widely used algorithms in supervised classification are minimum-distance-to-means, parallelepiped, and maximum likelihood.

The *minimum-distance-to-means-classification algorithm* begins with calculating a mean vector (i.e., a set of mean spectral values, one value for each of the n

Fig. 3.17 Unsupervised classification of Oxnard, coastal California, based on the ISODATA method. (Image courtesy of Verkere and Khorram 2015)

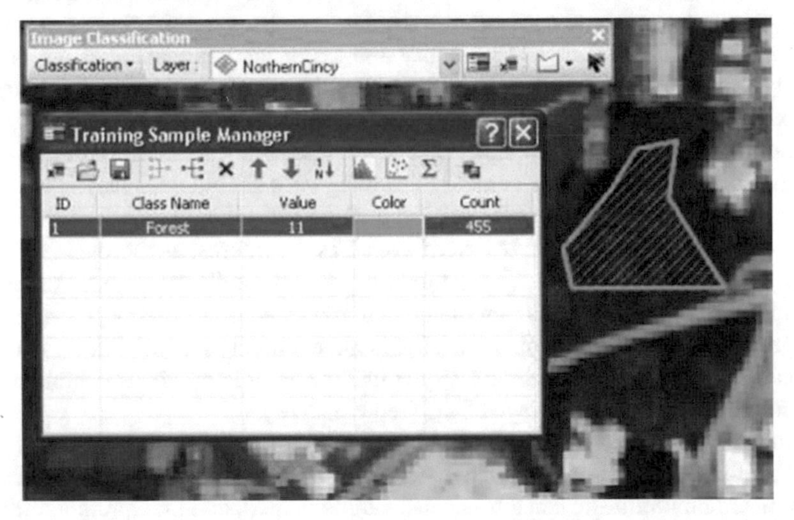

Fig. 3.18 Delineation of a training site representing the class "forest" for use in supervised image classification. The training site is shown with a false color composite of the input image as a backdrop. (Credit: Environmental Systems Research Institute, ESRI)

image bands being analyzed) for each class in the classification scheme based on the training site data. Each pixel in an image is then assigned to the class that exhibits the shortest distance between its mean vector and the pixel's spectral values. With respect to determining distance, measures such as Euclidean distance, which can be calculated straightforwardly in two dimensions, can further be extended to *n*-dimensional space. The minimum-distance-to-means algorithm is computationally efficient, but by utilizing only mean values, it ignores the variability in the spectral data (Lillesand et al. 2008; South et al. 2004). This omission could potentially decrease classification accuracy.

The *parallelepiped classification algorithm* is also computationally efficient, while accounting for some of the variation in the spectral data. Each class in the classification scheme is represented by a decision region, or parallelepiped, defined by a lower bound (one standard deviation below the mean spectral value for the class) and an upper bound (one standard deviation above the mean spectral value for the class), which are again calculated using the training site data. A distinct decision region is defined for each of the *n* image bands used in the analysis, which, taken together, form an *n*-dimensional box representing the class of interest (Lillesand et al. 2008). Individual image pixels can then be classified according to the box in which they fall. However, there is a possibility that the defined parallelepipeds will overlap, so a candidate pixel could legitimately be assigned to more than one class. In such cases, a minimum-distance-to-means or maximum likelihood algorithm may be used to assign the pixel to a single class.

The *maximum likelihood classification algorithm* assumes that the training site data (i.e., the spectral values of the associated pixels) for each class are normally distributed (Blaisdell 1993). Under this normality assumption, the distributions of spectral values (from each band) for a class of interest can be described by a mean vector as well as a covariance matrix, which depicts the variation of values both within and between image bands. In turn, the statistical probability that any pixel in an image belongs to a particular class of interest can be calculated based on the mean vectors and covariance matrices. In short, the pixel is assigned to the class in which it has the highest probability, or maximum likelihood, of membership. Although the maximum likelihood classification algorithm is widely used, it is computationally intensive as compared with the minimum-distance and parallelepiped algorithms (Jensen 2005). Moreover, it depends heavily on an accurate covariance matrix, which necessitates sufficient, high-quality training site data (South et al. 2004). Figure 3.19 illustrates an application of the maximum likelihood algorithm for image classification as compared with a TCC of the classified area.

A more sophisticated implementation of the maximum likelihood classifier is the *Bayesian classifier,* in which weights are assigned to the probability estimates for each class based on prior knowledge of the expected likelihood of class occurrence (Hester et al. 2008; Hord 1982). For example, a historical map of the study area could provide an initial estimate of the proportion of the area represented by each class. In other words, the classes are not assumed to have equal probabilities of occurrence, as is the case with the traditional maximum likelihood classifier.

Fig. 3.19 The true color composite *(top)* and a supervised classified image of an urban area *(bottom)*. (Based on QuickBird satellite data)

At more detailed level, the results of a supervised classification for an urban subdivision under construction in Raleigh, North Carolina, based on QuickBird satellite data are shown in Fig. 3.20. In addition, Fig. 3.20 represents an example of a supervised classification of the San Francisco Bay Area of California using the maximum likelihood classifier.

In LULC classification from satellite data, one must carefully choose the spatial and spectral resolution of the data to be utilized depending on the size of the study

Fig. 3.20 The results of a supervised classification using the maximum likelihood classifier for San Francisco Bay Area

area and the objectives of the project. As an example, while Landsat ETM+ data are very useful for a large variety of LULC applications, for smaller area within the cities, higher-resolution data are preferred. In Fig. 3.21, the Landsat ETM+ data on the left are compared with the QuickBird data on the right. The overall classification accuracy for ETM+ was 86% for ETM+ and 89% for QuickBird (Khorram et al. 2005).

3.3.5 *Fuzzy Logic Classification*

As hard-logic classifiers, the unsupervised and supervised classification algorithms described above assign the pixels of an image to mutually exclusive categories. For many image classification projects, this is an acceptably accurate approach. However, real-world conditions do not always conform to definitive boundaries between classes of interest. Because classification is done based on spectral response data, and despite the best effort of an analyst to define classes uniquely, cases of overlap are probably inevitable. This is especially true of low-to-moderate-spatial-resolution multispectral imagery, where image pixel values may be generated by more than one ground phenomenon, resulting in *"mixed pixels"* (Fisher

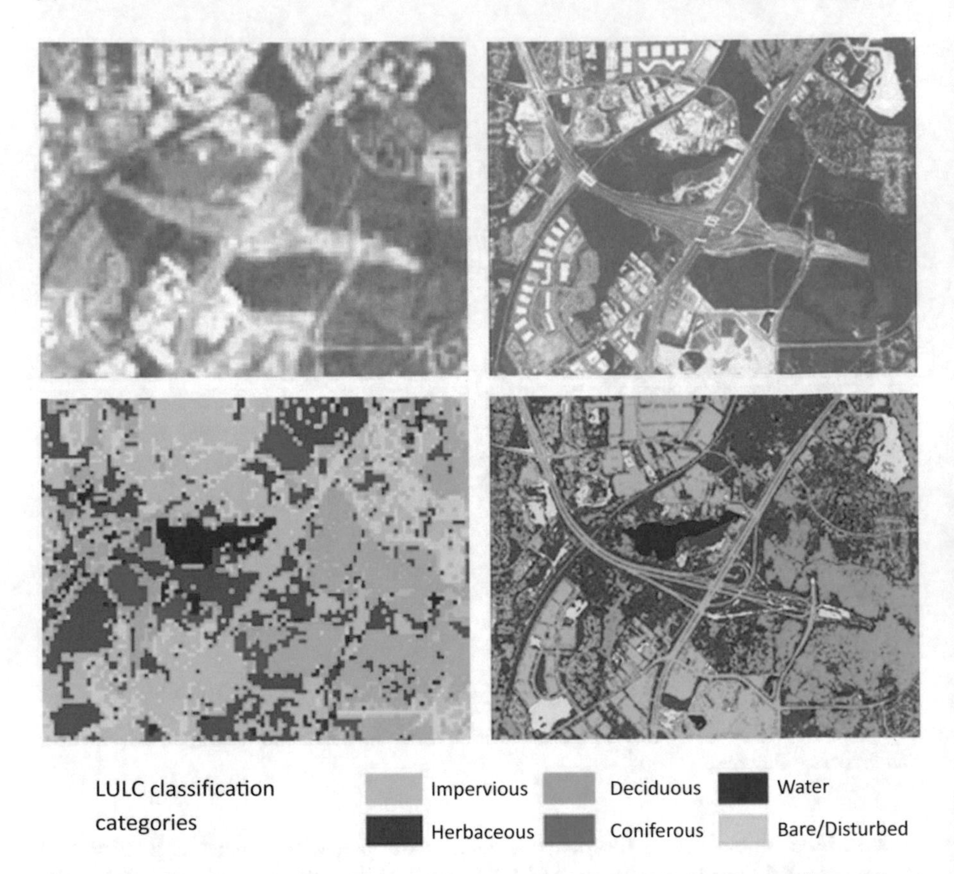

Fig. 3.21 Comparison of Landsat Enhanced Thematic Mapper *(ETM)+ (upper left image)* and QuickBird *(upper right image)* of the same area for land use and land cover *(LULC)* mapping leading to the determination of impervious surface areas. This image also demonstrates the increased distinguishable classes when comparing the lower-resolution Landsat ETM+-derived LULC classification product *(lower left image)* to those created with the higher-resolution QuickBird data *(lower right image)*

and Pathirana 1990). *Fuzzy logic* classifiers circumvent the definitive boundaries of hard-logic classifiers by establishing transitional, or "fuzzy," regions between classes in a scheme, which come into play when a pixel potentially belongs to two or more classes (Jensen 2005; Zadeh 1965). For instance, for any pixel in a multispectral image, it is possible to measure the distance between the pixel's spectral values and the mean vectors (i.e., the sets of mean spectral values) for all classes in the classification scheme. These measured distances may then be translated to the pixel's likelihood of membership in each class.

There are additional ways to incorporate fuzzy logic into a classification effort, and fuzzy classifiers can resolve other issues besides the problem of mixed pixels. For example, the application of a fuzzy classifier in a change detection study in an urban watershed (Hester et al. 2010) helped to address a major source of error: misregistration between images of the study area collected on different dates (and,

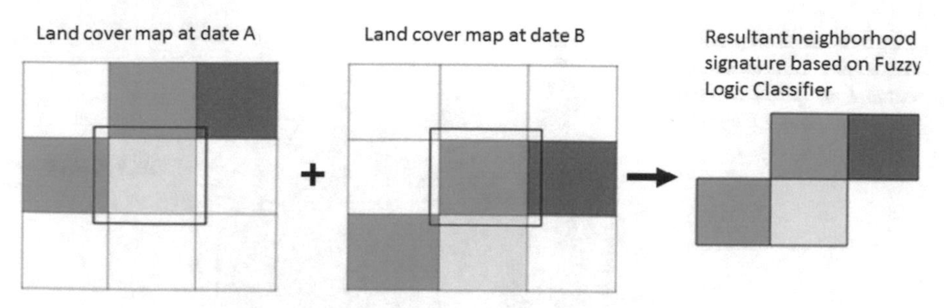

Fig. 3.22 Comparison of fuzzy neighborhood vectors for a target pixel (outlined in *red*) in classifications collected on different dates

in turn, apparent disagreement in the resulting classified maps). One conceptual approach for such an application is illustrated in Fig. 3.22. Each pixel in a classified image can be described by a "fuzzy neighborhood vector" (Hagen 2003), which incorporates the class memberships of neighboring pixels based on their (spatial) distance from the target pixel. When comparing the fuzzy neighborhood vectors for a target pixel location in two different maps, it is possible to find a shared neighborhood signature despite different class values for the specific pixel in question.

3.3.6 Other Classification Approaches

The supervised and unsupervised classification algorithms described above are well-accepted standards in remote sensing. However, there are many other approaches developed from research in fields such as artificial intelligence and pattern recognition. For cxample, *artificial neural network (ANN)* approaches have been applied for remote sensing image classification since the early 1990s (Dai and Khorram 1999; Qiu and Jensen 2004; Yang and Chung 2002). The essential processing unit of an ANN is called a neuron, which is analogous to a neuron in the human brain (Jensen et al. 2009) . Essentially, ANNs attempt to mimic the performance of the human brain for the purpose of data processing (Mohanty and Majumdar 1996). They resemble the brain in two major ways. First, the ANN acquires knowledge through a learning process. Second, this knowledge is stored in the connection strengths between neurons, which are known as synaptic weights (Haykin 1994; Jensen et al. 2009). In what is known as *supervised learning,* these synaptic weights may be adapted through the application of training samples. Thus, an ANN adaptively relates inputs to outputs through its set of neurons, rather than depending on statistical procedures. One popular type of ANN, a *multilayer perceptron* (Fig. 3.23), is characterized by a high degree of network interconnectivity, and contains one or more hidden layer of nodes, distinct from the layers of input and output nodes, that enable the network to learn especially complex tasks (Jensen et al. 2009; Mohanty and Majumdar 1996).

There are numerous advantages to ANNs for classification of remotely sensed images. Most prominently, they are not bound by statistical assumptions such as the

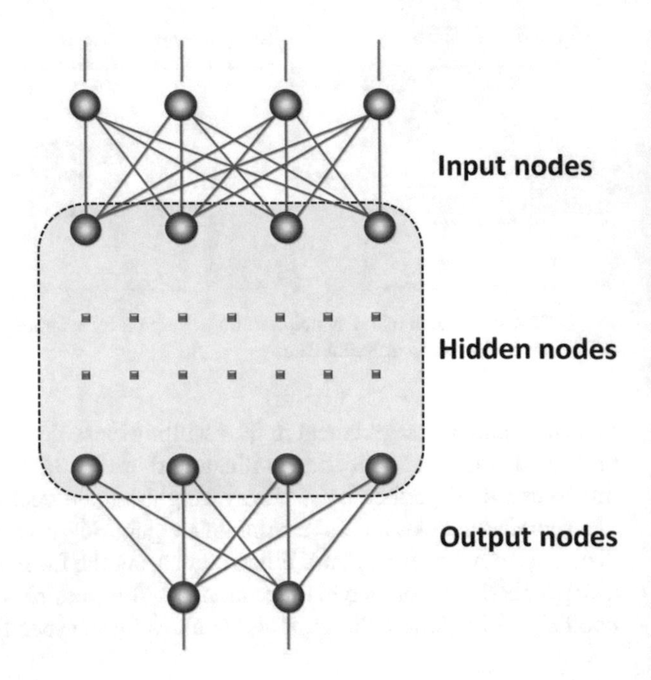

Fig. 3.23 A conceptual model of a multilayer perceptron. (From Mohanty and Majumdar 1996)

assumption of normality. They are also well suited to the incorporation of ancillary data sets (e.g., GIS layers such as digital elevation models (DEMs) or soil maps), which is a limitation of more traditional classification algorithms such as maximum likelihood. However, the supervised learning process usually requires a great deal of computation time, and the relationships between inputs and outputs in an ANN may not be as readily discernable as with more traditional classifiers (Jensen et al. 2009). An application of this methodology for LULC classification is illustrated in Fig. 3.24, using an example in coastal North Carolina, US.

Related to the topic of ANNs are simulated annealing (SA) classifiers, which offer an alternative to the commonly used *k*-means clustering method. The *k*-means method assigns each observation in a data set (e.g., the pixels in a remotely sensed image) to one of *k* clusters based on which cluster centroid, or mean, is closest to the observation in multidimensional Euclidean space. This iterative procedure is fundamentally similar to the minimum-distance-to-means algorithm used in supervised classification. Unfortunately, *k*-means clustering is known to produce good results only if the clusters are well separated in the multidimensional feature space, similar in size, and hyperspherical in shape. Clustering is more problematic when the observations (e.g., pixels) are distributed across the feature space in a complex way. In such situations, as the clusters iteratively converge to their local minima, it is possible that one or more clusters will include observations that are not especially close to each other in the feature space; yet sit at similar distances from the cluster mean. In short, the spatial arrangement of the clusters (and their means) might appear to contradict the underlying spatial structure of the data.

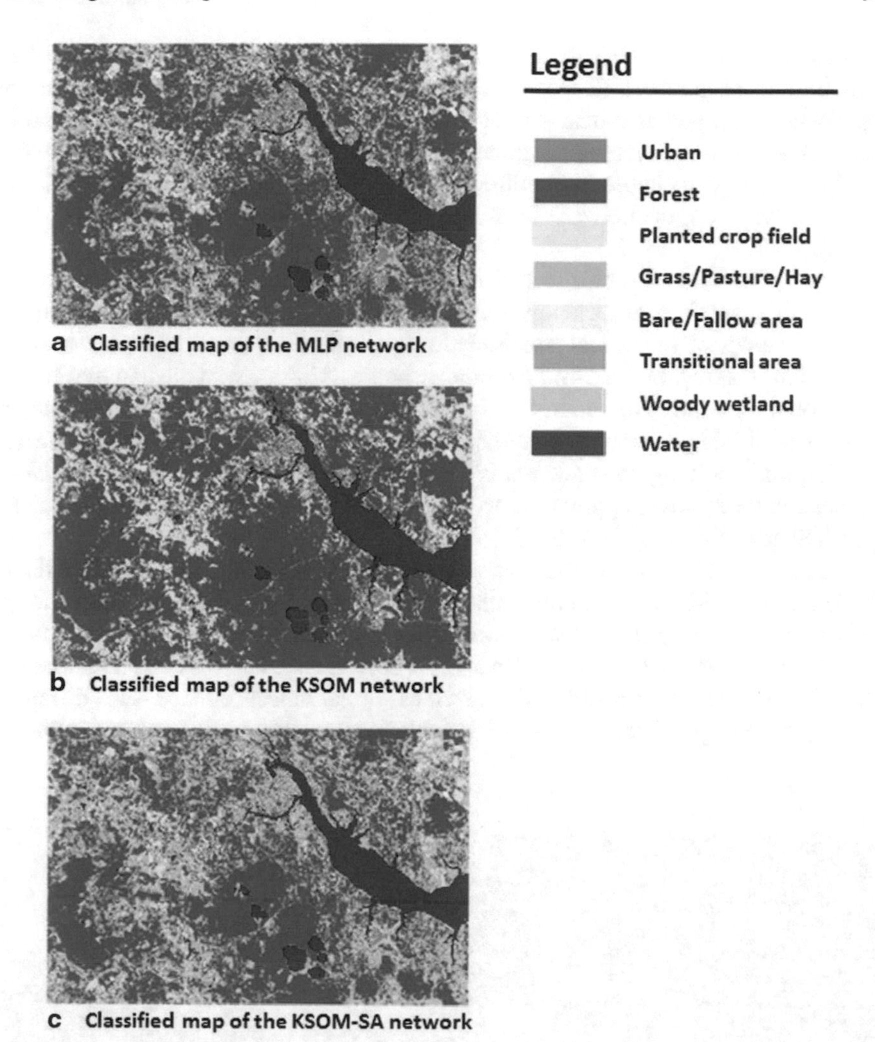

Fig. 3.24 Classified images of coastal North Carolina, US, based on three versions of the artificial neural network *(ANN)* classification method. *MLP* multilayer perceptron, *KSOM* Kohonen self-organizing map, *KSOM-SA* Kohonen self-organizing map-simulated annealing

SA was developed on the basis of an analogy between the physical annealing process of solids and large combinatorial optimization problems (Cerny 1985; Das and Chakrabarti 2005; De Vincente et al. 2003; Kirkpatrick et al. 1983; Khorram et al. 2011; Yuan et al. 2009). SA was proven to have great potential to find or approximate the global or near global optimal in a combinatorial optimization problem (Geman and Geman 1984). The premise of SA is to incorporate some randomness in the assignments of cluster labels to pixels in the clustering procedure, thus reducing the limitation of local minima. As a result, using the SA-based approach to classification has the potential to improve the accuracy for land cover classification.

Geman and Geman (1984), Pang et al. (2006), Wang and Wang (2009), Yang et al. (2010), and Zheng et al. (2008) demonstrated that SA has the potential to find or approximate the global or near-global optimal in a combinatorial optimization problem. Although SA often requires greater computational time, the improved network could self-adapt to choose momentum parameters according to annealing temperature, thus enabling the network to escape from local minimum spots and converge stably.

Finally, traditional multispectral image classification methods have relied on the use of supervised or unsupervised classification algorithms that operate on the spectral values of individual pixels. However, with the advent of higher-spatial-resolution imagery, in the 1-m to submeter range, it has proved challenging to use traditional algorithms to obtain highly accurate classifications because the images tend to exhibit high spectral variability in the target classes (Yu et al. 2006). Newer techniques, involving *object-oriented classification* and *image segmentation algorithms* represent novel approaches to multispectral image classification (Blaschke 2010; Shackelford and Davis 2003).

The object-oriented classification procedure varies from the traditional methods in that, rather than broadly applied classifications based on groups of pixels identified as training samples, instead entire objects that make up surface features to be classified are selected (Fig. 3.25). Once these objects have been established, rules are then defined to model the objects', as well as similar objects', shape, size, distance, texture, spectral pixel values, position, orientation, and relationship between objects,

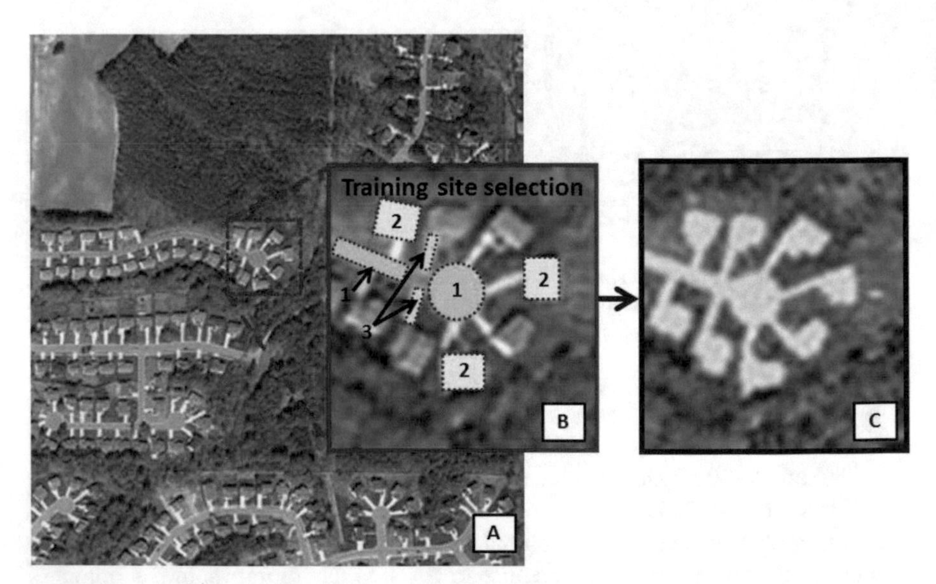

Fig. 3.25 **a** An unclassified 2003 true color, digital orthoimagery with a spatial resolution of 33 cm (USGS 2006a). **b** Three manually selected impervious surface features to be used as training samples: *(1)* roads, *(2)* houses, and *(3)* driveways. **c** The results of an object extraction classification based on the training samples of the selected impervious surface features

as well as the distribution of the objects throughout the image, along with a number of additional user-defined parameters. Each object may represent a particular cover type. The applied object-oriented procedure then utilizes algorithms such as nearest neighbor analysis, neural networks, and decision tree analyses to complete the classification operation (Herold et al. 2003; Thomas et al. 2003).

The object-oriented classification represents a type of image segmentation algorithm. In general, the image segmentation algorithm operates by partitioning or segmenting an image into multiple segments (or sets of pixels) based on spectral pixel values, the shape of merged pixels, and an iterative threshold limit, as well as other user-defined parameters. The results of the segmentation process produces an image that, while originally a complex mosaic of pixels, is now composed of regions of similar pixels that are easier to analyze.

Many studies have applied the object-oriented image segmentation process to studies of land use analyses. Miller et al. (2009) classified 111 high-resolution images using an object-oriented approach based on the analysis of spatial context in relation to spectral data and achieved an overall total accuracy of 92 % in regard to the classification of impervious surface.

Mitri and Gitas (2004) evaluated the performance of an object-based classification model in comparison to topographically and non-topographically corrected Landsat TM imagery developed for burned area mapping. The object-based classification model provided an increase in accuracy over the Landsat-based imagery.

Al Fugara et al. (2009) developed an innovative object segmentation approach to compare classification accuracies in land cover types between a pixel-based and the object-oriented image classification approaches. Image segmentation parameters used for the object-oriented approach included scale, color, shape (smoothness and compactness), as well as a fuzzy dimension function to assign class memberships to the identified or segmented objects. The classification results for the object-oriented approach achieved higher accuracies for most of the land cover classes examined compared with those of the pixel-based method with an overall accuracy that exceeded 85 %.

3.4 Post-processing Image Data

A number of post-processing operations may subsequently be applied to classified multispectral images. The most typical operations involving remotely sensed data include filtering, accuracy assessment, change detection, and spatial modeling.

3.4.1 Spatial Filters

Spatial filters are applied to classified images in order to either highlight or downplay features in an image based on their spatial frequency. Spatial frequency is defined as the number of changes in pixel brightness value per unit distance for any particular part of an image. High variation in brightness values translates into high spatial frequency, or what is commonly known as a salt-and-pepper effect, in classified images. Spatial filtering is the process of dividing the image into its constituent spatial frequencies, and selectively altering certain spatial frequencies to emphasize or deemphasize some image features. Filters are applied to reduce these high frequently occurring categories and thus smoothing the image or simply transferring the image to a lower frequency image and therefore removing the salt-and-pepper effects.

A number of algorithms can be applied to a classified image for better presentation and visualization. Kernels, which can be thought as moving windows, are applied to classified images to reduce noise and filter unwanted information. An example of the application of a 3 × 3 and a 5 × 5 median filter to a classified map is shown in Fig. 3.26. In short, median values are calculated within a moving 3-pixel

Fig. 3.26 A filtering technique has been applied to classified land use and land cover of Oxnard, CA, as shown in Fig. 3.16. The original classified image was subjected to a 3 × 3 median filter *(top left)*, a 5 × 5 median filter *(top right)*, and a 7 × 7 median filter *(bottom)*. (Images courtesy of Verkere and Khorram 2015)

by 3-pixel or 5-pixel by 5-pixel window and assigned into the central pixel. Mean and median filters usually have a smoothing effect. Other filters that have been frequently used in image processing employ functions that highlight differences between pixel values. They are usually used for sharpening the edges of objects in an image. These edge detection filters are also known as directional filters.

3.4.2 Accuracy Assessment

Image accuracy assessment has been a key component and the focus of a significant number of remote sensing studies (Congalton 1991; Congalton and Green 1999; Goodchild et al. 1992; Khorram et al. 1992, 1999a; Paine and Kiser 2003; Van Genderen and Lock 1977). Without assessing the accuracy of the classified data, the reliability and repeatability of the output products are in question.

Sophisticated statistical procedures in the analysis of error matrices are developed for the accuracy assessment of land cover classifications for a single date (Aaronoff 1985; Congalton et al. 1983; Foody 2002; Khorram et al. 1999b; Lunetta and Lyons 2003; Morisette and Khorram 1998; Pal and Mather 2003; Stehman 2001). However, the accuracy assessment of change detection procedures is not in operational stage quite yet and involves issues such as not yet widely accepted sampling techniques, image registration, boundary problems, and reference data. The error sources involved in the accuracy assessment include registration differences between reference data and remotely sensed data, delineation errors in digitizing, data entry errors, errors in image classification and delineation, and errors involved in sampling, collection, and interpretation of reference data.

The most commonly used procedure for accuracy assessment is error matrix analysis. As shown in Table 3.2, an error matrix can be constructed by the results based on reference data (e.g., data collected on the ground or from substantially higher-spatial-resolution image) on one side of the table, and the results based on the classified image on the other side of the table. An adequate number of samples are identified on the classified image and corresponding reference data are collected using various sampling strategies. The accuracy is determined in terms of percent correctly classified sample sites, as compared with their corresponding reference data, for each category of interest as well as the overall classification accuracy involving all categories.

Traditionally, the total number of correct samples in a given category is divided by the total number of samples in that category based on reference data. This accuracy measure indicates omission errors and is often referred to as "producer's accuracy" because the producer of image classification is interested in how well he has classified a certain category. If the total number of correctly identified samples in a given category is divided by the total number of samples based on classified data, then this indicates the commission error. This measure is called the "user's accuracy" or reliability because the user is interested in the probability that a classified

Table 3.2 The accuracy assessment performed on the land use and land cover classification shown in Fig. 3.19

Category	1	2	3	4	5	6	Row total	User's accuracy (%)	Kappa (κ)
1. Impervious	45	0	4	0	0	1	50	90.0	0.88
2. Water	0	50	0	0	0	0	50	100	1.00
3. Bare/disturbed	4	0	45	0	0	1	50	90.0	0.88
4. Deciduous	0	0	0	46	2	2	50	92.0	0.90
5. Evergreen	0	0	0	5	44	1	50	88.0	0.86
6. Herbaceous	6	0	0	3	3	38	50	76.0	0.72
Column total	55	50	49	54	49	43	300		
Producer's Accuracy (%)	81.8	100	91.8	84.9	89.8	88.4		89.3	

Overall accuracy = 89.3 % and the overall kappa = 0.87

sample representing the actual category on the ground (Story and Congalton 1986). Multivariate statistical procedures have also been used for accuracy assessment. The most commonly used is a discrete multivariate technique, called KAPPA, which is a measure of agreement or accuracy by KHAT statistics (Cohen 1960). Kappa is computed from the error matrix table constructed from both reference and classified datasets.

$$K_{\text{hat}} = \frac{N \sum_{i-1}^{r} x_{ii} - \sum (x_{i+} \times x_{+1})}{N^2 - \sum_{i-1}^{r} (x_{i+} \times x_{+i})},$$

where r is the number of rows in the matrix, x_{ii} is the number of observations in row i and column i, x_{i+} and x_{+i} are the marginal totals for row i and column i, respectfully, and N is the total number of observations.

Table 3.2 is an example of the accuracy assessment performed on the LULC classification shown in Fig. 3.19

3.4.3 Change Detection

Change detection using remote sensing data is an attempt to record natural and anthropogenic transitions of LULC on Earth's surface that have occurred over time (e.g., see Fig. 3.19). This type of analysis is usually completed by using either manual or automated methods, or a combination of both (see studies by Cakir et al. 2006; Hester et al. 2010; Khorram et al. 1996; Nelson et al. 2002). Appropriate spatial and temporal resolutions to provide the proper base data for sound change detection must be considered prior to change detection or delineation of study area

(Kennedy et al. 2009; Townsend et al. 2009). Spatial, temporal, scale, and atmospheric conditions must be considered in change analysis. Change detection typically starts with image classification for all dates in process. It is necessary to account for the compatibility of all datasets and future data continuity and availability of remote sensing data when choosing datasets (Gross et al. 2009). Change detection is conducted using both manual (by visual inspection) and automated change detection methodologies.

The spatial scale and information content of satellite-derived remote sensing data has inspired the development of automated change detection algorithms and methodologies, especially for evaluating and recording LULC change. These automated methods can be categorized as pre-classification, post-classification, or even constitute more advanced procedures.

Pre-classification methods develop change maps from multi-temporal data (i.e., data captured over the same area on different dates) without first generating classified LULC maps from that data. The algorithms used in pre-classification procedures may transform or simplify the original data before creating a change map, but they do not rely first on the generation of meaningful of LULC classifications of the individual image dates. One of the most important aspects of any pre-classification change detection algorithm is the specification of a change threshold. This parameter, derived either by the analyst using the initial algorithm output or by the analyst using only a priori knowledge, represents the interpretative mechanism by which the algorithm judges whether a change has occurred.

Some pre-classification methods that rely on user-defined change thresholds include image differencing, image regression, image rationing, vegetation index differencing, principal component analysis (PCA), and change vector analysis (Lu et al. 2004). In each of these methods, the analyst's major responsibility is to interact with the algorithm, through an iterative or automated process, in defining the change threshold. If input imagery is suitably matched in terms of geometric, radiometric, and atmospheric quality, then the major advantage of pre-classification change detection is that these approaches do not introduce a great deal of analyst subjectivity as they evaluate the quantitative differences in image dates, despite being computationally intensive.

One limitation to pre-classification change analysis is that geometric and radiometric alignment between image dates may be difficult or even impossible to achieve. This limits the ability to develop a complete *"from–to"* representation of change (Jensen 2005). Pre-classification methods are primarily equipped to identify either binary or scaled indicators of *"change"* or *"no change"* (Fig. 3.27). As such, they are particularly robust in measuring in-category and landscape-level change, as in the forest defoliation study by Muchoney and Haack (1994) and the analysis of continent-scale land cover change in Africa by Lambin and Ehrlich (1996).

As an example of visual inspection/analysis method, deforestation, and conversion of forests to agricultural and urban lands between 1975 and 1992 in Rondonia, Brazil, and the Athens, Greece, stadium are shown in Figs. 3.28 and 3.29, respectively.

Post-classification change detection can generally provide a detailed matrix of *"from–to"* change and requires a great deal of preparatory analyst input (Table 3.3).

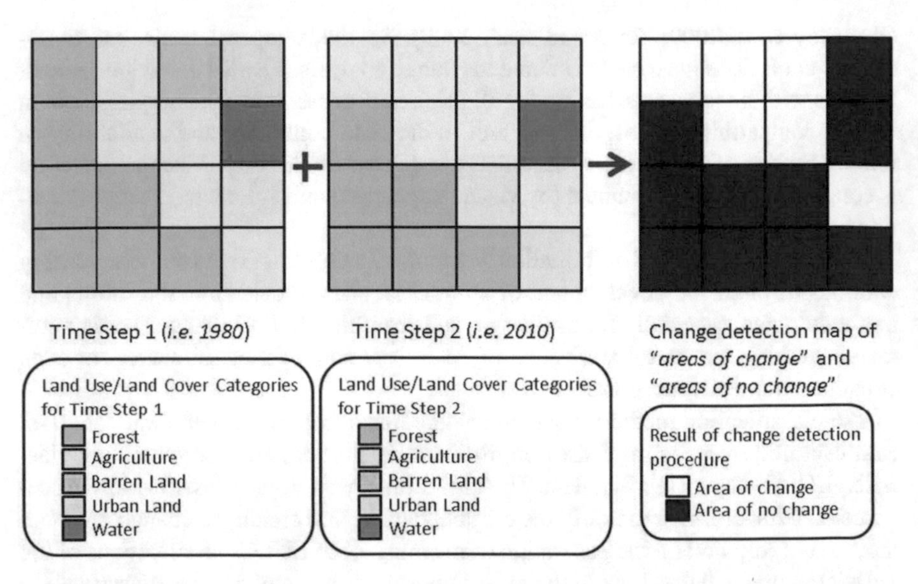

Fig. 3.27 Depicts the change occurring between time step 1 and time step 2 (i.e., 30 years change from 1980 to 2010)

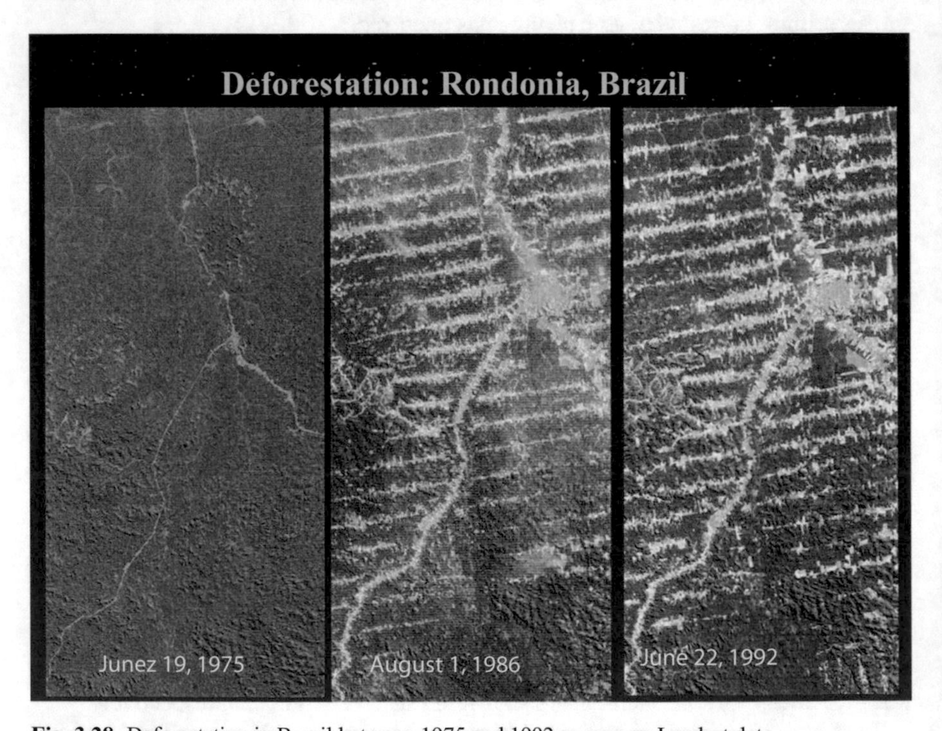

Fig. 3.28 Deforestation in Brazil between 1975 and 1992 as seen on Landsat data

Fig. 3.29 Two QuickBird satellite images of a section of the Athens, Greece, stadium, showing the changes in the stadium under construction *(left)* and completed *(right)* while preparing for 2004 Summer Olympics

Table 3.3 Post-classification *"from–to"* change matrix representing the classified figure (see Fig. 3.30) of the Upper Barataria Basin Estuary, LA (Nelson et al. 2002). The table values denote areas of land (expressed in percentage) converted from 1972 to 1992. The areas of no change are in italic font within parentheses. The total column represents the land use areas in the earlier 1972 time step. Changes within this matrix show the total percentage of each land cover in the study area that changed into other categories. For example, of the total (100 %) Urban land in 1972, by 1992, 64 % remain unchanged, however, *"from"* Urban, 33 % use converted *"to"* Agriculture, 2 % *"to"* Forest, 1 % *"to"* Swamp, and <1 % *"to"* Marsh. This change represents a *"From–To"* land conversion

		1992					
		Urban	Agriculture	Forest[a]	Swamp[b]	Marsh[c]	*1972 Total*
1972	Urban	*(64)*	33	2	1	<1	*100*
	Agriculture	40	*(54)*	5	<1	<1	*100*
	Forest[a]	6	4	*(43)*	46	1	*100*
	Swamp[b]	1	<1	5	*(93)*	1	*100*
	Marsh[c]	<1	<1	16	26	*(57)*	*100*

[a] Bottomland forest
[b] Swamp forest
[c] Freshwater marsh

In this approach, the original study imagery (image dates 1 and 2, or more) is thematically classified. The change analysis occurs when the classified maps are compared by one or more post-classification methods (Fig. 3.30). These include the expectation-maximization algorithm, ANNs, unsupervised change detection, and hybrid change detection (Lu et al. 2004). In addition to the advantage of the

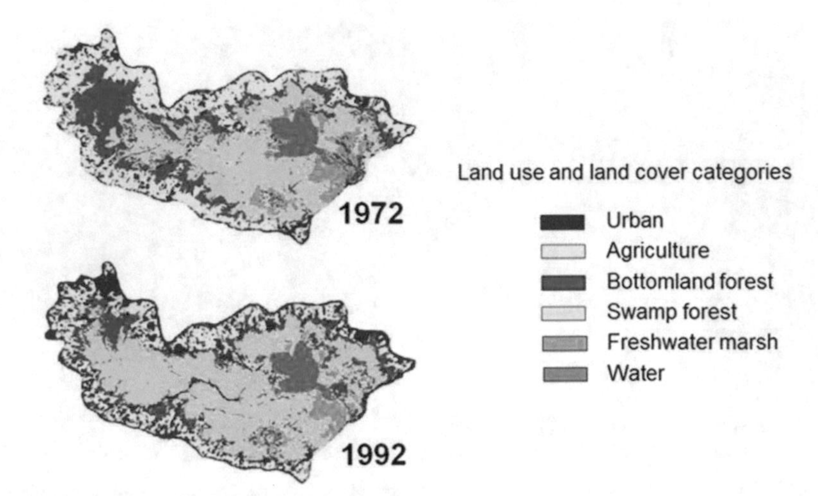

Fig. 3.30 Classified land use and land cover from Landsat Multispectral Scanner System *(MSS)* images for 1972 and 1992 time steps used to generate post-classification *"from–to"* change matrix seen in Table 3.3

"from–to" output generated by these approaches, post-classification change detection can be used to minimize the effect of atmospheric and environmental disparity between input images by identifying unusually high feature values that may result from these conditions. In addition, the computational simplicity of post-classification change analysis method have made this type of analysis a very popular procedure in comparison to the pre-classification methods (Jensen 2005). However, it is important for the analyst to be aware that each classification input introduces error into the post-classification change analysis, and this error is magnified in the final change map by the aggregate classification error from each input image.

As an example of automated change detection, the changes in the San Francisco Bay/Delta region between 1986 and 1993 along with the changed image are shown in Figs. 3.31 and 3.32.

More advanced change analysis methods integrate ancillary GIS data, statistical models, spatial analysis, textural metrics, or other alternative to pre- and post-classification approaches (Lu et al. 2004). These methods are too diverse to befit many general characterizations, but they are usually developed in response to specific research area issues or LULC categories of particular interest. Examples of advanced methods include those using measures of spatial dependence (Henebry 1993), generalized linear models (Morisette et al. 1999), spatial-spectral structure (Zhang et al. 2002), and spatial statistics (Read and Lam 2002).

The automated change detection can also be further analyzed to reflect various LULC objectives. Figure 3.33 shows an example of changes in impervious surface areas due to additional urban development in Raleigh, North Carolina between 1992 and 2001 (Khorram et al. 2005) based on National Land Cover Data (NLCD) data.

The dynamic nature of terrestrial environments presents unique challenges to remote sensing-based change detection. Concern over the rate, permanence, and

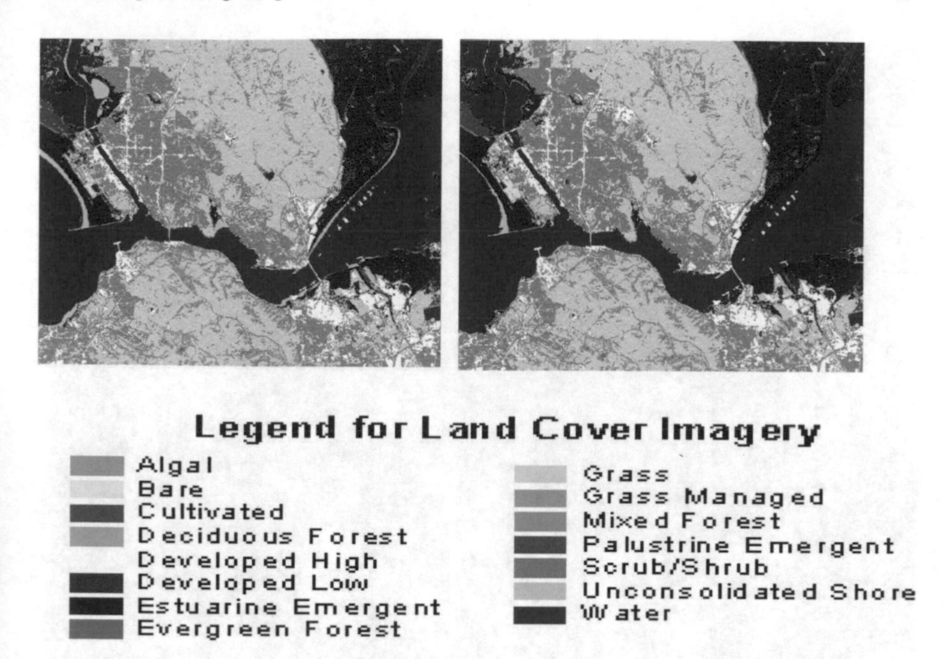

Fig. 3.31 Classified images of San Francisco Bay/Delta area in 1986 *(left)* and 1993 *(right)*. (Image courtesy of C-CAP and Khorram)

impact of certain land cover changes, such as forest conversion to agriculture or human developed land cover types, has spawned a diverse research interest in mapping and understanding LULC change (Arnold and Gibbons 1996; Stone 2004; Wilson and Lindsey 2005). The next-generation of high spatial resolution satellites such as QuickBird and IKONOS provide high-quality, 11-bit data with submeter ground pixel size and have shown land cover mapping promise. However, spectral challenges still remain due to the heterogeneity of terrestrial landscapes. For example, within the urban environment, challenges are imposed due to the high degree of this imagery's spatial detail along with a complex spectral response. The QuickBird and IKONOS sensors provide four discrete spectral bands in the visible and near-infrared wavelengths at a high spatial resolution. However, sunlit and shaded sides of the same feature are shown to have different spectral responses (Carleer et al. 2005). Ancillary information, such as GIS data and aerial photography, may be used to provide additional information to improve change detection classification accuracies of features obscured by shadows, clouds, or other obstructions during the post-classification process.

In selecting an appropriate change detection algorithm, it is important to understand there is no one most appropriate algorithm for any image type. The right solution will ultimately be the method that extracts the desired information content necessary to answer the research question at hand. An analyst should always be mindful of his or her study goals. Carefully defined study questions will directly inform the appropriate LULC categorization scheme. In addition, the temporal scale

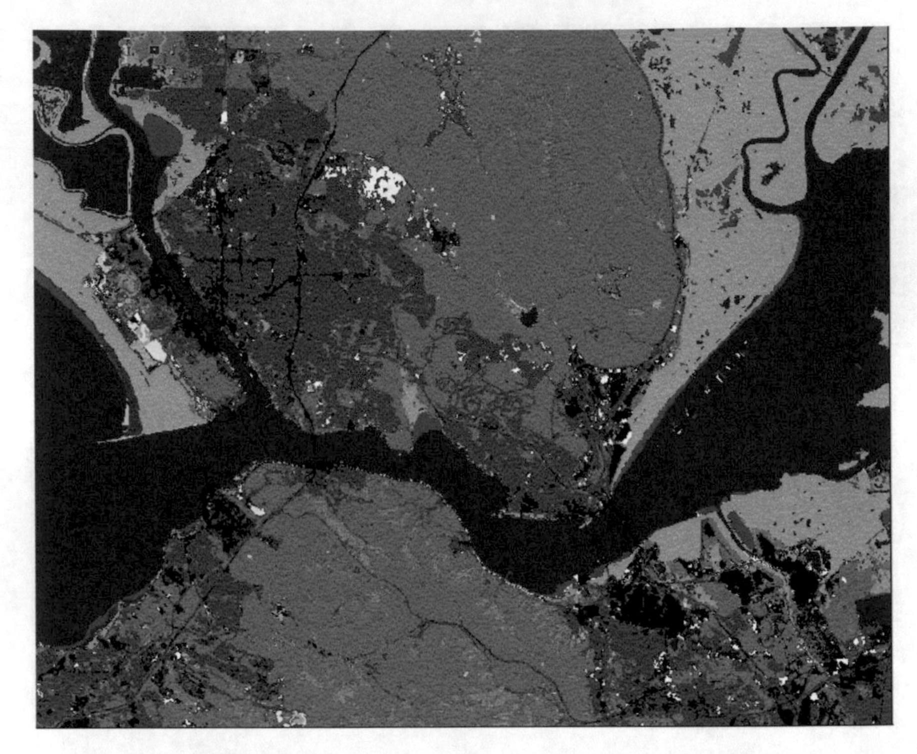

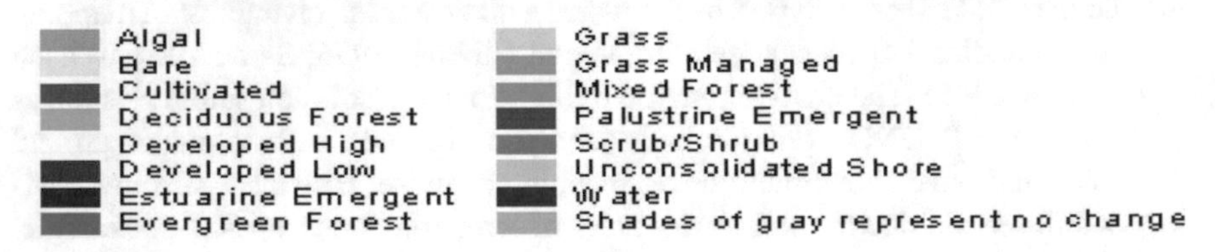

Fig. 3.32 Change image San Francisco Bay/Delta area between 1986 and 1993. (Image courtesy of C-CAP and Khorram)

of the change detection evaluation is also critical. So care should also be used when determining the appropriate interval, or time steps, selected to use between images.

3.4.4 Data Integration and Geospatial Modeling

The integration of remotely sensed data and the advent of increases in remote sensing data resolution have symbiotically led to new applications and new abilities to model data outputs. This trend is likely to continue as higher-quality data emerges, new uses for the data are discovered, new methods are developed to process them,

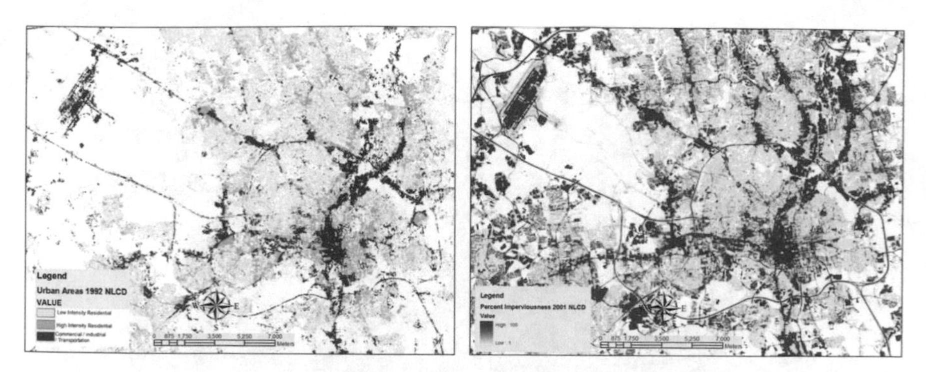

Fig. 3.33 Changes in impervious surface areas in Raleigh, North Carolina, US, between 1992 and 2001. *NLCD* National Land Cover Data

new computing power is designed to analyze, and future missions and platforms are created to capture even higher-quality data. In addition, new advancements in image processing and data integration also include our ability to incorporate additional sources of data into our analyses. An example of integrating various data layers composed of remotely sensed and other data types within a GIS is shown in Fig. 3.34.

A GIS makes it possible to manage, combine, and model various data layers and their interactions to identify spatial patterns within the dataset. Integration of remotely sensed (raster) data with vector data within the GIS, allows for the combination and modeling of these data with other available data to provide a vast amount of information for a variety of applications.

As described in the Handbook of Satellite Applications (Khorram et al. 2013b), there are many potential ways remotely sensed data can be integrated with a large datasets in areas such as public health information (hospitals, pharmacies, urgent care, veterinary care, etc.); weather data, potential hazards, and disaster management information (flooding, hurricane, earthquake, wildfires, tsunami, tornadoes, etc.); housing (residential, industrial, commercial, hotels, etc.); transportation (highways, streets, bridges, railway and bus stations, airports, etc.); and utilities (power, communication facilities, gas stations, water and sewer networks, etc.).

An additional useful strategy for incorporating ancillary data into classification is the implementation of an *expert system* (Goodenough 1987; Stefanov et al. 2001). The basic principle behind expert systems is that use of ancillary data sources can help to distinguish classes that are difficult to identify from the image spectral data alone, particularly if those classes exhibit high spectral variability (Jensen 2005; McIver and Friedl 2002). Both the image data to be classified and the supporting geospatial data follow a set of decision rules in order to classify each pixel in the image. These decision rules can be applied to an unclassified image (i.e., incorporating the original spectral values in the rule set) or instead, be applied to an image that has already been classified using a supervised or unsupervised approach (i.e., the class assignments of pixels are used in the rule set). In the expert system approach, there

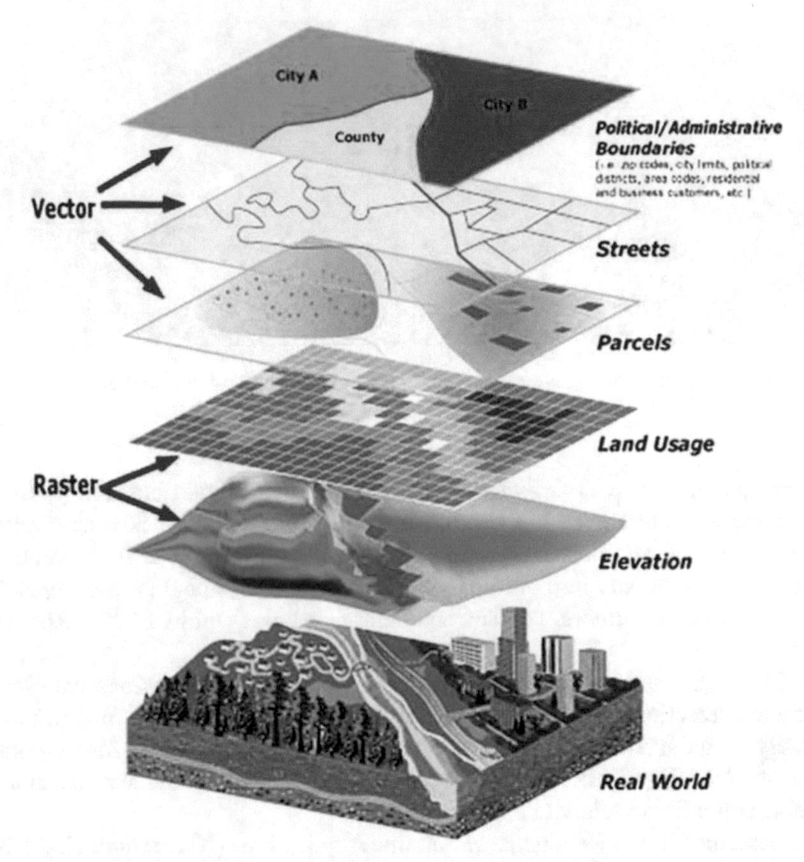

Fig. 3.34 A diagram displaying the integration of remote sensing and other data types in the form of geographic information systems. (Image courtesy of NOAA)

are a number of statistical procedures for constructing the set of decision rules in order to classify multifactor data, with decision tree methods perhaps most prominent. For examples of the use of decision tree methods in a remote sensing context see work by Pal and Mather (2003), and Friedl and Brodley (1997).

Hyperspectral remote sensing has been an evolving new technology in the world of remote sensing. In this process, hyperspectral sensor collects and processes spectral information across a broad range of the electromagnetic spectrum, at typically very narrow bandwidths. The narrower bandwidths of the hyperspectral sensor increases the ability to detect specific features on the Earth's surface, such as certain minerals or even subtle variances between similar structures, that may have strong reflectances within very narrow ranges in the electromagnetic spectrum. In addition, these types of sensors are usually able to capture data using a large number of bands. For example, in comparison to the Landsat program sensors, the AVIRIS (Airborne Visible/Infrared Imaging Spectrometer) hyperspectral sensor records 224 different bandwidths that range between 400 and 2500 μm.

Imaging spectroscopy is an additional technique for gaining information from hyperspectral data sources. In this technique, the spectral profiles within the data are analyzed to distinguish features, such as discriminating between different types of minerals or identifying stressed vegetation. However, before these data can be used for feature analysis, unique spectral "endmembers" must be generated to represent the diverse reflectance or emittance behavior of potential scene features (Lillesand et al. 2008). Additional challenges in the use of hyperspectral data sets include the large size of the dataset, as well as processing requirements. These challenges are not unique to remotely sensed or space-derived hyperspectral data, but also to the scale and diversity of the landscapes studied with these data that result in particularly intense computations. Despite these challenges, the increased availability and affordability of space-derived hyperspectral data have spawned broader interests in its application. Whereas hyperspectral data's widespread use had previously been limited to relatively specific fields such as mineralogy, the advent of hyperspectral sensors on space-borne platforms is opening up new research areas for which spectral profiles are not currently well-defined. For example, recent work has been attempted to develop a set of hyperspectral profiles for the complex urban landscape (Herold et al. 2004). Other new applications of hyperspectral data include studies of marine and estuarine water quality in Australia (Brando and Dekker 2003), invasive species in China (Galvao et al. 2005), sugarcane species in Brazil (Ramsey et al. 2005), and forest classification in the Pacific Northwest of the US (Goodenough et al. 2003). Because these are all novel applications of hyperspectral data, the required developmental work is still forthcoming.

Other satellite-derived remote sensing data quality improvements have occurred recently in the form of increased radiometric resolution. This is a measure of the range of possible data values to which reflectance intensity can be assigned, and it is expressed as digital bit depth. Higher radiometric resolution means that a sensor is more sensitive to differences in brightness values (Jensen 2005). Conventionally, satellite imagery has been stored as 8-bit data, although AVHRR and Geostationary Operational Environmental Satellite (GOES) data are stored in at least 10-bit format. These satellites, however, have coarse spatial resolutions (1000–4000 km) and are primarily used for monitoring global phenomena such as weather patterns. The relegation of higher radiometric resolution data to spatially coarse data has been a function of digital storage and processing requirements. As a result, some applications which would have required both high spectral and spatial resolutions have been hindered. Kutser et al. (2005) found that, while 8-bit Landsat 7 imagery was not adequate to map colored dissolved organic matter (CDOM) in boreal lakes in Finland, 11-bit Advanced Land Imager (ALI) data from the EO-1 platform were sensitive enough to map CDOM at a wide range of concentrations.

The most recent generation of high spatial resolution satellites, such as WorldView-2 and WorldView-3, all collect imagery in 11-bit radiometric depth (see Chap. 2). This aspect of data quality is often overshadowed by spatial resolution, but analysts using these data have found that their radiometric quality is an extremely important element of their significance (Elvidge et al. 2004; Kutser et al. 2005). High levels of radiometric resolution continue to present data storage and processing concerns, but Rogan and Chen (2004) pointed out that recent advances in remote sensing data quality have been matched with advances in computer hardware and image processing software.

As has been the case with high radiometric resolution, the highest temporal resolutions in satellite data have been limited to sensors with relatively low spatial resolution like GOES and AVHRR, whose revisit periods are 15 min to 12 h, respectively (NASA 2006). This is appropriate to their functional purposes in monitoring ever-changing weather and oceanic conditions. Sensors such as the Landsat and SPOT series have revisit periods from 16 to 26 days. In the past decade, satellites have been producing higher-spatial-resolution data by traveling at lower orbital altitudes (705 km for Landsat 7 vs. 450 km for QuickBird), but this translates into much narrower at-nadir ground swaths (185 km for Landsat 7 vs. 16.5 km for QuickBird). More than compensating for this loss of revisit frequency, however, these sensors are pointable by as much as 30° off-nadir, which means that their effective revisit period is from 1 to 3 days. Thus, recent increases in spatial and temporal resolution have been coupled in some sensors. This off-nadir acquisition capability also enables stereo-pair image capture, which can be used for automated feature and elevation extraction. The disadvantage of off-nadir image capture is that spatial resolution varies in the imagery captured by these pointable sensors, and geometric distortions can be severe. Both of these complications can require extensive preprocessing time and can ultimately limit map accuracy. Furthermore, these stated revisit periods depend on the availability of cloud-free acquisition windows.

The increases in satellite spatial resolutions have also been exhibited in imagery provided by a number of American commercial ventures, a direct result of the US government's 1994 decision to allow firms to market this type of data (Rogan and Chen 2004). In 1999, Space Imaging's IKONOS satellite (now owned by the DigitalGlobe Corporation) was launched to acquire landmark 1-m (panchromatic) and 4-m (multispectral) image products. Subsequently, sensors that have matched or exceeded the 1-m pixel size threshold (see Chap. 2) have qualified as a part of the new class of high spatial resolution platforms. At the same time, aerial-based systems have steadily increased in spatial resolution, and many now capture imagery with pixel sizes as small as 30 cm (Thomas et al. 2003). Nevertheless, high-resolution satellite imagery has the potential to be a more consistent and affordable source of data of similar quality. In addition, satellite-derived satellite imagery can capture large areas under identical temporal and atmospheric conditions, a task for which aerial sources are not ideally suited (Carleer and Wolff 2004). This is true even in areas that are difficult to service by aircraft.

The existing meter- or submeter spatial resolution satellite imagery is captured in panchromatic or, at its spectral best, in blue-green-red-near-infrared multispectral format. This represents a potential limitation of these data for applications that are highly sensitive to spectral response, such as automated species-level tree mapping. In general, the minimal spectral detail accompanying high-spatial-resolution imagery may make automated image land cover classification difficult because the high level of spatial detail will result in increased scene heterogeneity (Carleer et al. 2005). Spatially coarse data, in contrast, have a smoothing effect on variable spectral responses. In the same way, high spatial detail presents added classification complexity because this imagery can be replete with shadows (Dare 2005; Fisher and Goetz 2001; Thomas et al. 2003). Methods to deal with this image complexity are in their infancy, and data storage and processing requirements are a concern with this kind of data as well.

As is true for any data type, cost is an important factor in the use of high-spatial-resolution imagery. Its price has decreased and varied, but multispectral QuickBird imagery (DigitalGlobe corporation) may cost as much as US\$58/km^2 since its initial offering in 2000 (Read et al. 2003). This is more than 27,000 times the per-kilometer cost of acquiring Landsat ETM+ imagery in 2003, US\$0.00213/km^2 (Rogan and Chen 2004). This type of imagery expense may be limiting for most large-area mapping applications, a conclusion that demonstrates the trade-offs forced during remote sensing data selection.

High-spatial-resolution data present a particular set of methodological challenges in its use for terrestrial land cover classification, change detection, and spatial modeling. For example, urban areas are spatially complex, and they represent a mosaic of heterogeneous LULC. High image spatial detail offers analysts the option of mapping land cover directly without deference to land use, a fundamental mapping element for users of coarse spatial data. For instance, McCauley and Goetz (2004) used Landsat TM data and a decision tree classifier to map residential land use along a low-to-high gradient of development intensity. Such an approach acknowledges that urban land use is diverse at scales smaller than 30 m TM data, but it does not attempt to directly map this diversity. This is similar to a convention used in the recent Multi-Resolution Land Characteristics Consortium 2001 National Land Cover Database (Homer et al. 2004) in which development classes were defined by the percentage of pixel impervious surface coverage.

Under the direct mapping of land cover schemes made possible by high-spatial-resolution data, there is no need to aggregate cover types into specific land use categories. As discussed earlier, the effective aggregation of landscape features (such as shadows or small features such as individual, isolated trees) by spatially coarse data may actually be desirable for LULC mapping purposes. The strategies that have previously been successful in LULC classification for coarse data, in turn, may not necessarily translate into classification success with spatially fine data (Thomas et al. 2003). This reality has motivated a whole new range of textural (Johansen and Phinn 2006), morphological (Cablk and Minor 2003), and image segmentation (Carleer et al. 2005) classification approaches for use with high-spatial-resolution imagery. Many of these new methods are inspired by the mapping of urban places.

LULC change detection is also both enhanced and challenged by increases in image spatial resolution. A fundamental error source in change analysis is image misregistration (Dai and Khorram 1998). Co-registration of multi-date high-spatial-resolution satellite data is limited both by the positional accuracy of image registration reference data, which can be as erroneous as 1 m even in differentially corrected Global Positioning System (GPS) data, and by the likelihood that two images captured by a pointed high-spatial-resolution satellite sensor at different times will have different degrees of topographic distortion. As a result, change detection error due to image misregistration is a particular problem with high-spatial-resolution data (Wang and Ellis 2005). Furthermore, just as the minimum mapping unit is difficult to define in classifying LULC with these data, meaningful change units will require careful post-processing of high-spatial-resolution change maps. Nonetheless, high-spatial-resolution imagery is rich in information content and has already been used in a wide variety of unique spatial modeling and environmental monitoring applications. The diverse application areas of these data have already included rain forest tree mortality (Clark et al. 2004), coral reef damage (Philipson and Lindell 2003), lake water clarity (Sawaya et al. 2003), prediction of submersed aquatic vegetation (Nelson et al. 2006) precision agriculture (Seelan et al. 2003), impervious surface mapping (Goetz et al. 2003), and prairie dog habitat (Sidle et al. 2002).

Current advancements in remote sensing sensor technology (i.e., data integration with GIS, hyperspectral data, increased temporal, radiometric, and spatial resolutions) provide better sensing and classification capabilities of our environment. Satellite-based sensors, over the past decade, have seen an unprecedented period of increasing data abundance, diversity, and quality. This has been a function of increases in research demand as well as the commercialization of satellite data. Inevitably, new research and processing challenges have emerged alongside these data improvements, and the remote sensing community has responded with continued innovation and growth.

3.4.5 Processing of Airborne LiDAR Data

The processing of an airborne LiDAR data set is substantially different from multispectral or hyperspectral image processing primarily due to the nature of the data. (Airborne LiDAR data are also qualitatively different from space-borne LiDAR data captured by sensors such as CALIOP, although some of the processing issues are similar; see Winker et al. 2009). An airborne LiDAR data set consists of a dense (often greater than one pulse per square meter), three-dimensional, and geospatially referenced point cloud, where each point documents a single elevation estimate for a reflective object scanned by the system's laser. The primary challenge for analyzing this topographic point cloud is to separate ground, or "bare-earth," points from those corresponding to vegetation or to man-made structures such as buildings and vehicles (Chen 2007; Zhang et al. 2003). This filtering of the cloud into ground and non-ground returns facilitates the development of a bare-earth digital terrain model (DTM)—also sometimes called a DEM —which is necessary to estimate the height of vegetation and other objects (Chen 2007). Various approaches have been used to remove non-ground returns. For example, iterative linear interpolation applies a low-degree polynomial trend to reduce the original elevation data estimates so that they conform to a relatively smooth surface representing the ground. Other options include slope-based and curvature-based filters that compare the measurements of neighboring points. Morphological filters treat a LiDAR data set as a grayscale image, such that the shapes of trees and other objects can be straightforwardly identified by a shift in gray tone between neighboring points (Zhang et al. 2003). In general, bare-earth extraction filters can be categorized in two ways: neighborhood-based or directional (Meng et al. 2009). Neighborhood-based filters preserve the shapes of larger non-ground objects (e.g., buildings, trees) but are not very sensitive to small elevation changes caused by low-lying features like shrubs. In contrast, directional filters are more sensitive to small objects but because of this sensitivity have a tendency to generate artificial edge-lines across the ground (or even across large objects). Some of these limitations can be overcome by more complex algorithms (e.g., multidirectional filtering; see Meng et al. 2009), but all ground filters struggle with complex terrain, such as in highly urbanized areas.

After ground and non-ground returns have been distinguished in a LiDAR data set, the next step is to classify non-ground measurements according to the category of object they represent. Typically, most non-ground measurements in a LiDAR data set correspond to either buildings or vegetation (Zhang et al. 2006). Automated extraction of building footprints has been an especially active area of research (e.g., Awrangjeb et al. 2010; Rottensteiner 2003; Sohn and Dowman 2007), driven by the need of such data for applications such as the estimation of flood insurance rates (Zhang et al. 2006). Similar to efforts to identify and characterize individual trees using LiDAR data (Chen 2007; Holmgren et al. 2008), multispectral images are often used in conjunction with the building extraction algorithms, thereby representing another prominent example of data fusion to improve the accuracy and quality of output products.

As a final note, the use and implementation of LiDAR-derived models is continuing to be explored by the research and resource management communities. For example, many functions of hydrologic modeling depend on accurate calculation of the associated watershed's topography (i.e., slope gradient, etc.). Recently, DEMs produced from LiDAR elevation point 2 data have been replacing older types of DEMs produced from stereo photography or the Shuttle Radar Topography Mission. LiDAR DEMs were found to be significantly more accurate than Level I and II USGS DEMs (Hodgson et al. 2003). Due to the high resolution at which LiDAR DEMs can be produced and their superior accuracy in representing elevation, LiDAR DEMs are also frequently used to delineate possible flood inundation areas for disaster management planning and have proven to be significantly more reliable than traditional datasets (Miller et al. 2004; Raber 2003). LiDAR DEMs are capable of representing small changes in topography that influence the accuracy of flood inundation models. In addition, many of the LiDAR processing techniques described above can also be applied to data sets generated using radar interferometry (see Chap. 2). Radar interferometric data sets are functionally similar to LiDAR data sets in that they consist of elevation estimates for the ground and other reflective objects (Sithole and Vosselman 2004). However, they are typically coarser in resolution, limiting their applicability for fine-scale operations such as identifying individual tree crowns.

3.5 Summary

This chapter provides just a few of the many data processing tools currently being used in the field of remote sensing. While not exhaustive, as is always the case, technologies and the research continues to evolve, thus expanding our opportunities to extract increased amounts of data from current and future remote sensing platforms. The next chapter will provide examples of how remote sensing is used to better understand many terrestrial applications.

3.6 Review Questions

1. What are the three stages that the processing of multispectral image data are typically divided into?
2. A false color composite (FCC) image displays the combination of any three bands from a multispectral image other than the true-color (blue/green/red) combination. In the standard FCC, the green band of the input image is displayed using the blue color ramp, the red image band is displayed using the green color ramp, and a near-infrared band from the input image (e.g., Band 4 from a Landsat TM image) is displayed using the red color ramp. Why might this FCC display be of interest to vegetation studies? The reason that this particular FCC is so popular is because it highlights the presence of healthy green vegetation.
3. Why might image data preprocessing applications involve operations such as, geometric correction, atmospheric correction, radiometric correction, and the calculation of various spectral band combinations and ratios, and data fusion?
4. The process of pan-sharpening an image requires the integration of a multispectral image and a corresponding, higher-resolution panchromatic image. Why might this procedure be of value for an image analyst?
5. Image classification of multispectral data is a very important process in which each pixel in an image is assigned to a particular category in a set of categories of interest. The classification process is generally based on the pixels' values (i.e., their spectral values) in each band of the image. What should be an image analyst's main goal for developing a final classification output product?
6. In a supervised classification procedure, the analyst selects multiple training sites during the initial stage of the classification project. Why do you think the selection of training sites in this type of procedure may or may not have an advantage over an unsupervised classification procedure?
7. An image accuracy assessment may be performed in a variety of ways and has been a key component and the focus of a significant number of remote sensing studies. Why should this procedure be an important part of any land use and land cover classification study?
8. Change detection using remote sensing data is an attempt to record natural and anthropogenic transitions of land use/land cover on Earth's surface that have

occurred over time. Why is it necessary to consider appropriate spatial and temporal scale when developing a land use and land cover change detection study?

References

Aaronoff, S.G. 1985. The minimum accuracy value as an index of classification accuracy. *Photogrammetric Engineering and Remote Sensing*. 57(5):501–509.

Al Fugara, A.M., B. Pradhan, and T. A. Mohamed. 2009. Improvement of land-use classification using object-oriented and fuzzy logic approach. *Applied Geomatics*. 1:111–120.

Alparone, L., B. Alazzi, S. Baronti, A. Garzelli, F. Nencini, and M. Selva. 2008. Multispectral and panchromatic data fusion assessment without reference. *Photogrammetric Engineering and Remote Sensing*. 74(2):193–200.

Anderson, J.R., E. Hardy, J. Roach, and R. Witmer. 1976. *A Land Use and Land Cover Classification System for Use with Remote Sensing Data*. Washington, DC: US Geological Survey Professional Paper 964. 28 p.

Arnold, C. and C. Gibbons. 1996. Impervious surface coverage: The emergence of a key environmental indicator. *Journal of American Planning Association*. 62:243–258.

Awrangjeb, M., M. Ravanbakhsh, and C.S. Fraser. 2010. Automatic detection of residential buildings using LIDAR data and multispectral imagery. *ISPRS Journal of Photogrammetry and Remote Sensing*. 65(5):457–467.

Blaisdell, E.A. 1993. *Statistics in Practice*. New York: Harcourt Brace Jovanovich. 653 p.

Blaschke, T. 2010. Object based image analysis for remote sensing. *ISPRS Journal of Photogrammetry and Remote Sensing*. 65:2–16.

Brando, V.E. and A.G. Dekker. 2003. Satellite hyperspectral remote sensing for estimating estuarine and coastal water quality. *IEEE Transactions on Geoscience and Remote Sensing*. 41:1378–1387.

Cablk, M.E. and T.B. Minor. 2003. Detecting and discriminating impervious cover with high-resolution IKONOS data using principal component analysis and morphological operators. *International Journal of Remote Sensing*. 24:4627–4645.

Cakir, H.I. and S. Khorram. 2008. Pixel level fusion of panchromatic and multispectral images based on correspondence analysis. *Photogrammetric Engineering and Remote Sensing*. 74(2):183–192.

Cakir, H.I., S. Khorram, X.L. Dai, and P. de Fraipont. 1999. Merging SPOT XS and SAR imagery using the wavelet transform method to improve classification accuracy. In: Proceedings of the IEEE 1999 International Geoscience and Remote Sensing Symposium (IGARSS), 28 June-2 July 1999, Hamburg, Germany, pp. 71–73.

Cakir, H. I., S. Khorram, and S.A.C. Nelson. 2006. Correspondence analysis approach for detecting land use land cover changes. *Remote Sensing of Environment*. 102:306–317.

Carleer, A.P., O. Debeir, and E. Wolff. 2005. Assessment of very high spatial resolution satellite image segmentations. *Photogrammetric Engineering and Remote Sensing*. 71:1285–1294.

Carleer, A. and E. Wolff. 2004. Exploitation of very high resolution satellite data for tree species identification. *Photogrammetric Engineering and Remote Sensing*. 70:135–140.

Celik, T. 2009. Unsupervised change detection in satellite images using principal component analysis and *k*-means clustering. *IEEE Geoscience and Remote Sensing Letters*. 6(4):772–776.

Cerny, V. 1985. A thermodynamic approach to the traveling salesman problem: An efficient simulation. *Journal of Optimization Theory and Applications* 45: 41–51.

Chavez, P.S. and A.Y. Kwarteng. 1989. Extracting spectral contrast in Landsat Thematic Mapper image data using selective principal component analysis. *Photogrammetric Engineering and Remote Sensing*. 55(3):339–348.

Chavez, P.S., S.C. Sides, and J.A. Anderson. 1991. Comparison of three different methods to merge multiresolution and multispectral: Landsat TM and SPOT Panchromatic. *Photogrammetric Engineering and Remote Sensing*. 57(3):295–303.

Chen, Q. 2007. Airborne Lidar data processing and information extraction. *Photogrammetric Engineering and Remote Sensing*. 73(2):109–112.

Clark, D.B., C.S. Castro, L.D.A. Alvarado, and J.M. Read. 2004. Quantifying mortality of tropical rain forest trees using high-spatial-resolution satellite data. *Ecology Letters*. 7:52–59.

Cohen, J.A. 1960. A coefficient of agreement for nominal scales. *Educational and Psychological Measurements*. 20:37–46.

Congalton, R.G. 1991. A review of assessing the accuracy of classifications of remotely sensed data. Remote Sens. Environ. 37, 35–46.

Congalton, R.G., R.G. Oderwald, and R.A. Mead. 1983. Assessing Landsat classification accuracy using discrete multivariate statistical techniques. *Photogrammetric Engineering and Remote Sensing*. 49(12):1671–1678.

Congalton, R.G., and K. Green, Assessing the Accuracy of Remotely Sensed Data: Principles and Practices. (Boca Raton, Lewis Publishers 1999), p. 137.

Cowardin, L.M., V. Carter, F.C. Golet, and E.T. LaRoe. 1979. *Classification of Wetlands and Deepwater Habitats of the United States*. Washington: U.S. Fish and Wildlife Service, FWS/OBS-79/31. Pp. 103.

Dai, X.L. and S. Khorram. 1998. The effects of image misregistration on the accuracy of remotely sensed change detection. *IEEE Transactions on Geoscience and Remote Sensing*. 36:1566–1577.

Dai, X.L. and S. Khorram. 1999. Data fusion using artificial neural networks: a case study on multitemporal change analysis. *Computers, Environment and Urban Systems*. 23:19–31.

Dare, P.M. 2005. Shadow analysis in high-resolution satellite imagery of urban areas. *Photogrammetric Engineering and Remote Sensing*. 71:169–177.

Das, A. and B.K. Chakrabarti, eds. 2005. Quantum Annealing and Related Optimization Methods. Lecture Notes in Physics 679. Berlin: Springer.

De Vincente, J., J. Lanchares, and R. Hermida. 2003. Placement by thermodynamic simulated annealing. *Physics Letters A* 317: 415–423.

Elvidge, C.D., J.B. Dietz, R. Berkelmans, S. Andrefouet, W. Skirving, A.E. Strong, and B.T. Tuttle. 2004. Satellite observation of Keppel Islands (Great Barrier Reef) 2002 coral bleaching using IKONOS data. *Coral Reefs*. 23:123–132.

Fisher, J. and S. Goetz. 2001. Considerations in the use of high spatial resolution imagery: an applications research assessment. In: Proceedings of the American Society for Photogrammetry and Remote Sensing (ASPRS) Conference, April 23–27, 2001, St. Louis, MO.

Fisher, P.F. and S. Pathirana. 1990. The evaluation of fuzzy membership of land cover classes in the suburban zone. *Remote Sensing of Environment*. 34:121–132.

Fonseca, L.M.G. and B.S. Manjunath. 1996. Registration techniques for multisensory remotely sensed imagery. *Photogrammetric Engineering and Remote Sensing*. 62:1049–1056.

Foody, G. M. 2002. Status of land cover classification accuracy assessment, *Remote Sensing of Environment*. 80:185–201.

Forster, B.C. 1984. Derivation of atmospheric correction procedures for Landsat MSS with particular reference to urban data. *International Journal of Remote Sensing*. 5:799–817.

Friedl, M.A. and C.E. Brodley. 1997. Decision tree classification of land cover from remotely sensed data. *Remote Sensing of Environment*. 61:399–409.

Galvao, L.S., A.R. Formaggio, and D.A. Tisot. 2005. Discrimination of sugarcane varieties in Southeastern Brazil with EO-1 hyperion data. *Remote Sensing of Environment*. 94:523–534.

Gao, B.C., M.J. Montes, C.O. Davis, and A.F.H. Goetz. 2009. Atmospheric correction algorithms for hyperspectral remote sensing data of land and ocean. *Remote Sensing of Environment*. 113:S17–S24.

Geman, S. and D. Geman. 1984. Stochastic relaxation, Gibbs' distribution and the Bayesian restoration of images. *IEEE Transactions on Pattern Analysis and Machine Intelligence* 6: 721–741.

Goetz, S., R. Wright, A. Smith, E. Zinecker, and E. Schaub. 2003. IKONOS imagery for resource management: Tree cover, impervious surfaces, and riparian buffer analyses in the mid-Atlantic region. *Remote Sensing of Environment*. 88:195–208.

Goodchild, M.F., G.Q. Sun, and S. Yang. 1992. Development and test of an error model for categorical data. *International Journal of Geographical Information Systems*. 6(2):87–104.

Goodenough, D.G., Goldberg, M., Plunkett G., and, J. Zelek, 1987. An expert system for remote sensing. IEEE. Trans. Geosci. Remote Sens. GE-25, 349–359.

Goodenough, D.G., A. Dyk, O. Niemann, J.S. Pearlman, H. Chen, T. Han, M. Murdoch, and C. West. 2003. Processing Hyperion and ALI for forest classification. *IEEE Transactions on Geoscience and Remote Sensing*. 41:1321–1331.

Gross, J.E., Goetz, S.J., Cihlar, J., 2009. Application of remote sensing to parks and protected area monitoring: introduction to the special issue. *Remote Sensing of Environment*. 113(7):1343–1345.

Hagen, A. 2003. Fuzzy set approach to assessing similarity of categorical maps. *International Journal of Geographical Information Science*. 17:235–249.

Haykin, S. 1994. *Neural Networks: A Comprehensive Foundation*. Englewood Cliffs, NJ: Prentice Hall.

Henebry, G.M. 1993. Detecting change in grasslands using measures of spatial dependence with Landsat TM Data. *Remote Sensing of Environment*. 46:223–234.

Herold, M., D.A. Roberts, M.E. Gardner, and P.E. Dennison. 2004. Spectrometry for urban area remote sensing—Development and analysis of a spectral library from 350 to 2400 nm. *Remote Sensing of Environment*. 91:304–319.

Herold, M., X.H. Liu, and K.C. Clarke. 2003. Spatial metrics and image texture for mapping urban land use. *Photogrammetric Engineering and Remote Sensing*. 69:991–1001.

Hester, D.B. 2008. Land Cover Mapping and Change Detection in Urban Watersheds Using QuickBird High Spatial Resolution Satellite Imagery. Ph. D. Dissertation, North Carolina State University, 148p.

Hester, D.B., H.I. Cakir, S.A.C. Nelson, and S. Khorram. 2008. Per-pixel classification of high spatial resolution satellite imagery for urban land cover mapping. *Photogrammetric Engineering and Remote Sensing*. 74:463–471.

Hester, D.B., S.A.C. Nelson, H.I. Cakir, S. Khorram, and H.M. Cheshire. 2010. High-resolution land cover change detection based on fuzzy uncertainty analysis and change reasoning. *International Journal of Remote Sensing*. 31:455–475.

Hodgson, M.E., J.R. Jensen, L. Schmidt, S. Schill and B. Davis. 2003. An evaluation of LIDAR- and IFSAR-derived digital elevation models in leaf-on conditions with USGS Level 1 and Level 2 DEMs. *Remote Sensing of Environment*. 84(2):295–308.

Holmgren, J., Å. Persson, and U. Söderman. 2008. Species identification of individual trees by combining high resolution LiDAR data with multi-spectral images. *International Journal of Remote Sensing*. 29(5):1537–1552.

Homer, C., C.Q. Huang, L.M. Yang, B. Wylie, and M. Coan. 2004. Development of a 2001 National Land-Cover Database for the United States. *Photogrammetric Engineering and Remote Sensing*. 70:829–840.

Hord, R.M. 1982. Digital Image Processing of Remotely-Sensed Data. New York: Academic Press. Pp. 256.

Huete, A., K. Didan, T. Miura, E.P. Rodriguez, X. Gao, and L.G. Ferreira. 2002. Overview of the radiometric and biophysical performance of the MODIS vegetation indices. *Remote Sensing of Environment*. 83:195–213.

Jain, A.K. 1989. Fundamentals of Digital Image Processing. Englewood Cliffs, NJ: Prentice Hall, pp. 418–421.

Jensen, J.R. 2005. Introductory Digital Image Processing, 3rd edition. Upper Saddle River, NJ: Pearson Prentice Hall. 316 p.

Jensen, R.R., P.J. Hardin, and G. Yu. 2009. Artificial neural networks and remote sensing. *Geography Compass*. 3(2):630–646.

Jiang, Z., A.R. Huete, K. Didan, and T. Miura. 2008. Development of a two-band enhanced vegetation index without a blue band. *Remote Sensing of Environment*. 112:3833–3845.

Johansen, K., and S. Phinn. 2006. Mapping structural parameters and species composition of riparian vegetation using IKONOS and Landsat ETM plus data in Australian tropical savannahs. *Photogrammetric Engineering and Remote Sensing*. 72:71–80.

Kennedy, R.E., P.A. Townsend, J.E. Gross. 2009. Remote sensing change detection tools for natural resource managers: understanding concepts and tradeoffs in the design of landscape monitoring projects. *Remote Sensing of Environment*. 113(7):1382–1396.

Khorram, S. 2015, Landsat 8 Images of South San Francisco Bay, personal communications.

Khorram, S., S.A.C. Nelson, H.I. Cakir, and C.F. van der Wiele. 2013a. Digital image acquisition, preprocessing, and data reduction, in Handbook of Satellite Applications, 2nd ed. Editors Pelton, J. N., Madry, S., and Camacho-Lara, S. Springer-Verlag. New York City, New York, p. 250. ISBN: 978-1-4419-7670-3

Khorram, S., S.A.C. Nelson, H.I. Cakir, and C.F. van der Wiele. 2013b. Digital image processing, post- processing, and data integration, Handbook of Satellite Applications, 1st ed., ed. by J.N. Pelton, S. Madry, S. Camacho-Lara (Springer-Verlag, New York).

Khorram, S., C.F. van der Wiele, and H. Yuan. 2011. Development of a Modified Neural Network Based Land Cover Classification System Using Automated Data Selector and Multiresolution Remotely-Sensed Data. *Journal of Goecarto International*. 26(6):435–457.

Khorram, S. 1985. Development of water quality models applicable throughout the entire San Francisco Bay and delta. *Photogrammetric Engineering and Remote Sensing*. 51(1):53–62.

Khorram, S., S.A.C. Nelson, H.I. Cakir, D.B. Hester, and H.M. Cheshire. 2005. Cost effective assessment of land use practices influencing erosion and sediment yield, Technical report, Center for Earth Observation, North Carolina State University, submitted to North Carolina Water Resources research Institute, 46 p.

Khorram, S. 1982. Coastwatch: water quality mapping of the entire San Francisco Bay and delta from Landsat Multispectral Scanner data. R.N. Colwell, PI, Space Sciences Laboratory, Series 23, Issue 6, University of California, Berkeley. Pp. 34.

Khorram, S., G.S. Biging, N.R. Chrisman, D.R. Colby, R.G. Congalton, J.E. Dobson, R.L. Ferguson, M.F. Goodchild, J.R. Jensen, and T.H. Mace. 1999a. Accuracy Assessment of Remote Sensing-Derived Change Detection. Bethesda, MD: American Society of Photogrammetry and Remote Sensing Monograph. Pp. 64.

Khorram, S., G.S. Biging, N.R. Chrisman, D.R. Colby, R.G. Congalton, J.E. Dobson, R.L. Ferguson, M.F. Goodchild, J.R. Jensen, and T.H. Mace. 1999b. "Accuracy Assessment of Remote Sensing-Derived Change Detection," American Society of Photogrammetry, 78 p.

Khorram, S., H.M. Cheshire, X. Dai, and J. Morisette. 1996. Land cover inventory and change detection of coastal North Carolina using Landsat Thematic Mapper data. In: Proceedings, ASPRS/ACSM—Annual Convention and Exposition, Vol. 1: Remote Sensing and Photogrammetry. pp. 245–250.

Khorram, S., H.M. Cheshire, K. Siderellis, and Z. Nagy. 1992. Mapping and GIS Development of Land Use/Land Cover Categories for the Albemarle-Pamlico Drainage Basin. North Carolina Department of Environmental, Health, and Natural Resources, No. 91–08. Pp. 55.

Kirkpatrick, S., C.D. Gelatt, and M.P. Vecchi. 1983. Optimization by simulated annealing. *Science* 220: 671–680.

Kutser, T., D. Pierson, L. Tranvik, A. Reinart, S. Sobek, and K. Kallio. 2005. Using satellite remote sensing to estimate the colored dissolved organic matter absorption coefficient in lakes. *Ecosystems*. 8:709–720.

Lambin, E.F., and D. Ehrlich. 1996. The surface temperature-vegetation index space for land cover and land-cover change analysis. *International Journal of Remote Sensing*. 17:463–487.

Lee, D.H., K.M. Lee, and S.U. Lee. 2008. Fusion of lidar and imagery for reliable building extraction. *Photogrammetric Engineering and Remote Sensing*. 74(2):215–225.

Lillesand, T., R. Kiefer, and J. Chipman. 2008. Remote Sensing and Image Interpretation, 6th edition. New York: John Wiley and Sons. Pp. 763

Lu, D., P. Mausel, E. Brondizio, and E. Moran. 2004. Change detection techniques. *International Journal of Remote Sensing*. 25:2365–2407.

Lunetta, R. L. and J.G. Lyons, eds. 2003. Geospatial Data Accuracy Assessment. Las Vegas: US Environmental Protection Agency, Report No. EPA/600/R-03/064. Pp. 335.

Maktav, D., F. Erbek, and C. Jurgens. 2005. Remote sensing of urban areas. *International Journal of Remote Sensing*. 26:655–659.

Mausel, P.W. and J.K. Lee. 1990. Optimum band selection for supervised classification of multispectral data. *Photogrammetric Engineering and Remote Sensing*. 56:55–60.

McCauley, S., and S. Goetz. 2004. Mapping residential density patterns using multi-temporal Landsat data and a decision-tree classifier. *International Journal of Remote Sensing*. 25:1077–1094.

McIver, D.K. and M.A. Friedl. 2002. Using prior probabilities in decision-tree classification of remotely sensed data. *Remote Sensing of Environment*. 81:253–261.

Meng, X., L. Wang, J.L. Silván-Cárdenas, and N. Currit. 2009. A multi-directional ground filtering algorithm for airborne LIDAR. *ISPRS Journal of Photogrammetry and Remote Sensing* 64(1):117–124.

Miller, J.E., S.A.C. Nelson, and G.R. Hess. 2009. A new object-oriented method of impervious surface classification using Feature Analyst. *The Professional Geographer*. 61:250–264.

Miller, S.N., S.R. Shrestha, and D. Semmens. 2004. Semi-automated Extraction and Validation of Channel Morphology from LIDAR and IFSAR Terrain Data. In: Proceedings of the ASPRS Annual Conference, Denver, CO.

Mitri, G. H. and I. Z. Gitas. 2004. A performance evaluation of a burned area object-based classification model when applied to topographically and nontopographically connected TM imagery. *International Journal of Remote Sensing*. 25:2863–2870.

Mohanty, K.K. and T.J. Majumdar. 1996. An artificial neural network (ANN) based software package for classification of remotely sensed data. *Computers & Geosciences*. 22:81–87.

Morisette, J.T. and S. Khorram. 1998. Exact binomial confidence interval for proportions. *Photogrammetric Engineering and Remote Sensing* 64(4):281–283.

Morisette, J.T., S. Khorram, and T. Mace. 1999. Land-cover change detection enhanced with generalized linear models. *International Journal of Remote Sensing*. 20:2703–2721.

Muchoney, D.M. and B.N. Haack. 1994. Change detection for monitoring forest defoliation. *Photogrammetric Engineering and Remote Sensing*. 60:1243–1251.

National Aeronautics and Space Administration (NASA). 2006. ASTER Instrument Characteristics [web site]. http://asterweb.jpl.nasa.gov/characteristics.asp (last accessed 27 May 2011).

National Oceanic and Atmospheric Administration (NOAA). 2004. Coastal Change Analysis Program (C-CAP) [web page]. Charleston: NOAA Coastal Services Center. http://www.csc.noaa-gov/crs/lca/ccap_program.html

Nelson, S.A.C., K.S. Cheruvelil, and P.A. Soranno. 2006. Remote sensing of freshwater macrophytes using Landsat TM and the influence of water clarity. *Aquatic Botany*. 85:289–298.

Nelson, S.A.C., P. A. Soranno, and J. Qi. 2002. Land cover change in the Upper Barataria Basin Estuary, Louisiana, from 1972 to 1992: increases in wetland area. *Environmental Management*. 29:716–727.

Núñez, J., X. Otazu, O. Fors, A. Prades, V. Palà, and R. Arbiol. 1999. Multiresolution-based image fusion with additive wavelet decomposition. *IEEE Transactions on Geoscience and Remote Sensing*. 37(3):1204–1211.

Paine, D.P., and J.D. Kiser. 2003. Aerial Photography and Image Interpretation (2nd ed.). John Wiley, inc. New York. 632 pages.

Pal, M. and P.M. Mather. 2003. An assessment of the effectiveness of decision tree methods for land cover classification. *Remote Sensing of Environment*. 86:554–565.

Pang, X., L. Chen, and W. Chen. 2006. Application of neural network based on simulated annealing to classification of remote sensing image. In: Proceedings of the 6th World Congress on Intelligent Control and Automation, 21–23 June 2006, Dalian, China, 2874–2877.

Philipson, P., and T. Lindell. 2003. Can coral reefs be monitored from space? *Ambio*. 32:586–593.

Pohl, C. and J.L Van Genderen. 1998. Multisensor image fusion in remote sensing: concepts, methods and applications. *International Journal of Remote Sensing*. 19(5):823–854.

Qiu, F. and J.R. Jensen. 2004. Opening the black box of neural networks for remote sensing image classification. *International Journal of Remote Sensing*. 9:1749–1768.

Raber, G. 2003. The Effect of Lidar Posting Density on DEM Accuracy and Flood Extent Delineation. In Proceedings of the UCGIS Summer Assembly 2003, Pacific Grove, California.

Ramsey, E., A. Rangoonwala, G. Nelson, and R. Ehrlich. 2005. Mapping the invasive species, Chinese tallow, with EO1 satellite Hyperion hyperspectral image data and relating tallow occurrences to a classified Landsat Thematic Mapper land cover map. *International Journal of Remote Sensing*. 26:1637–1657.

Read, J., D. Clark, E. Venticinque, and M. Moreira. 2003. Application of merged 1-m and 4-m resolution satellite data to research and management in tropical forests. *Journal of Applied Ecology*. 40:592–600.

Read, J.M. and N.S.N. Lam. 2002. Spatial methods for characterising land cover and detecting land-cover changes for the tropics. *International Journal of Remote Sensing*. 23:2457–2474.

Rocha, A.V. and G.R. Shaver. 2009. Advantages of a two band EVI calculated from solar and photosynthetically active radiation fluxes. *Agricultural and Forest Meteorology*. 149:1560–1563.

Rogan, J. and D.M. Chen. 2004. Remote sensing technology for mapping and monitoring land-cover and land-use change. *Progress in Planning*. 61:301–325.

Rottensteiner, F. 2003. Automatic generation of high-quality building models from lidar data. *IEEE Computer Graphics and Applications*. 23(6):42–50.

Sabins, M.J. 1987. Convergence and consistency of fuzzy c-means/ISODATA algorithms. *IEEE Transactions on Pattern Analysis and Machine Intelligence*. 9:661–668.

Sawaya, K., L. Olmanson, N. Heinert, P. Brezonik, and M. Bauer. 2003. Extending satellite remote sensing to local scales: land and water resource monitoring using high-resolution imagery. *Remote Sensing of Environment*. 88:144–156.

Schnur, M.T., H. Xie, and X. Wang. 2010. Estimating root zone soil moisture at distant sites using MODIS NDVI and EVI in a semi-arid region of southwestern US. *Ecological Informatics*. 5:400–409.

Schott, J.R., C. Salvaggio, and W.J. Volchok. 1988. Radiometric scene normalization using pseudoinvariant features. *Remote Sensing of Environment*. 26:1–16.

Seelan, S.K., S. Laguette, G.M. Casady, and G.A. Seielstad. 2003. Remote sensing applications for precision agriculture: A learning community approach. *Remote Sensing of Environment*. 88:157–169.

Shackelford, A.K. and C.H. Davis. 2003. A combined fuzzy pixel-based and object-based approach for classification of high-resolution multispectral data over urban area. *IEEE Transactions on Geoscience and Remote Sensing*. 41:2354–2363.

Sidle, J.G., D.H. Johnson, B.R. Euliss, and M. Tooze. 2002. Monitoring black-tailed prairie dog colonies with high-resolution satellite imagery. *Wildlife Society Bulletin*. 30:405–411.

Simone, G., A. Farina, F.C. Morabito, S.B. Serpico, and L. Bruzzone. 2002. Image fusion techniques for remote sensing applications. *Information Fusion*. 3:3–15.

Sithole, G. and G. Vosselman. 2004. Experimental comparison of filter algorithms for bare-Earth extraction from airborne laser scanning point clouds. *ISPRS Journal of Photogrammetry and Remote Sensing*. 59:85–101.

Solberg, A.H.S., T. Taxt, and A.K. Jain. 1994. Multisource classification of remotely sensed data: fusion of Landsat TM and SAR images. *IEEE Transactions on Geoscience and Remote Sensing*. 32(4):768–778.

Sohn, G. and I. Dowman. 2007. Data fusion of high-resolution satellite imagery and LiDAR data for automatic building extraction. *ISPRS Journal of Photogrammetry and Remote Sensing*. 62:43–63.

South, S., J. Qi, and D.P. Lusch. 2004. Optimal classification methods for mapping agricultural tillage practices. *Remote Sensing of Environment*. 91:90–97.

Stefanov, W.L., M.S. Ramsey and P.R. Christensen. 2001. Monitoring urban land cover change: An expert system approach to land cover classification of semiarid to arid urban centers. *Remote Sensing of Environment*. 77:173–185.

Stehman, S.V. 2001. Statistical rigor and practical utility in thematic map accuracy assessment. *Photogrammetric Engineering and Remote Sensing*. 67:727–734.

Stone, B. 2004. Paving over paradise: how land use regulations promote residential imperviousness. *Landscape and Urban Planning*. 69:101–113.

Story, M. and R.G. Congalton. 1986. Accuracy assessment: a user's perspective. Photogrammetric Engineering and Remote Sensing. 52: 397–399.

Thomas, N., C. Hendrix, and R. Congalton. 2003. A comparison of urban mapping methods using high-resolution digital imagery. *Photogrammetric Engineering and Remote Sensing*. 69:963–972.

Tou, J.T., and R.C. Gonzalez. 1977. Pattern Recognition Principles. Reading, MA: Addison-Wesley, pp. 377.

Townsend, P.A., Lookingbill, T.R., Kingdon, C.C., et al., 2009. Spatial pattern analysis for monitoring protected areas. *Remote Sensing of Environment*. 113 (7), 1410–1420.

Treuhaft, R.N., B.E. Law, and G.P. Asner. 2004. Forest attributes from radar interferometirc structure and its fusion with optical remote sensing. *BioScience*. 54(6):561–571.

Turner, R.E. and M.M. Spencer. 1972. Atmospheric model for correction of spacecraft data. In: Proceedings of the Eighth Annual Symposium on Remote Sensing of Environment, Ann Arbor, MI: ERIM, pp. 895–934.

U.S. Geological Survey (USGS). 2006a. EO-1 Home—Earth Observing Mission 1 [web page]. http://eo1.gsfc.nasa.gov/ (last accessed 27 May 2011).

U.S. Geological Survey (USGS). 2006b. Seamless data distribution system [web page]. National Center for Earth Resources Observation and Science. http://seamless.usgs.gov (last accessed 2 June 2011).

U.S.Geological Survey (USGS). 2004. USGS National Land Cover Data [web page]. Sioux Falls: EROS Data Center. http://landcover.usgs.gov/prodescription.html

van Aardt, J.A.N., and R.H. Wynne. 2001. Spectral separability among six southern tree species. *Photogrammetric Engineering and Remote Sensing*. 67:1367–1375.

Van Genderen, J.L., and B.F. Lock. 1977. Testing land use map accuracy. Photogrammetric Engineering and Remote Sensing. 43, 9:1135–1137.

Verkere, J. and S. Khorram. 2015. Personal communication

Wald, L. 1999. Some terms of reference in data fusion. *IEEE Transactions on Geoscience and Remote Sensing*. 37(3):1190–1193.

Wang, H.Q., and E.C. Ellis. 2005. Image misregistration error in change measurements. *Photogrammetric Engineering and Remote Sensing*. 71:1037–1044.

Wang, S. and X. Wang. 2009. Local search algorithm for K-means clustering based on minimum sub-cluster size. In: Proceedings, CCPR 2009, Chinese Conference on Pattern Recognition, 4–6 November 2009, Nanjing, China, 1–4.

Wardlow, B.D., S.L. Egbert, and J.H. Kastens. 2007. Analysis of time-series MODIS 250 m vegetation index data for crop classification in the U.S. Central Great Plains. *Remote Sensing of Environment*. 108:290–310.

Wilson, J. and G. Lindsey. 2005. Socioeconomic correlates and environmental impacts of urban development in a central Indiana landscape. *Journal of Urban Planning. D-ASCE* 131:159–169.

Winker, D.M., M.A. Vaughan, A. Omar, Y. Hu, K.A. Powell, Z. Liu, W.H. Hunt, and S.A. Young. 2009. Overview of the CALIPSO mission and CALIOP data processing algorithms. *Journal of Atmospheric and Oceanic Technology*. 26(11):2310–2323.

Xiao, X., S. Boles, J. Liu, D. Zhuang, S. Frolking, C. Li, W. Salas, and B. Moore III. 2005. Mapping paddy rice agriculture in southern China using multi-temporal MODIS images. *Remote Sensing of Environment*. 95:480–492.

Yang, C. and P. Chung. 2002. Knowledge-based automatic change detection positioning system for complex heterogeneous environments. *Journal of Intelligent and Robotic Systems*. 33:85–98.

Yang, W., Z. Zeng, and S. Zhang. 2010. Application of combining watershed and fast clustering method in image segmentation. In: Proceedings of the 2nd International Conference on Computer Modeling and Simulation, 22-24 January 2010, Sanya, China, 170–174.

Yu, Q., P. Gong, N. Clinton, G. Biging, M. Kelly, and D. Schirokauer. 2006. Object-based detailed vegetation classification with airborne high spatial resolution remote sensing imagery. *Photogrammetric Engineering and Remote Sensing*. 72:799–811.

Yuan, H., C.F. van der Wiele, and S. Khorram. 2009. An automated artificial neural network system for land use/land cover classification from Landsat TM imagery. *Remote Sensing* 1: 243–265.

Zadeh, L.A. 1965. Fuzzy sets. *Information and Control*. 8(3):338–353.

Zhang, K., S.C. Chen, D. Whitman, M.L. Shyu, J. Yan, and C. Zhang. 2003. A progressive morphological filter for removing nonground measurements from airborne LIDAR data. *IEEE Transactions on Geoscience and Remote Sensing*. 41(4):872–882.

Zhang, K., J. Yan, and S.C. Chen. 2006. Automatic construction of building footprints from airborne LIDAR data. *IEEE Transactions on Geosciences and Remote Sensing*. 44(9):2523–2533.

Zhang, Q., J. Wang, X. Peng, P. Gong, and P. Shi. 2002. Urban built-up land change detection with road density and spectral information from multi-temporal Landsat TM data. *International Journal of Remote Sensing*. 23:3057–3078.

Zheng, J., Z. Cui, A. Liu, and Y. Jia. 2008. A K-means remote sensing image classification method based on AdaBoost. In: Proceedings of the 4th International Conference on Natural Computation, 18-20 October 2008, Jinan, China, 4: 27–32.

Zitová, B., and J. Flusser. 2003. Image registration methods: a survey. *Image and Vision Computing*. 21:977–1000.

Suggested Reading

Dai, X. and S. Khorram. 1997. "A New Automated Land Cover Change Detection System for Remotely-Sensed Imagery based on Artificial Neural Networks," Proc. IEEE/IGARSS'97 International Geoscience and Remote Sensing Symposium, Singapore.

Jain, A.K. 1989. "Fundamentals of Digital Image Processing," Englewood Cliffs, NJ: Prentice Hall, pp. 418–421.

Jensen, J.R. 2005. Introductory Digital Image Processing, 3rd edition. Upper Saddle River, NJ: Pearson Prentice Hall. Pp. 316.

Lillesand, T., R. Kiefer, and J. Chipman. 2008. Remote Sensing and Image Interpretation, 6th edition. New York: John Wiley and Sons. pp. 763.

Lunetta, R. L. and J.G. Lyons (Eds.), 2003, Geospatial Data Accuracy Assessment, Las Vegas: US Environmental Protection Agency, Report No. EPA/600/R-03/064, pp. 335.

Zadeh, L.A. 1965. Fuzzy sets. Information and Control 8(3): 338–353.

Chapter 4
Terrestrial Applications of Remote Sensing

Each day, millions of individual images and observations collect an enormous variety of information about the Earth's surface and subsurface. This routine surveillance enables the monitoring and modeling of ecosystem health, detecting seismic activity, identifying surface vegetation, promoting sustainable agriculture, and characterizing the physical and social vulnerability of human settlements.

Continuous advances in satellite remote sensing and geospatial technology have dramatically accelerated our understanding of environmental phenomena—extracting data which is often invisible (or less visible) from the ground—allowing us to develop innovative solutions to ever more complex social, environmental, and economic challenges. Decreasing costs of computing technology and satellite imagery; increasingly robust proprietary and open source software; rapid availability of geospatial data easily acquired online; higher quality image resolutions (e.g., under 1 m); and the ever-greater number of satellite sensors have all contributed to transformational changes in every field of study and professional practice.

Terrestrial applications of remote sensing include mapping, change detection, monitoring, modeling, and other observations of land uses stemming from human activities. For example, remote sensing has shown great promise to map and monitor land use and land cover change (LULCC) in an efficiently and cost-effective manner (see Gong et al. 2013; Hansen et al. 2008; Hester et al. 2009; Khorram et al. 2011 2012; Townsend et al. 2012). Utilizing remotely sensed data may be the only viable mechanism for tracking LULCC changes at the subregional, regional, and global scales. An exhaustive review of terrestrial remote sensing applications is beyond the scope of this chapter. However, this chapter does provide examples of demonstrative case studies of the diverse and exciting ways that remote sensing has been put into use in tackling difficult local, regional, and global challenges.

S. Khorram et al., *Principles of Applied Remote Sensing*,
DOI 10.1007/978-3-319-22560-9_4

4.1 Classifying Land Use and Land Cover

One of the most frequently used applications of remote sensing is in the creation of land use and land cover (LULC) maps (Khorram et al. 2012). LULC maps are created by taking an image and categorizing natural and artificial features into representative groups or "classes". Resource managers and researchers use LULC maps to study everything from plant composition to human settlement patterns, as well as to answer questions such as "how is vegetation changing and why?" Several large-scale LULC studies include the European Environmental Agency's classification of 25 European Union member states and other European countries (Fig. 4.1) and the UN Food and Agricultural Organization's (UNFAO) LULC mapping of the Sahel region of West Africa (Fig. 4.2). In addition, new initiatives are increasingly being

Fig. 4.1 The European Environmental Agency's CORINE 44-category land use land cover map of European countries. (Image courtesy of EEA)

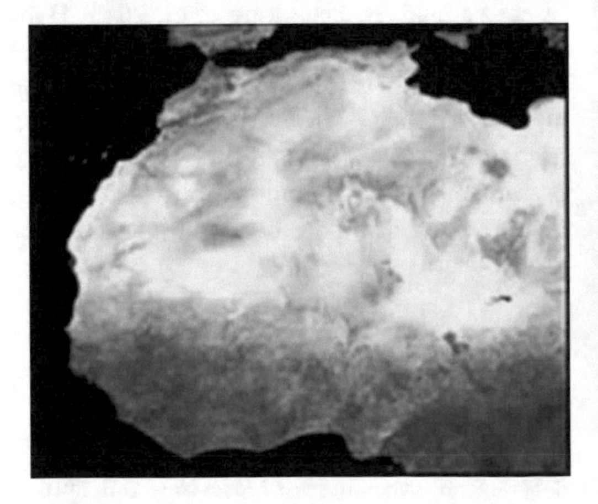

Fig. 4.2 The UNFAO's Africover Project monitors land use and land cover changes in the Sahel region of West Africa. (Image courtesy of UNFAO)

taken by agencies acting at national, regional, or global level in an effort to establish a base data layer for LULC to enable the monitoring of LULCC at all scales (Ioannis and Braun 2014).

In 2010, China initiated the Global Land Cover (GLC) mapping project, the aim of which was to produce a 30 m GLC data product (GlobeLand30) with ten land cover classes for the years 2000 and 2010 within a 4-year period (Chen et al. 2011). Over 10,000 Landsat-like satellite images were combined to cover the entire Earth at 30 m resolution. The project thus far has produced many interesting datasets relating to various aspects of land cover and its relation with the availability of water and other natural resources. As always, accuracy is an important consideration. According to Gong et al. (2013), the first attempt to conduct an accuracy assessment on the GLC dataset was an experimental evaluation on the usability of existing automated classification techniques (Gong et al. 2013). Four classifiers, that is, Maximum Likelihood Classifier (MLC), J4.8 Decision Tree Classifier (DT), Random Forest Classifier (RF), and Support Vector Machine (SVM) were tested with more than 8000 images captured during the year 2000. It was found that the highest overall classification accuracy (OCA) was produced by SVM, at 64.9% as shown in the following Fig. 4.3.

There are other major programs of interest at national level in the US involving LULCC mapping. A good example is the Multi-Resolution Land Characteristics (MRLC) Consortium involving a number of the US Federal agencies. Starting in the mid-1990s, MRLC has grown into a group of ten agencies that coordinate the production of five different products, including the National Land Cover Database (NLCD), the Coastal Change Analysis Program (C-CAP), the Cropland Data Layer (CDL), the Gap Analysis Program (GAP), and the Landscape Fire and Resource Management Planning Tools (LANDFIRE). As described by Wickham et al. (2014), these products include almost every aspect of land cover from impervious surface to detailed crop and vegetation types to fire fuel classes. Some products can be used for land cover change assessments because they cover multiple time periods. The MRLC Consortium has become a collaborative forum, where members share research, methodological approaches, and data to produce products using established protocols, and we believe it is a model for the production of integrated land cover products at national to continental scales.

Another example is the North American Landscape Characterization (NALC) project that has produced LULC data for the entire US. The NALC project was sponsored by the US Environmental Protection Agency's (EPA) and the US Geological Survey, EROS Data Center. This project is a component of the National Aeronautics and Space Administration's (NASA) Landsat Pathfinder program of experiments to study global change issues (Lunetta et al. 1998). Figure 4.4 shows the land cover classification and accuracy assessment for Region 5 Great Lakes; Khorram et al. 2004).

At the regional scale, monitoring urban growth is one of the questions social scientists, urban planners, and decision makers deal with most frequently. The direct impacts of urban expansion on physical, ecological, and social resources have made research on urban sprawl an issue of increased interest. Traditional census sources are extremely useful in that they capture changes in the socioeconomic and demographic structure of cities, but they lack spatial details and are not frequently updated.

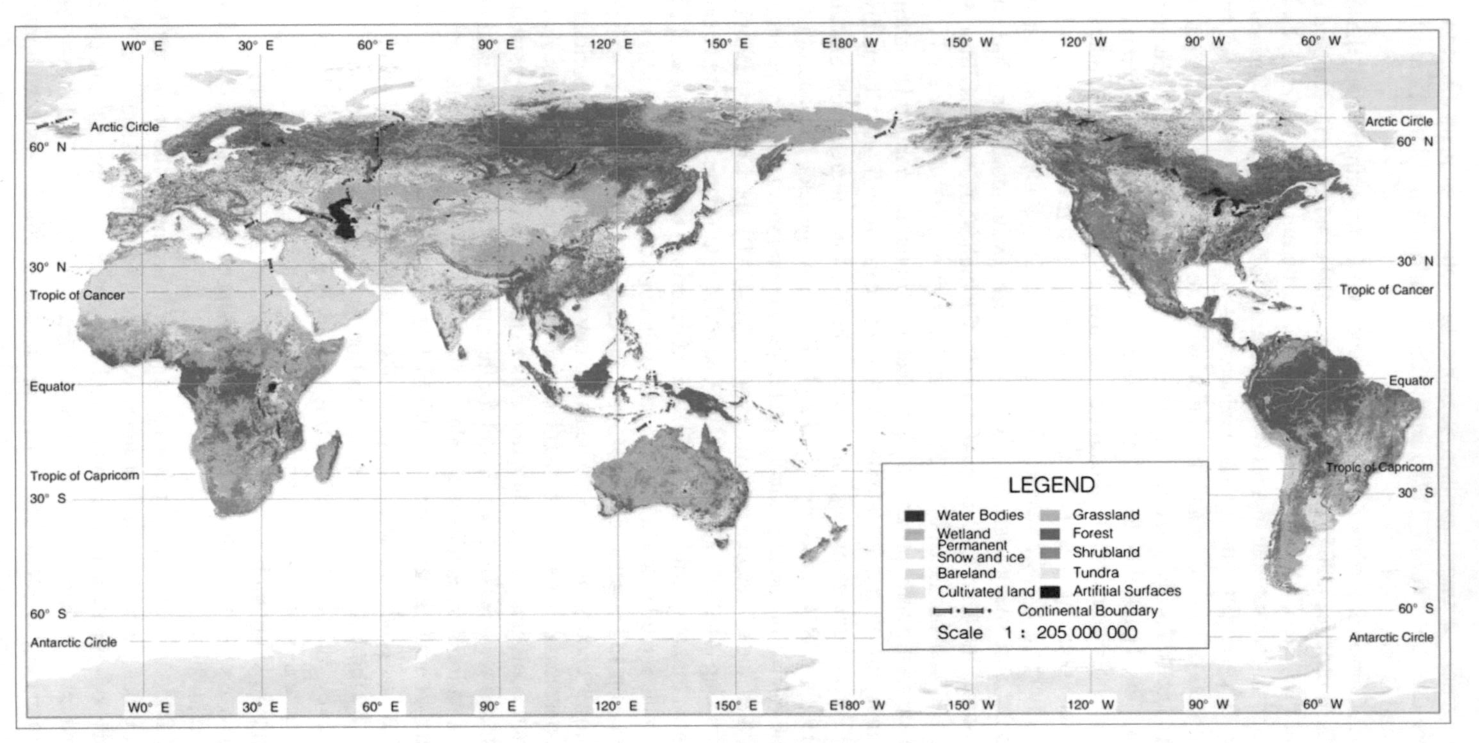

Fig. 4.3 The global land cover map with the accuracy of 64.0%. (Gong et al. 2013)

Fig. 4.4 The accuracy assessment of the Region 5 LULC map

Remote sensing, in contrast, makes available a vast amount of data with continuous temporal and spatial coverage and can, therefore, provide a successful means for monitoring urban growth and changes. Current research is oriented toward the use of remotely sensed data in conjunction with demographic and socioeconomic data to study urban change dynamics or to better understand the spatial distribution of population and socioeconomic phenomena.

This has been particularly important in developing countries where many cities are experiencing a rapid increase in population and consequential urban expansion, and where remote sensing may provide fundamental observations of urban growth that are not available from other sources.

Most of the research on urban growth is still based on traditional land cover classifications and change detection methodologies, but recently there has been a shift toward the development of integrated datasets that can be used in planning and urban monitoring efforts. Such relatively new trends include, for example, the development of urban environmental quality and quality-of-life indices to monitor physical and social changes in cities.

Turning to the local scale, Hester et al. (2009) produced LULC maps from high-resolution satellite (Quickbird) data of an area in Raleigh, North Carolina that was subsequently used for impervious area estimation, land use change detection, and water quality studies. The impervious surface area maps of this study are shown in Fig. 4.5.

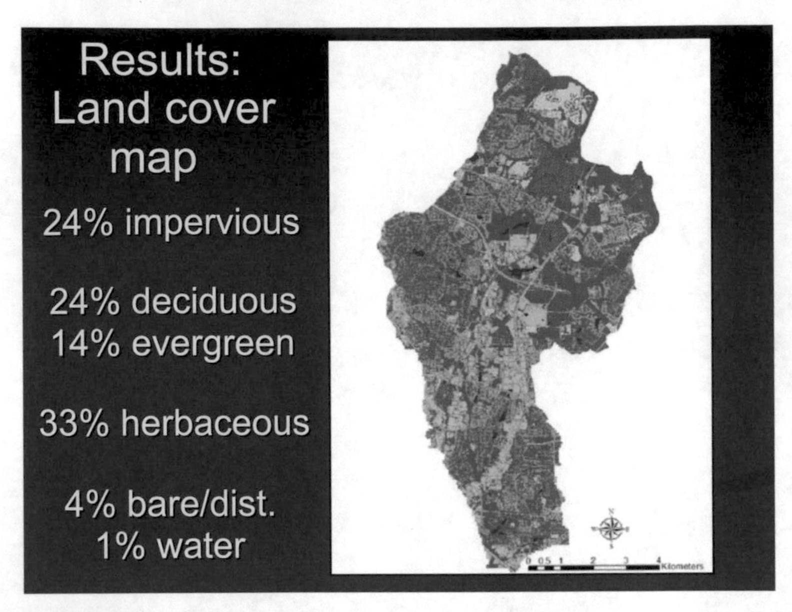

Fig. 4.5 The impervious surface area map based on the land use land cover map from Quickbird satellite data

4.2 Understanding and Protecting Biodiversity Through Wildlife Tracking

Our understanding of large animal movement, migration patterns, and habitat utilization patterns are largely tied to the landscape. As these animals move over large geographic areas, our understanding of their movement patterns can often be quite limited. Common approaches have involved tagging and recapturing animals in different locations, whereas more modern approaches involve monitoring active animal movement with radio telemetry applications. Newer approaches include the process of "*bio-logging*". Bio-logging, which uses small animal-attached tags for recording and/or relaying of data about an animal's movements, behavior, physiology, and/or environment, has led to important new discoveries about migratory patterns, habitats, breeding and wintering grounds, speed of migration, orientation, the influence of weather conditions and climate change, and an enormous host of other factors.

Studies conducted in the field offer unique opportunities to observe natural phenomena; however, there are many challenges—from being prohibitively expensive to logistically difficult. Wildlife biology has advanced through an influx of technology in which audio and video recordings, GPS, and satellite remote sensing are important aids in gaining biological and physical information from free-ranging species. The ARGOS satellite system, created in 1978 by the French Space Agency (CNES) to collect and relay meteorologic and oceanographic data, is one of the most established satellite-based systems used to track mammals, marine animals, and birds across the globe.

Direct remote sensing from aerial platforms has also been applied to larger animal movement studies. Often, all of these approaches are limited to observing only a few animals at a time, and within a manageable/limited spatial scale, due to cost, the labor involved, and the time necessary to complete the study. Satellite remote sensing, despite its potential to provide data over a large spatial scale and at a relatively low cost, has been applied a lot less to larger animal inventories and tracking animal movement studies. However, large, region-scale, satellite imagery has been useful in monitoring associated vegetation and landscape usage of animals.

Verlinden and Masogo (1997) used Normalized Difference Vegetation Index (NDVI) data products generated from the Advanced Very-High-Resolution Radiometer (AVHRR-NOAA) sensor to correlate habitat greenness estimates with species density distributions of ostrich populations (*Struthio camelus*) and large, grazing herbivores: wildebeest (*Connochaetes taurinus*), hartebeest (*Alcelaphus buselaphus*), eland (*Taurotragus oryx*), and gemsbok (*Oryx gazella*). This study was carried out in the semi-arid region of the Kalahari Desert, Africa. The Kalahari Desert comprises an area of over 900,000 km^2 and encompasses much of Botswana, and parts of Namibia and South Africa. Larger animal species in this region must travel over large distances in search of grass "greenness" patches in order to fulfill their water and nutrient requirements. The green patches of grasses result from rainfall, which may be highly variable both temporally and spatially. For this study computer-generated density and distribution models were developed for each species, coupled with rainfall data and field observations made from aircraft (Verlinden and Masogo 1997). The generated density and distribution estimates were compared to NDVI data. The NDVI data used consisted of 10-day composite data and were derived from corrected and georeferenced Global Area Coverage (GAC) data with a resolution of approximately 4×4 km. The GAC dataset is comprised of a reduced resolution image data that is processed while onboard the AVHRR satellite platform. The AVHRR-GAC data are resampled by extracting one line out of every three captured and averaging four of five adjacent samples along the sensor's scan line. This process results in a reduced resolution image apart from the original 1.1 km resolution image data.

The Verlinden and Masogo (1997) NDVI model results showed low, but significant, correlations to aerial survey greenness scores (Spearman rank correlation was 0.43, $p < 0.001$). NDVI model results also showed low, but significant, correlations to ground-sampled greenness scores (Spearman rank correlation was 0.33, $p < 0.001$). These significant positive relationships between NDVI and grass greenness (both aerial survey and ground sampled) suggested that, with some caution, NDVI estimates could be used in this area to monitor greenness. Relationships between NDVI and animal distribution proved more difficult to determine. However, the NDVI-greenness correlations may show a means of assessing habitat suitability, on a regional scale in the Kalahari, for a few of the species examined (i.e., wildebeest, hartebeest, and ostrich).

Researchers and environmental managers may be beginning to view image analysis applications derived from remote sensing as an efficient and promising means of confronting data gaps in inventory challenges, status and trend monitoring, recovery planning, and predictive assessments of species biodiversity and wildlife population management (see Franklin 2010; Kerr and Ostrovsky 2003; Turner 2003). Current limitations of spaceborne remote sensing and the development of effective management approaches in tracking and monitoring animal movement include identifying an appropriate scale of study, as well as the availability and appropriateness of the type and resolution of the remotely sensed data or remotely sensed data product required. However, with the continuing evolution of sensor technologies ever expanding, as well as higher sensor ground spatial resolutions becoming more commonly available, the ability to use spaceborne sensors may provide an efficient means of investigating animal movements over larger spatial scales (Nelson 2010). The potential of integrating remote sensing and geospatially derived data-driven analyses may additionally have a huge impact on the public, as well as on professionals within the field, and further increase the possibility for the use of these technologies to create opportunities for greater public and scientific discussion (Franklin 2010).

4.3 Water Resources

Surface water supply, reservoir mapping and monitoring, water quality mapping and modeling, runoff forecast modeling, drainage network mapping, hydrologic research, watershed characterization, and surface water impairment are all water-resources-related candidates for remotely sensed data applications (Khorram 1982).

Changes in land use and land cover may have severe impacts on aquatic ecosystems. Activities such as agriculture, urbanization, and erosion are primary sources of sediment and nutrient input into water bodies (Dahl 1990; Feierabend and Zelazny 1987; Sasser 1986). A few examples of the consequences of increased sediment and nutrient loading on water bodies include the loss of littoral or coastal zone habitat, eutrophication, depletion of oxygen, and changes in food webs dynamics (Carpenter et al. 1998; Petersen 1991; Wear et al. 1998).

Nonpoint source pollution (NPS) is the term used to describe loadings to water bodies from diffuse sources within the landscape. These sources make up nutrient and chemical pollutants from developed areas, construction sites, paved surfaces, cultivated fields and pasture lands, and atmospheric deposition. However, despite the variety of pollutants classified as NPS, sediments are the largest amounts of material transported to surface waters (Loehr 1974; Tim and Jolly 1994). Approximately 60 % of all sediment and sediment-associated contaminants delivered to the US's rivers, lakes, and coastal regions are from agricultural soil loss (Madden et al. 1987; Petersen et al. 1991).

Remote sensing provides a unique technological advantage in monitoring aquatic systems due to its ability to capture data over large areas, and assorted spatial and temporal resolutions. However, satellite remote sensing of aquatic systems has been less studied as a result of the difficulties inherent in interpreting reflectance values of water (Verbyla 1995). Clear water provides little spectral reflectance because longer wavelengths are absorbed and the reflected shorter

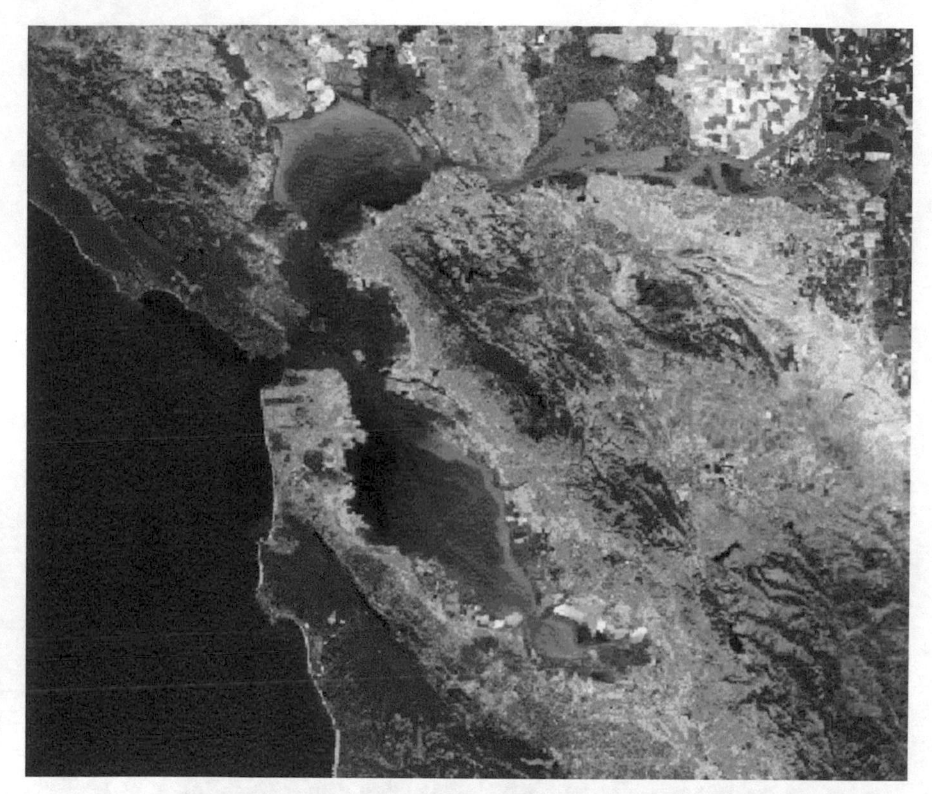

Fig. 4.6 Due to the absorption of near-infrared wavelengths, most deep and clear water bodies appear *dark* and depending on the turbidity levels in various shades of *blue* on this unclassified Landsat imagery of San Francisco Bay Area

wavelengths, which sensors rely on for surface feature detection, are subject to higher atmospheric scattering. Due to the absorption of longer wavelengths, most deep and clear water bodies appear dark on unclassified satellite imagery (Fig. 4.6). Sensors, such as the Landsat Multispectral Scanner (MSS) and Thematic Mapper (TM), were primarily designed for land studies, and thus are not completely applicable to aquatic studies (Nelson et al. 2006). Recent improvements in sensor technologies have allowed for better spatial and spectral resolutions than previously available, which may improve their usefulness to aquatic studies (Kloiber et al. 2000).

In a study by Nelson et al. (2003), Landsat-7 data were used to investigate the relationship between spectral values extracted from Landsat Enhanced Thematic Mapper (ETM+) data and water clarity using Secchi disk transparency (SDT) measurements The SDT measurements incorporated the range of SDT values found within a geographic region comparable to 80 % of the lakes within the lower peninsula of Michigan, US. The Secchi depth is inversely proportional to the absorbance of light by water and dissolved or particulate substances, such as algal biomass.

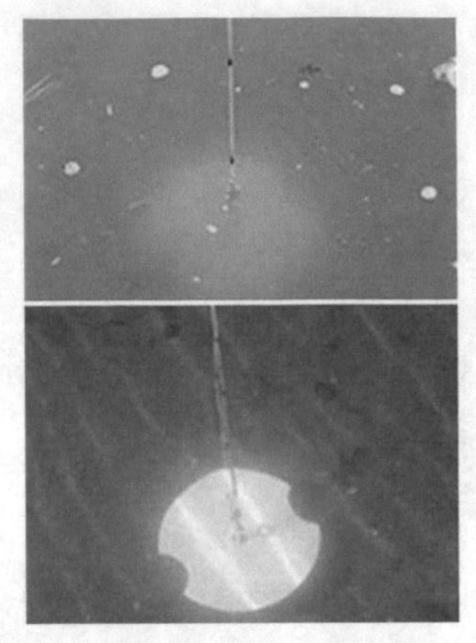

Fig. 4.7 The physical appearance of a water body with a shallow Secchi disk transparency (SDT) value versus deep SDT value can be very different. The image on the *top* represents a shallow SDT measurement where the water may be discolored due to high algal biomass content, sediment, etc. The image on the *bottom* represents a deep SDT measurement where the water appears clearer. Thus, the measurement of SDT values can provide important information as an indication of water clarity

This study represented an assessment of water clarity on a regional scale using remotely sensed data and identified an important consideration necessary for modeling water clarity on such a large scale—models developed for large regions must incorporate the full range of lake SDT values found throughout the region in the model calibration (Nelson et al. 2003). Models that incorporate this complete range of SDT values, such as the one developed in this study, may not produce statistical relationship results as high as models that are developed for groups of lakes that do not incorporate the full regional range of SDT values (Dekker and Peters 1993; Kloiber et al. 2000; Lathrop and Lillesand 1986). In addition, these results suggest that Landsat may be better suited for predicting shallow SDT values (i.e., eutrophic-higher algal biomass) than clearer water containing deeper SDT values (Fig. 4.7). These types of models, based on remotely sensed data, have the capacity to provide managers with a method to inventory shallow SDT lakes across the entire state where in situ data are difficult to obtain. In particular, regulatory and management agencies would be able to monitor private lakes without spending resources on these lakes and without requiring the permission of the property owner. Additionally, models developed from remotely sensed data may be useful in the historical analysis of watershed changes that may have caused lake eutrophication to occur over time. This application is a very attractive use of remote sensing since historical imagery may be obtained for aquatic systems that have become eutrophic over time.

Beyond this specific study, despite current limitations, remote sensing does possess the potential to provide a valuable tool for monitoring freshwater systems. Several studies have successfully used remote sensing to measure characteristics of aquatic systems such as chlorophyll and SDT suspended sediments, and dissolved organic matter (Dekker and Peters 1993; Jensen et al. 1993; Khorram and

Cheshire 1985; Kloiber et al. 2000; Lathrop and Lillesand 1986; Lillesand et al. 1983; Narumalani et al. 1997; Nelson et al. 2003). Other studies have used remote sensing to map aquatic plant abundance in large areas of homogenous cover or for species detection in smaller areas they have used high-resolution hand or aircraft sensors (Ackleson and Klemas 1987; Armstrong 1993; Lehmann and Lachavanne 1997; Nelson et al. 2006; Penuelas et al. 1993). While remote sensing has been used in large-scale land change studies for many years, few studies have used currently available remotely sensed data to develop standardized methods for measuring water quality parameters across a wide range of aquatic environments, over large regions, or covering multiple images.

Accessing water quality parameters on a regional scale provides a challenge in that regional remote sensing requires that multiple remotely sensed images be combined from multiple dates or regions of varying atmospheric influences (i.e., haze, cloud cover, etc.). Multitemporal image-to-image differences are a result of atmospheric absorption and light scattering, which can be highly variable from one period of time to the next (Moore 1980). This difference can have a pronounced effect on the interpretation of imagery for trend analyses when using multiple images that are collected on different days, months, or years. Song et al. (2001) found that atmosphere corrections are necessary when using two or more satellite images collected on different dates, such as imagery used in land use/cover change studies. For example, for aquatic studies, Brivio et al. (2001) found that without the appropriate corrections for atmospheric scattering and transmittance effects, successful remote sensing of water quality parameters could not be accurately achieved from four Landsat TM images. Brivio et al. (2001) were able to develop two image-based rectification models which were used to correct regional atmospheric effects in multi-seasoned Landsat images of Lake Lseo and Lake Garda, Italy. However, accurate and simple procedures to correct for these effects are still an area of future research (Rahman et al. 2001; Song et al. 2001).

These atmospheric limitations may be reduced by many recent technological advances. A few of the newest sensors include IKONOS, the Earth-Observing System (EOS) Terra sensors, and the Landsat ETM+. These sensors will ultimately employ more advanced data calibrations than was possible with some of the older sensors. However, the advantage of the older Landsat platforms is that they provide consistent historical data not yet available from new sensors, and the temporal and spatial resolutions provided by Landsat MSS and TM are still more effective in monitoring landscape changes. For example, since the early 1980s, remote sensing techniques have become commonly used to detect wetland change (Ricketts 1992). However, few studies have taken advantage of the wealth of historical data available from the satellite platforms that have been operating across several decades, and that are becoming widely available and inexpensive (Munyati 2000). Images obtained using the Landsat MSS sensor presently represent our longest, continuous satellite remotely sensed historic record.

The availability of newer data will, however, allow for the integration of cross-platform data in developing band and image sharpening algorithms which will serve to fill in the gaps present in the older system data acquisitions (clouds, haze, etc.)

and enhance the ability to develop multi-image calibrations for cross referencing. Furthermore, three of the main Terra sensors (MODIS—Moderate Resolution Imaging Spectrometer, ASTER—Advanced Spaceborne Thermal Emission and Reflectance Radiometer, and MISR—Multi-angle Imaging Spectroradiometer) have been identified as being particularly useful in the remote sensing of large-area land cover and land use changes and vegetation dynamics studies (Jensen 2000; NASA 1998). However, aquatic remote sensors have yet to see the production of sensors specifically designed to monitor aquatic environments.

Currently, Landsat provides one of our richest data sources for the remote sensing of historical changes in large areas of wetlands, water quality, and aquatic plants. This is evident from the high revisit cycle, spatial resolutions, and 30 years of operation provided by these sensors.

4.4 Forest Resources

Applications of remotely sensed data to forest resources and environmental management are numerous. Examples may include any of the following applications:

- Forest canopy extent and structure mapping
- Forest health mapping and monitoring
- Timber inventory
- Forest-type mapping
- Timber stand conditions assessment
- Harvesting and procurement priority implementation
- Reforestation
- Deforestation

Deforestation, the clear-cutting of forests for subsequent conversion to other land uses, is common throughout much of the world, especially in tropical regions (Asner et al. 2006; Curran et al. 2004; Curran and Trigg 2006; Hansen et al. 2008). During the past decade, the five countries with the highest rates of forest loss (Brazil, Papua New Guinea, Gabon, Indonesia, and Peru) lost a combined 3.6 million hectares of primary forest each year (Koh et al. 2011). Large-scale deforestation has major implications for ecosystem function and biodiversity, as well as climate change, since it affects regional carbon storage and sequestration (Hansen et al. 2010; Kuemmerle et al. 2009; Laporte et al. 2007; Linkie et al. 2004). In addition, selective logging is widespread in rapidly deforesting areas such as the Amazon, further exacerbating the ecological impacts (Asner et al. 2006; Laporte et al. 2007).

Deforestation is typically driven by economic pressure (i.e., a perceived economic incentive for forest conversion rather than conservation) and is usually facilitated by limited governmental oversight in the affected regions (Curran et al. 2004; Koh et al. 2011; Kuemmerle et al. 2009; Laporte et al. 2007). Because of the lack of regulatory enforcement, illegal logging is often rampant, encroaching upon (and reaching into) supposedly protected forest areas (Curran et al. 2004; Linkie et al. 2004). Unfortunately, governments in those world regions with the highest deforestation rates typically lack resources for monitoring or predicting patterns of forest conversion (Linkie et al. 2004). However, a number of recent studies have shown that remote sensing can be a cost-effective approach for monitoring deforestation (due to both legal and illegal logging) and for identifying management priority areas. The typical remote-sensing-based strategy is to analyze a time series of Landsat images (or similar moderate-resolution imagery) captured over a period of many years (Curran and Trigg 2006). Time series analyses of moderate-resolution satellite images have been used to study deforestation in many regions of the world: Southeast Asia (Curran et al. 2004; Koh et al. 2011; Linkie et al. 2004), the Amazon (Asner et al. 2006), Central Africa (Laporte et al. 2007), and Eastern Europe (Kuemmerle et al. 2009). Below, we highlight a few such applications.

Instead of focusing on actual changes in forest cover, Laporte et al. (2007) mapped the expansion of logging roads over a 27-year period (1976–2003) in the humid forests of Central Africa. Despite the persistence of cloud cover in this region (an estimated 1.6 % of which was completely obscured by cloud cover), the researchers typically had multiple images available to delineate an area; in total, they used more than 300 Landsat images to map their four-million-square-kilometer study area. The images were contrast stretched to optimize road detection, and then the roads were manually digitized. Basically, the researchers found that this manual approach outperformed semiautomated approaches, particularly when delineating older roads or working from older imagery. The researchers also used IKONOS imagery of a 25 km^2 area in the northern Republic of Congo to examine the finer-scale impacts of logging activities, finding that canopy gaps in logged areas were five to six times larger than gaps in adjacent unlogged areas.

Laporte et al. (2007) mapped nearly 52,000 km of road in the study region (Fig. 4.8). Overall, logging roads accounted for 38 % of the length of all roads. The most rapidly changing region was the northern Republic of Congo, where the rate of road construction increased from 156 km per year for 1976–1990 to 660 km per year after 2000. The researchers estimated that approximately 5 % of the total forested area was disturbed. In summary, this effort yielded the first comprehensive dataset on the status of logging in Central Africa, and so will serve as a key resource for future planning and management.

Deforestation in the Amazon region of South America is a well-known problem, but actually less than 1 % of the region was deforested prior to 1975 (Guild et al. 2004). Deforestation rates increased exponentially in subsequent decades because of economic incentives for conversion to agriculture (e.g., favorable credit policies for cattle ranchers; Guild et al. 2004; Moran 1993). The state of Rondônia in western Brazil has become one of the most rapidly deforested parts of the Amazon (Guild et al. 2004; Pedlowski et al. 2005). Many migrants have been brought to the state through government colonization projects (Pedlowksi et al. 2005), placing a huge burden on the state's remaining forests. While only 4200 km^2 of forest had been cleared in Rondônia in 1978, by 2003 the amount of cleared forest had increased to nearly 68,000 km^2 (Pedlowski et al. 2005).

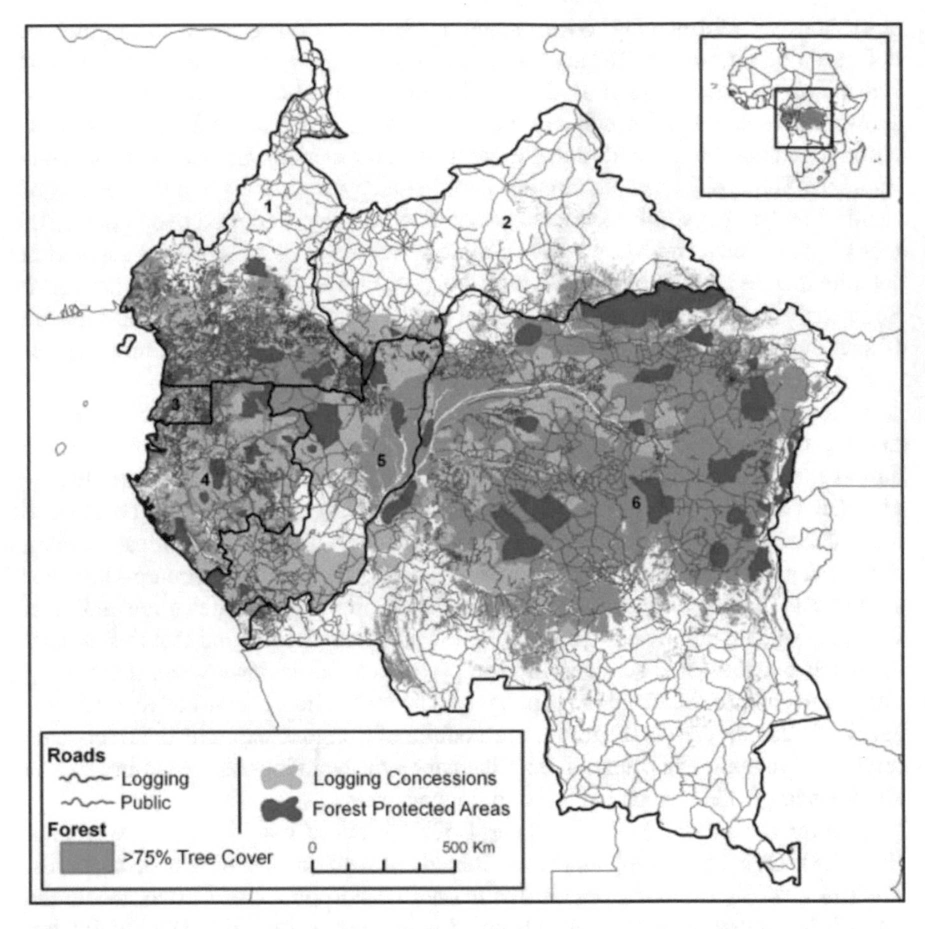

Fig. 4.8 Logging concessions and road distribution in Central Africa: Cameroon *(1)*, Central African Republic *(2)*, Equatorial Guinea *(3)*, Gabon *(4)*, Republic of Congo *(5)*, Democratic Republic of Congo *(6)*. (Image credit: Woods Hole Research Center; www.whrc.org)

The 2000 and 2010 MODIS true-color-composite images from the remote northwestern corner of Rondônia are shown in Fig. 4.9. The time series illustrates the typical pattern of initial deforestation along (both legal and illegal) roads, which permits small farmers to colonize previously inaccessible areas (NASA 2011). Over time, this leads to the establishment of large settlements and the clearing of most forest parcels. Figure 4.10 shows a map, developed using MODIS vegetation indices, of the intensity of vegetation change in this region between 2000 and 2008. Intensely red areas indicate the largest changes, typically the conversion of original rain forest into agricultural land (NASA 2011).

Like the state of Rondônia, the southeastern Asian nation of Papua New Guinea has similarly seen a large population increase in recent decades, combined with an increased demand for timber and other resources (Shearman et al. 2009). A team of researchers from the University of Papua New Guinea and Australian National

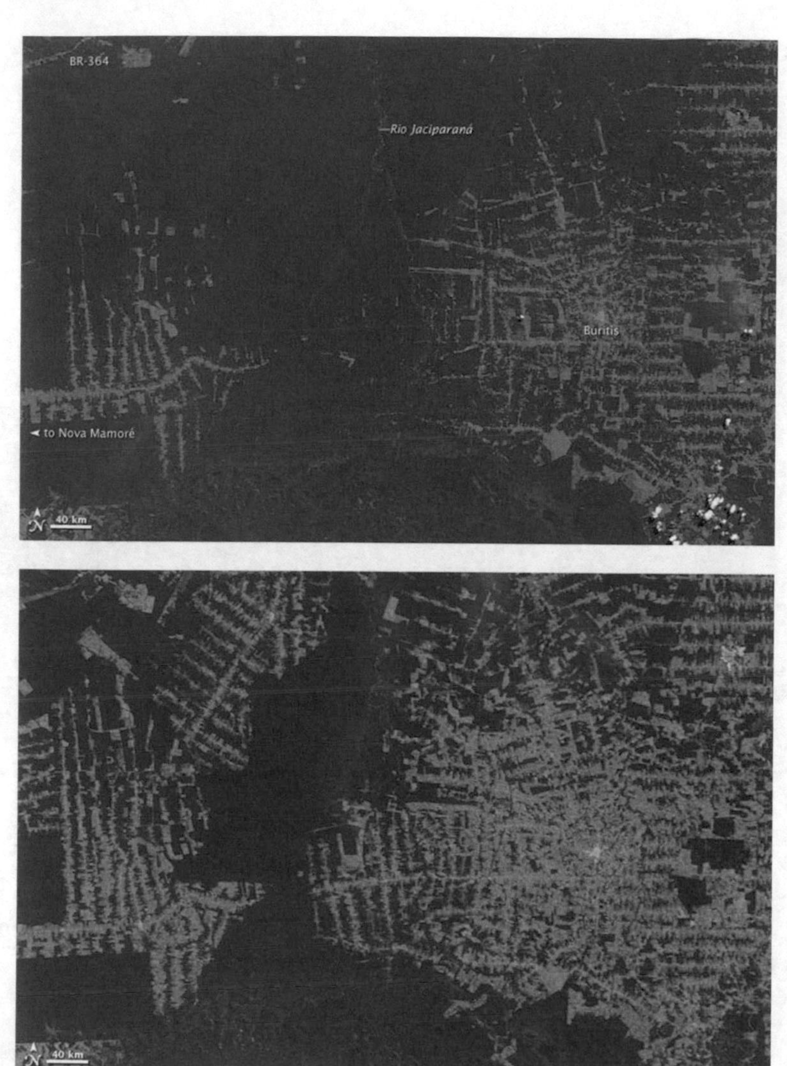

Fig. 4.9 MODIS true-color-composite images from northwestern Rondônia, Brazil, captured in 2000 (*top*) and 2010 (*bottom*). (Images courtesy of NASA)

University analyzed deforestation and forest degradation in the nation—home to the world's third largest rain forest (BBC 2008)—over a 30-year period. They used Landsat, Spot 4, and Spot 5 imagery to create a map of land cover as of 2002, which they then compared to a 1972 land cover map created by the Australian Army (Shearman et al. 2009). The researchers found that 15% of Papua New Guinea's forest had been cleared over the 30-year period, while an additional 9% was severely degraded by logging. They also found that the rate of forest conversion had generally risen since 1990; by 2001, the estimated deforestation rate was just over 3600 km^2 per year (BBC 2008; Shearman et al. 2009). The current rate of

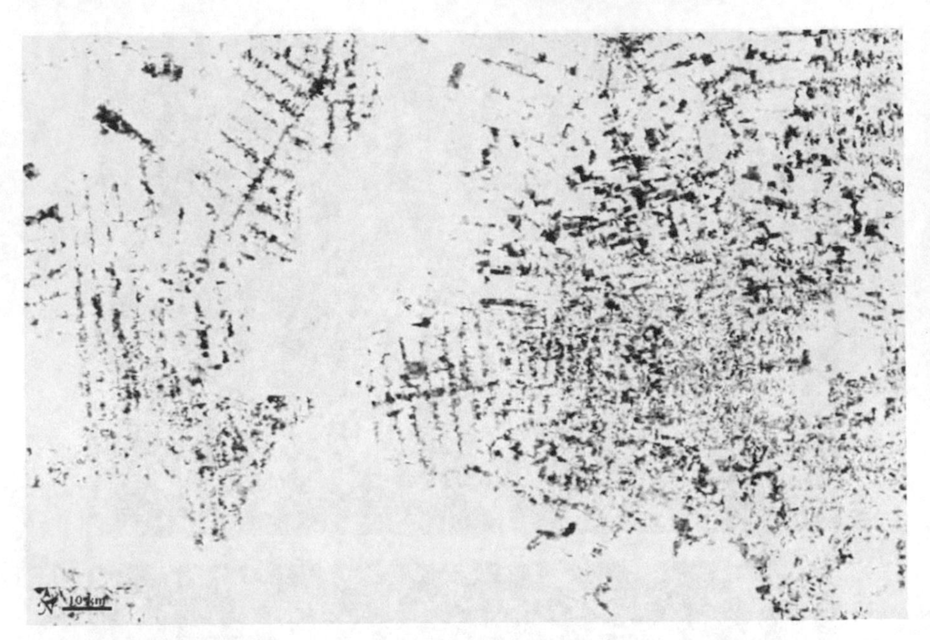

Fig. 4.10 Map of vegetation loss in northwestern Rondônia, Brazil, between 2000 and 2008. (Image courtesy of NASA, Robert Simmon, and Reto Stöckli)

conversion could result in the loss of more than half the nation's forest cover by 2021 (BBC 2008).

Two satellite images from Papua New Guinea's Gulf Province, the first captured in 1988 and the second in 2002, are shown in Fig. 4.11. Because Papua New Guinea is a mountainous island country, logging has mostly been limited to accessible areas (i.e., valleys and coastal lowlands). However, Shearman et al. (2009) estimated that 13 % of upper montane forests were also lost between 1972 and 2002.

4.4.1 Forest Health

In recent years, disturbances from extensive wildfire, drought, nonnative insect/pest species invasions, and climate change have threatened the health of forest throughout the globe. These threats, which can stretch across periods of time that last many years, may result in shifts in the dominant forest tree and animal species, as well as extensive forest mortality (Hargrove et al. 2012; Spruce et al. 2013, 2014). Satellite and aircraft remote sensing techniques have relied on the capture of spectral color and infrared data to access forest health. From acquired remotely sensed data, the NDVI has been used as a traditional tool for forest vegetation health assessments. The NDVI is an essential index of vegetation greenness that results from light interacting with the vegetation canopy. Chlorophyll pigments found in the leaves of healthy vegetation, that make up the forest canopy, interact with incoming solar

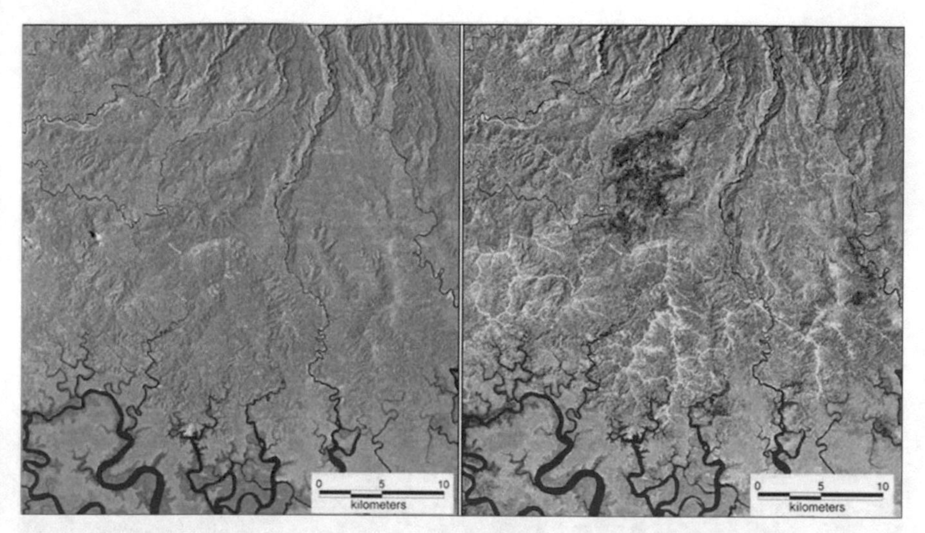

Fig. 4.11 Satellite images depicting the impact of logging in Papua New Guinea's Gulf Province. The 1988 image (*left*) shows intact rain forest, while the 2002 image (*right*) shows the impact of logging that began in 1995. (Images courtesy of the University of Papua New Guinea)

radiation by strongly absorbing visible light in the blue (0.45 μm) and red range (0.67 μm) of the electromagnetic spectrum, which is primarily used for the process of photosynthesis. Alternatively, light is reflected within the green range (0.5 μm) and even more so strongly reflected in the near-infrared portion (0.7–1.3 μm) of the electromagnetic spectrum. This strong reflection of light in the near-infrared portion of the electromagnetic spectrum is largely due to the leaf's internal structure. The response of the absorption and refection regions within the electromagnetic spectrum of the vegetation depicts the typical spectral signature curve for healthy vegetation when comparing the relationship of wavelength with percent reflectance (Fig. 4.12).

The spectral signature response curve for healthy vegetation may vary slightly over time as a plant canopy changes from early spring growth to late-season maturity and senescence. However, when vegetation becomes stressed as a result of disturbances, reflectance in the infrared range of the electromagnetic spectrum may also change, becoming significantly lower. In fact, stressed vegetation will typically reflect more light in the visible range of the spectrum, since less is being taken up for photosynthesis, and reflect less light in the near-infrared range. Thus assessing these differences (i.e., healthy vegetation vs. stressed vegetation) remote-sensing-derived NDVI results can reveal where vegetation is thriving and where it is under stress.

The US Forest Service, in collaboration with NASA-Stennis Space Center, DOE Oak Ridge National Laboratory, and the USGS EROS Data Center, developed a near-real-time regional monitoring system for forest disturbances (Hargrove et al. 2012; Spruce et al. 2013). This system is known as ForWarn and was developed in response to the 2003 US Healthy Forest Restoration Act, which mandated that a

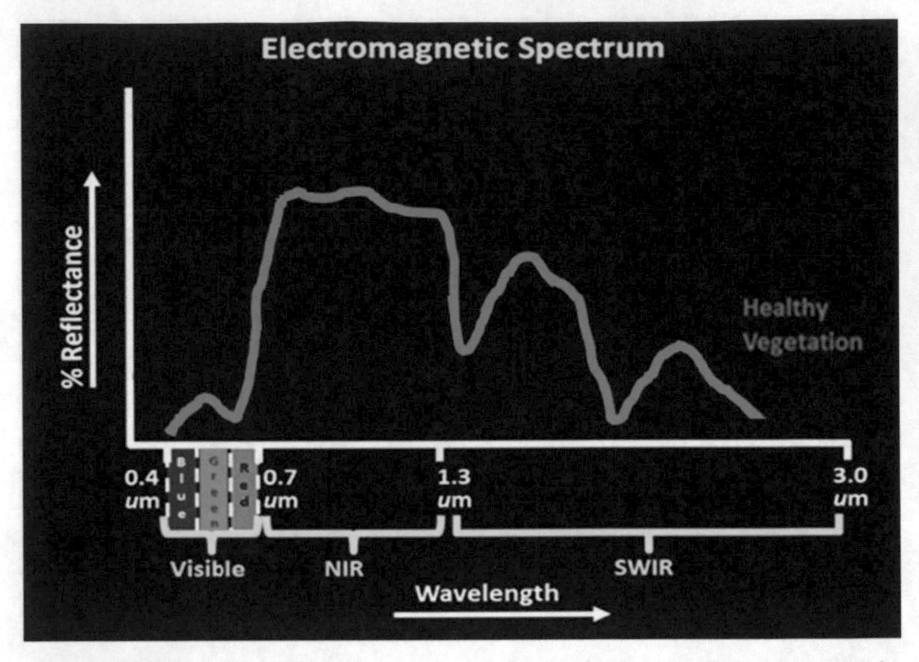

Fig. 4.12 The *green line* in the *above* graph represents a classic spectral signature curve for healthy vegetation

national forest threat Early Warning System (EWS) be enacted due to the increased extent and threat of forest disturbances across the country (Spruce et al. 2014). ForWarn started online operations in 2010 (http://forwarn.forestthreats.org/). This system uses multiple historical baseline datasets and satellite-derived NDVI values from the MODIS Aqua and Terra satellites (232 m ground spatial resolution) to derive weekly forest change products across the lower 48 United States (Fig. 4.13).

Data products are accessed through the ForWarn website, and ForWarn's geospatial data viewer provides user access to forest disturbance dynamics throughout the year of interest. ForWarn products have also been used to identify and track regional disturbances, including defoliation and mortality from insects, to drought, to wildfire, to severe storm events, and to human deforestation activities. Disturbance maps are available online through the ForWarn Change Assessment Viewer (http://forwarn.forestthreats.org/fcav; Fig. 4.14).

4.4.2 Biomass Estimation

The term biomass in the terrestrial environment, from an ecological perspective, often refers to the numbers, weight, or mass of living organisms. For vegetation, the term biomass typically refers to a dry weight calculation of the aboveground components of larger vegetation, such as grasses, shrubs, and trees, whereas smaller algae and fungi species are often less considered, particularly in satellite remote

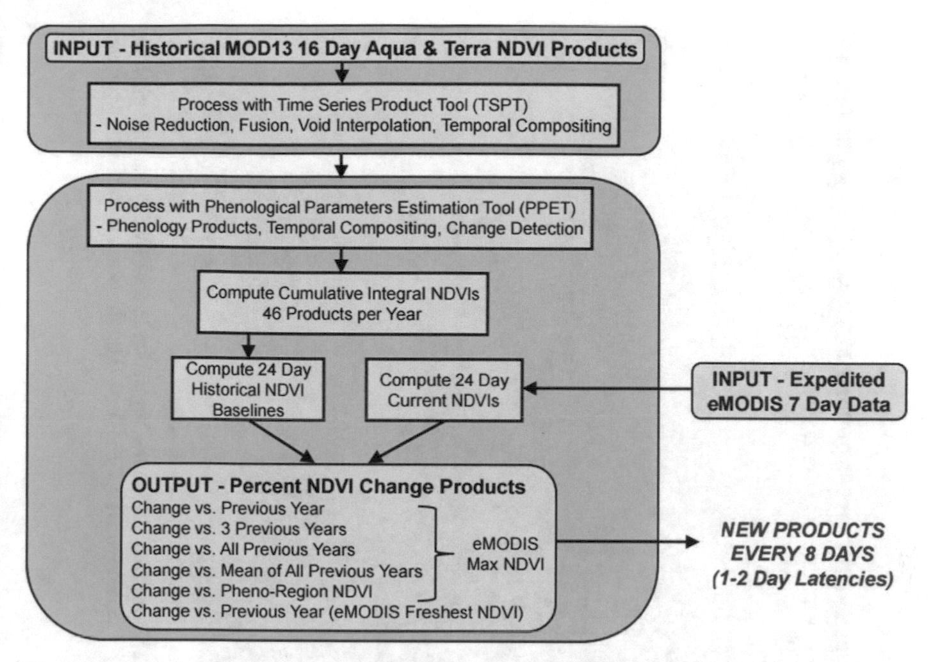

Fig. 4.13 Workflow procedure for developing ForWarn's MODIS Normalized Difference Vegetation Index (*NDVI*) change products. (NASA-Stennis Space Center)

sensing studies. For animals, few studies capture carnivores and herbivores with remote sensing. Thus, satellite remotely sensed estimates of terrestrial biomass largely include studies of the latter, macro-vegetation. However, estimates of vegetative terrestrial biomass can be ecologically very important and provide great capacities for assessing ecosystem biodiversity and implications for management strategies (Houghton et al. 2009). The definition of biomass typically implies the measurement of the mass of living biological organisms in a given area at a given time. This measurement may come in the form of weights, or even less commonly referred to as an in situ value, or natural state, such as standard vegetation indices (e.g., NDVI, etc.) as previously discussed. However, traditional methods of estimating of terrestrial biomass are usually made from field-based measurements collected over extremely small areas and usually nonrepresentative of a larger region in which synoptic sampling would be prohibitive (Houghton 2005; Houghton et al. 2009). Satellite remote sensing offers advantage in vegetative terrestrial biomass inventorying, monitoring, and analyses in that the platform's sensors have the unique ability for large-area, repetitive sampling.

TheNDVI (a ratio of red to near-infrared), a leaf area index (or LAI—a vegetation biophysical parameter that consists of a dimensionless variable and a ratio of leaf area to per unit ground surface area), and the fraction of vegetation cover (fCOVER) were used to evaluate the performance of SPOT imagery in estimating grassland biomass in a study by Dusseux et al. (2015). Linear regression models applied to 15 samples were developed to determine relationships between biomass measurements and variables estimated from SPOT sensor imagery. Biomass

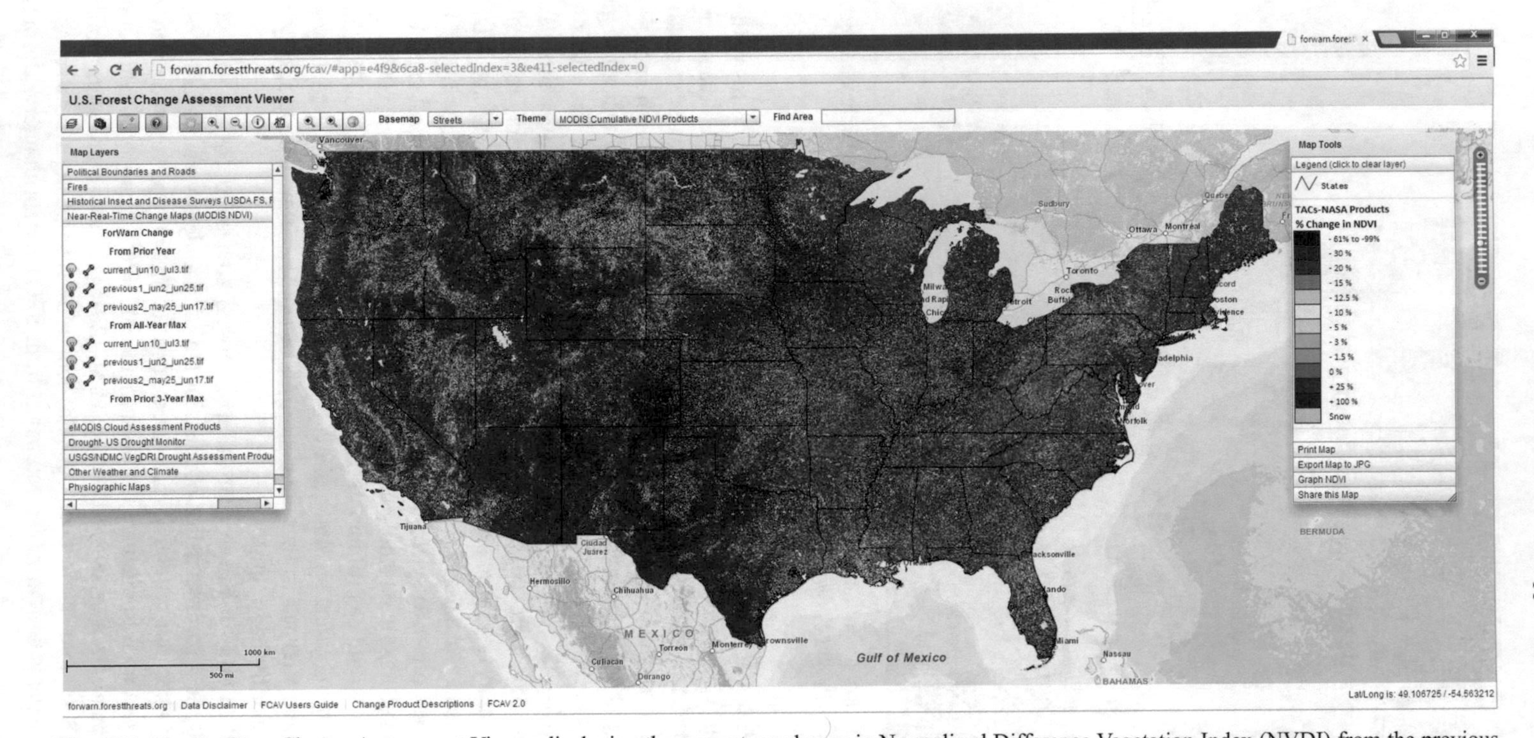

Fig. 4.14 The ForWarn Change Assessment Viewer displaying the percentage change in Normalized Difference Vegetation Index (NVDI) from the previous year (2014) to the current year (June 10-July 3, 2015). Derived from available MODIS Cumulative NDVI Products (http://forwarn.forestthreats.org/fcav). Extreme loss of vegetation is shown in shades of *orange, red,* to *darker red*

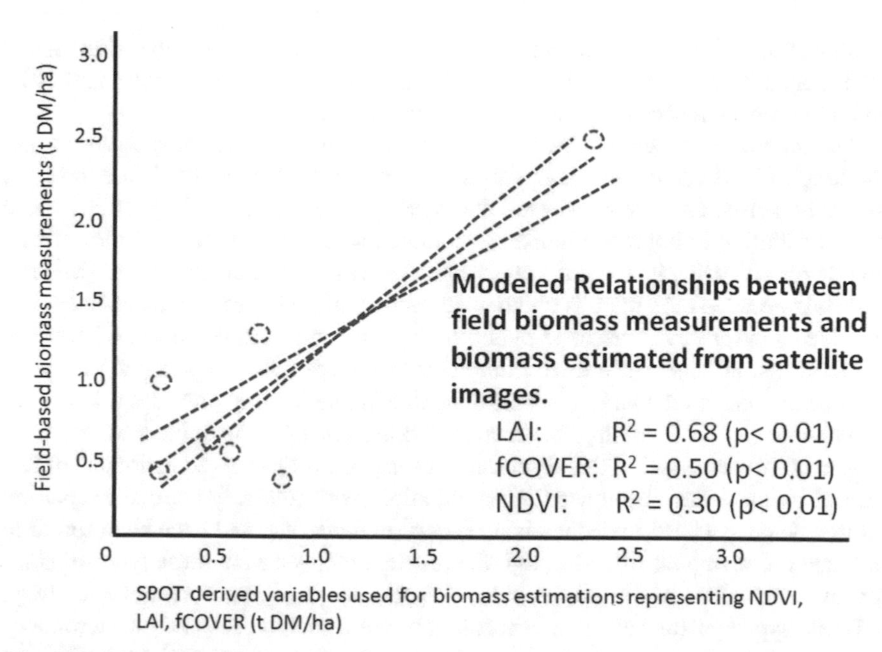

Fig. 4.15 Linear regression models showing relationship between SPOT-derived variables of biomass (Normalized Difference Vegetation Index (NDVI), leaf area index (LAI), fraction of vegetation cover (fCOVER); independent variables) and field-based measurements of biomass (dependent variables). Measurements expressed in tonnes of dry matter (t DM)/ha. (Graph adapted from Dusseux et al. 2015)

measurements were taken in the field from nine agricultural sites in the region of Brittany, in northwestern France, throughout the 2011 growing season. The remote sensing variables were developed from five cloud-free SPOT 4 images (20 m ground spatial resolution) that were acquired throughout the 2011 growing season. The image acquisition dates and field samples varied by a range of 2–11 days, with the imagery being corrected for both radiometric and atmospheric effects. The results of this study suggested that significant statistical relationships were possible between the observed biomass measurements and the remote-sensing-derived measurements (NDVI and LAI values; R^2 values of 0.30 and 0.68, respectively, at a level of significance $p<0.01$; see Fig. 4.15), thus providing improved estimates of grassland productivity corresponding to varying climate conditions.

Remote-sensing-derived estimates of aboveground biomass have also been implemented by other recent studies. Latifi et al. (2015) developed biomass estimate of stratified and non-stratified forest stands using a combination of data from LiDAR (2 m height threshold) and an airborne hyperspectral data scene (4 m pixel size, 125 spectral bands). For further details on the use of hyperspectral data and combined LiDAR data, please see Cocks et al. 1998; and Latifi et al. 2012. Finally, Barrachina et al. (2015) also used linear regression analyses based on field samples and vegetation indices (vegetation wetness index/VWI and NDVI) derived from Landsat-5 TM to estimate grass aboveground biomass in mountain meadows and pasture lands of the Pyrenees region of southwest France. Linear regression results (R^2) ranged between 0.76 (2008 and 2009 mid-summer) and 0.74 (2008 and 2009 end of summer), respectively, at a level of significance $p < 0.01$.

Estimates of aboveground biomass are critically important to enhance our understanding of biodiversity, species richness, and primary productivity. More recently, remotely sensed data, both airborne and satellite, are increasing in popularity and are contributing to improved biomass estimates for modeling the dynamics of carbon fluxes occurring between the atmosphere and the terrestrial ecosystem (Misbari and Hashim 2014). As more sophisticated models of global carbon balance are being developed, researchers need synoptic and conterminous estimates of aboveground terrestrial biomass that are limited by local-area field samples that may only provide estimates relative to a specific local. Remote sensing provides a larger regional capacity for sampling larger areas within certain geographic and temporal windows. As presented in this discussion, remote-sensing-derived estimates of terrestrial biomass have been used in combination with routinely applied vegetation indices, such as NDVI and statistical regression modeling, and have been found to produce promising results. Although the results produced from these types of studies are not perfect, newer methods involving emerging sensor technologies (e.g., LiDAR, hyperspectral data, higher resolution multispectral data, and interferometry synthetic aperture radar data, etc.) may help to improve current methodologies.

4.4.3 Carbon Estimation

Forests worldwide contain a large portion of terrestrial carbon stocks and depending on the decisions humans make in the face of continued climate and land use change, forests will be either net carbon sinks or sources. However, the exact amount of carbon contained in the earth's forests is not known to even one significant figure, ranging from 385 to 650 PgC (Goodale et al. 2002; Houghton et al. 2009; Saugier et al. 2001). Furthermore, there are unacceptably large variations and uncertainties in the estimates of terrestrial emissions, with a northern terrestrial uptake ranging from 0.5 to 4 Pg C year^{-1}, and tropical terrestrial emissions of -1 to $+4$ Pg C year^{-1} (Stephens et al. 2007). At least half of the uncertainty in carbon emission estimates from land use change results from uncertain estimates of biomass (Houghton 2005).

Recent multinational initiatives, such as the United Nations Reducing Emissions from Deforestation and Degradation (REDD+), are seeking to decrease tropical forest emissions and turning them from net carbon sources to net carbon sinks. In the Sierra Nevada forests of California, strategic placement of treatments (SPLATS) areas across the landscape are being implemented as part of the Sierra Nevada Forest Plan Amendment to interrupt potential wildland fire spread, to reduce the extent and severity of these fires, and to improve the continuity and distribution of old forests across landscapes (Bahro and Barber 2004). The United Nations Blue Carbon Program aims to increase carbon storage from coastal ecosystems by producing credits for mangrove and wetland conservation.

Fig. 4.16 Terrestrial Laser Scanner. **a** Picture of the Leica C10 terrestrial laser scanning (TLS) used in the Everglades study. **b** Point cloud image of intermediate height mangroves. The *black circle* marks the location TLS data acquisition, as no data were acquired beneath the TLS. (Photo courtesy of Emanuelle Feliciano, Feliciano et al. 2014)

The success of these new policy initiatives requires accurately quantifying and detecting changes in forest structure, biomass, and carbon stocks at spatio-temporal scales that are simultaneously relevant from an ecological, policy, and management perspective. Utilizing remote-sensed data provides the best path forward. Currently, three complementary remote-sensed technologies: (i) Terrestrial Laser/LiDAR Scanning (TLS), (ii) Airborne Laser/LiDAR Scanning, and (iii) Synthetic Aperture Radar (SAR) are being used in conjunction with field data to estimate forest structure, biomass, and carbon.

Necessity of ground truthing None of the remote sensing technologies discussed below directly measure biomass or forest carbon. Rather they all measure a variety of structural parameters of vegetation (e.g., canopy height, canopy volume, stem diameter) that can then be used with field-derived allometric equations to estimate vegetation volume, biomass, and carbon. Allometric relationships or allometry are mathematical equations describing the relationship between one or more parameters of an object and its shape. In the case of biomass estimation, information on wood specific density is needed and in the case of carbon both the wood-specific density and carbon content of tissue are needed. With regards to biomass and carbon estimation, most allometric equations estimate these using tree diameter and/or height. Chave et al. (2004) reviewed the literature on the use of allometric equations to estimate forest biomass and carbon. They found four types of uncertainty that could affect biomass estimates: (i) errors related to tree parameter estimates; (ii) errors related to the selection of allometric equations; (iii) sampling error due to sample plot size; and (iv) how representative the study plots were of the entire ecosystem. Overall, they concluded that the choice of allometric equations was the most important source of uncertainty. These findings point to the care that must be taken when using remote-sensed data to estimate forest biomass and carbon. Significant uncertainty can enter into the estimates both from the measurement uncertainty in the sensor and from the uncertainty embedded in the allometric measurements.

Terrestrial laser/LiDAR scanning (TLS) TLS is a relatively new technology that uses a rotating laser scanner to produce 3-D maps of various structures (Fig. 4.16). It has seen widespread application in a diversity of disciplines, including archaeology,

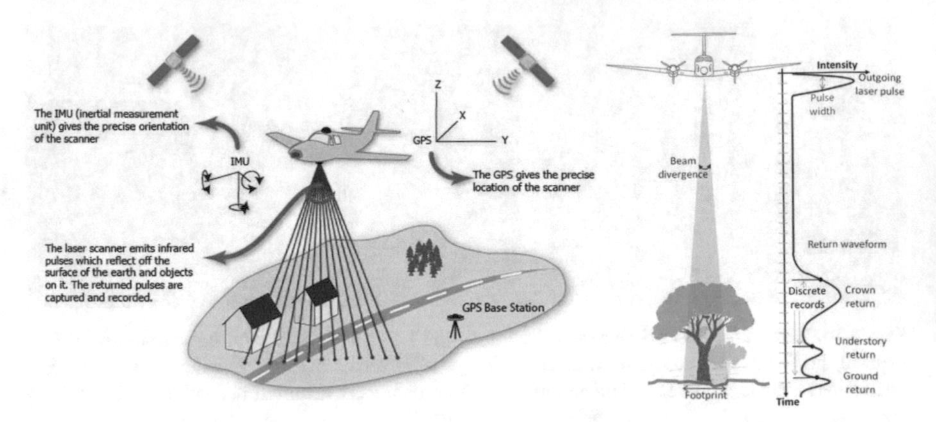

Fig. 4.17 Schematic illustrations of an airborne laser scanning (ALS) system. *IMU* Inertia Measurement Unit, *GPS* Global Position System. (Figure provided by Geospatial Modeling and Visualization & Imaging Notes)

criminology, engineering, geology, and forestry. TLS is capable of giving the coordinates (x, y, z) of any object with high resolution (mm level) and accuracy. The technology has been applied to vegetation structure characterization and monitoring since the mid-2000s (e.g., Watt and Donoghue 2007). TLS datasets provide high-resolution reconstructions (point clouds) of trunks, branches, and leafs distributions (Lucas et al. 2008). Furthermore, tree biomass can be estimated by multiplying the volume of its structure by its respective wood density.

Airborne laser/LiDAR scanning Airborne LiDAR systems are able to provide canopy height measurements over fairly large areas (Fig. 4.17). It fills the gap between high-resolution below the canopy data provided by TLS (discussed above) and coarser satellite-based SAR measurements (discussed below). The current airborne LiDAR system can estimate vegetation height with an accuracy of ~1 m (Hyyppä et al. 2004).

From a technical perspective, airborne LiDAR is similar to TLS in that the scanner emits a laser pulse that is reflected back from target objects and recorded by the scanner. In the case of airborne LiDAR the sensor is mounted on an aircraft with an Inertia Measurement Unit (IMU) and Global Position System (GPS). The IMU gives the precise orientation of the scanner and the GPS gives the precise location.

Synthetic aperture radar (SAR) Although airborne biomass and 3-D structure measurement methods (e.g., LiDAR) have significantly improved over the past decade, they are still too expensive to be conducted routinely on a global scale. Thus, the only current and near future available remote sensing technology capable of monitoring global vegetation structure, biomass, and carbon stocks is from the new generation of multiband and multi-polarization SAR satellites (TerraSAR-X, TanDEM-X, RADARSAT-2, ALOS-PALSAR). This method works because the SAR signal interacts with different vegetation structures depending on the signal wavelength (Fig. 4.18). A short wavelength signal (X-band) interacts mainly with

Fig. 4.18 Schematic illustration showing the interaction of the different Synthetic Aperture Radar (SAR) signal wavelengths with different vegetation structures. The 3-cm X-band signal interacts mainly with tree canopies, the 6-cm C-band signal interacts with canopies and branches, and the 24-cm L-band signal interacts with all structure levels. (Source: ESA website)

canopies, an intermediate wavelength signal (C-band) penetrates deeper into the vegetation and interacts also with branches, and a long wavelength signal (L-band) can penetrate throughout the vegetation and interact also with tree trunks. Thus, multiwavelength observations can provide useful estimates of vegetation structure and forest biomass (e.g., Bergen and Dobson 1999; Dobson et al. 1992; Kurvonen et al. 1999; LeToan et al. 1992). Another important advantage of SAR observations is the ability to acquire data in all weather conditions, which is extremely helpful in overcast tropical regions.

While promising early uses of SAR monitoring has been only partially successful mainly due to signal saturation problems. SAR-based biomass estimations are valid up to a certain threshold where saturation occurs (Lucas et al. 2007; Mitchard et al. 2009). The X- and C-band backscatter saturates at low biomass levels (30–50 Mg/ha), although a more recent study suggests the C-band estimate at a higher saturation level such as 250 Mg/ha (Pandey et al. 2010). An estimate for the L-band saturation level ranges from 40 to 600 Mg/ha (Lucas et al. 2007; Luckman et al. 1997 Mitchard et al. 2009). Most of the wide variation in biomass saturation estimates reflects different estimation methods as well as a different number of ground validation sites. A recent study by Englhart et al. (2011) that used both L-band ALOS and X-band TerraSAR-X data to measure aboveground biomass in tropical forests used about 400 validation points and obtained good biomass estimates up to 600 Mg/ha.

SAR data provide two main observables: amplitude and phase. Moving forward, we expect to see greater use of the Interferometric SAR (InSAR) technique to measure vegetation biomass and carbon. The InSAR technique is currently widely used to quantify changes and deformation of the earth due to earthquakes, volcanic activity, subsidence, and water level changes (Fig. 4.19). The InSAR technique uses information on phase differences through the creation of an interferogram between two SAR acquisitions of the same area. This may be done using either a repeat-pass (two-pass) or a single-pass bistatic acquisition (one-pass). For vegetation applications, single-pass bistatic InSAR is preferable as it avoids problems associated with coherence loss, temporal decorrelation, and wind decorrelation (Papathanassiou and Cloude 2003; Santoro et al. 2007).

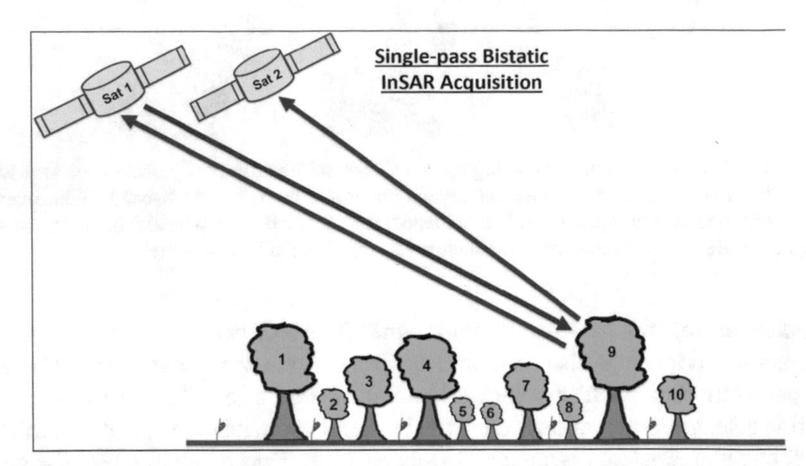

Fig. 4.19 Schematic illustration of single-pass bistatic Interferometric Synthetic Aperture Radar *(InSAR)*. Satelite 1 transmits a signal and Satellites 1 and 2 receive the reflected signal. Since both the signals (images) are acquired simultaneosly, temporal decorrelation is not present. (Figure courtesy of Emanuelle Feliciano)

4.4.4 Wildland Fire Risk Assessment

Wildland fire, defined as the relatively fast spreading of unintentional fire across a large areal extent of forested land, remains one of the greatest potentials to alter or disrupt ecosystem patterns and processes across the globe. These altered patterns and processes include terrestrial vegetation distribution and structure, LULCC dynamics, deforestation and desertification, global carbon cycle, and global climate change (Bowman et al. 2009; Keywood et al. 2013; Weise and Wright 2013). Whether this fire is caused by natural forces or human activity, the ecological consequences can be tremendous. Wildland fire is a significant source of harmful emissions worldwide with important implications for human health and carbon budget estimation. Many federal agencies have come to recognize the need for a unified approach to managing lands for prescribed and wildfires (Wildland Fire Executive Council 2009). However, the estimation of fire emissions is often based on the total replacement of forest or other ecosystems that, if the burn area is correctly estimated, undoubtedly overestimates the amount of emissions. In most circumstances, burn severity is not quantitatively estimated. (Khorram et al. 2014). In Fig. 4.20; the relative probability of wildland fire in California is shown as an example. Other parts of the US and most regions of the world also have significant fire risk.

Recent estimates suggest that wildland fires contribute to global greenhouse gas emissions at an alarming rate (approximately 6% of global carbon emissions; Langmann et al. 2009; Olivier et al. 2012). The current threat is changes in climate may be altering fire regimes and increasing wildfire contributions to global greenhouse gas emissions, further limiting humans' and the natural environment's capacity to reduce these impacts (Bowman et al. 2009; Stroppiana et al. 2010).

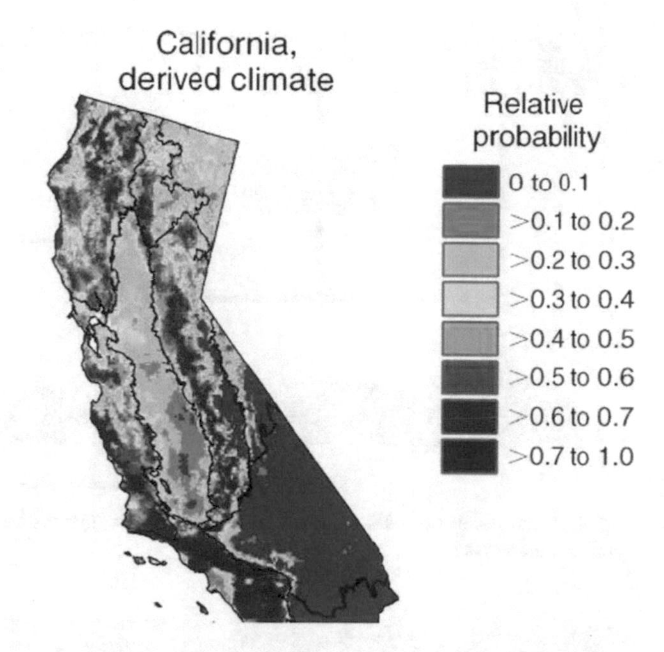

Fig. 4.20 Relative probability of fire occurrence in California

Advanced fire modeling for calculating wildland fire risk has relied on vegetation, weather, and meteorological inputs to calculate fire spread, fuel loads, and smoke production (Stratton 2006). However, despite numerous technological advances, our ability to manage fire and fire threat has been relatively limited. Additionally, the quantification of global greenhouse gas emissions as a result of fire is still poorly understood and inadequately represented in global emission models (Bowman 2009). However, remote sensing, beyond the commonly applied meteorological and weather applications (see Chuvieco 1996; Menzel 2006), provides greater capacity to model fire characteristics in that both aerial and spaceborne sensors have the ability to overcome costly field-collected data, which is often labor and time consuming, as well as requires lengthy processing intervals, thus minimizing the possibility of real-time model development (Arroyo et al. 2008; Keane et al. 2001).

(Khorram et al. 2014) have shown the applicability of airborne synthetic aperture (Unmanned Air Vehicle Synthetic Aperture Radar;UAVSAR) and Landsat TM multispectral data for modeling smoke emissions from wildland fires in northern California. The following images illustrate the utilization of the Berkeley Emission Estimation System (EES), which is used by several resource agencies as a decision support tool for their wildland fire management decision-making activities (Khorram et al. 2014). The EES uses First Order Fire Effects Model (FOFEM) as a framework and is based on the spatially specific inputs. The EES is based on a number of parameters, many of which can be estimated from remote sensing data (Fig. 4.21).

The results of this study based on Landsat TM and the UAVSAR are shown in Figs. 4.22 and 4.23.

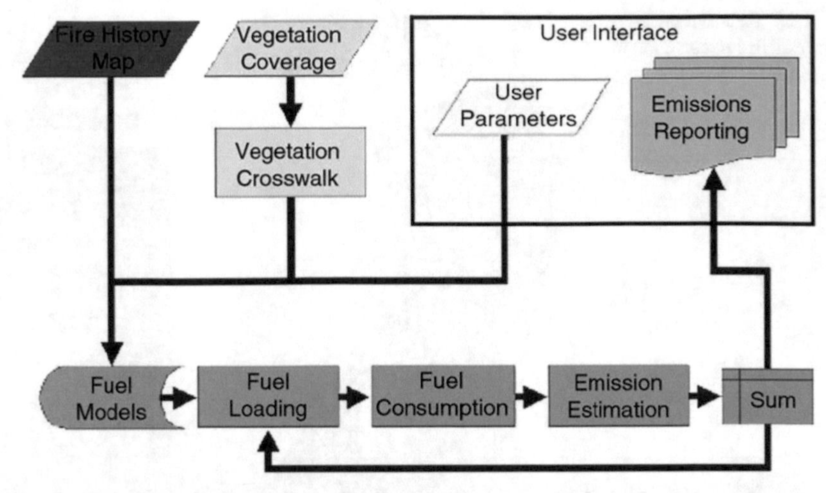

Fig. 4.21 Emission estimation system (EES) parameters, many of which can be estimated from remote sensing data

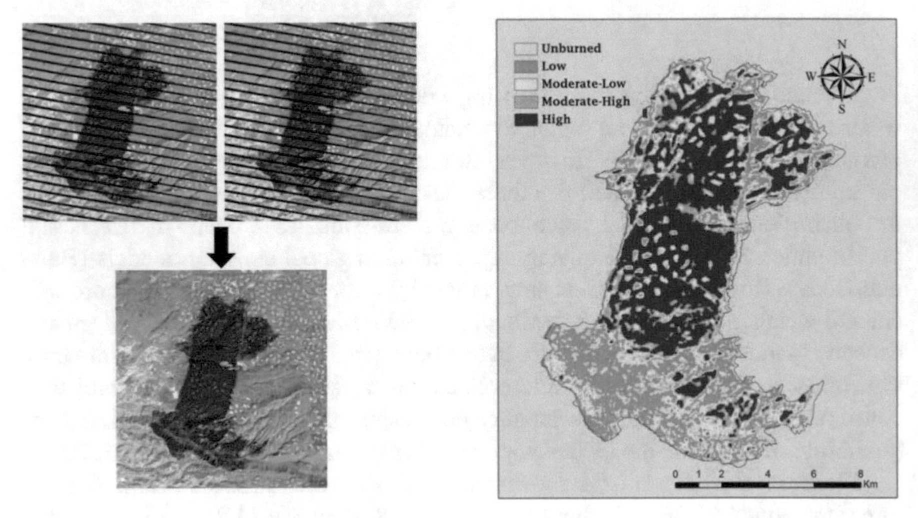

Fig. 4.22 Enhanced Thematic Mapper (ETM+)-based fire severity results and the scan lines removing *(left)* and burn severity map *(right)*

Another study of wildland fire was conducted by Hernandez-Leal et al. (2006), in which they used NDVI derived from the Advanced Very-High Resolution Radiometer (AVHRR-NOAA) sensor data over the Tenerife Island in Canary Islands, Spain to define a new wildland fire risk index. The new wildland fire risk index, which incorporated the NDVI values, is known as the Fire Risk Dynamic Index (FRDI; see Fig. 4.24). It is an extension of the Fire Risk Static Index (FRSI). The FRSI model is generated from land cover and environmental variables, which includes altitude and slope, solar insolation, proximity to main roads, and type of vegetation cover. The FRSI model output produces a static map of the probability model of

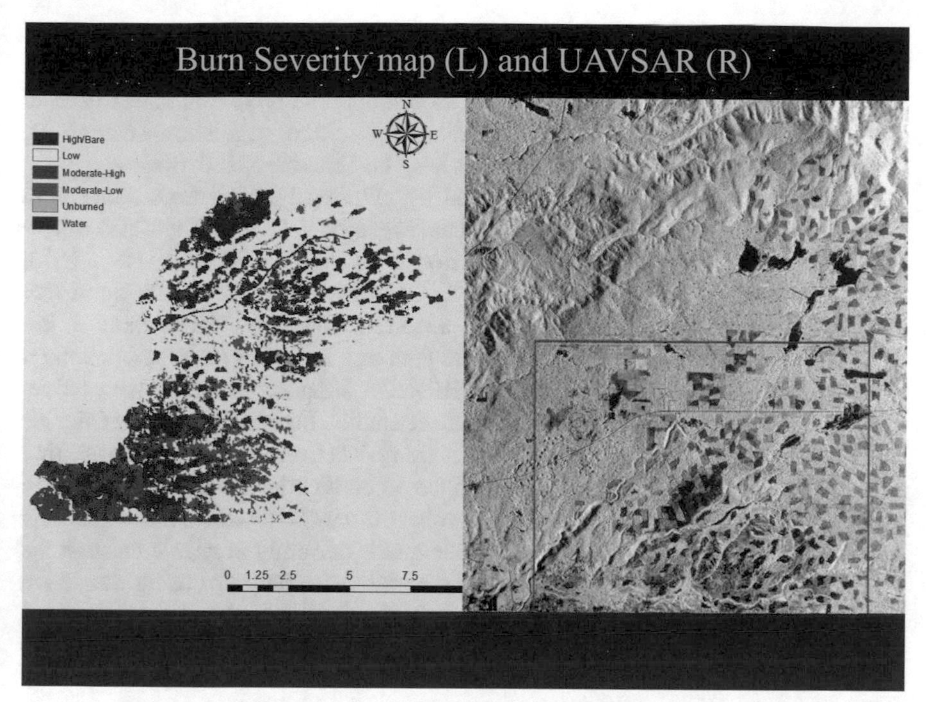

Fig. 4.23 The results based on the Unmanned Air Vehicle Synthetic Aperture Radar *(UAVSAR)*-processed data indicate burn severity *(left)* and the terrain-corrected raw UAVSAR data *(right)*

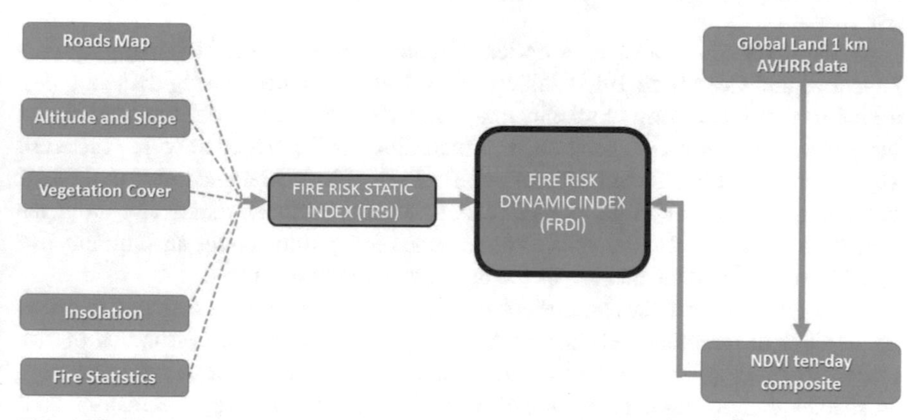

Fig. 4.24 Diagram of data used to develop the Fire Risk Static Index *(FRSI)* and Fire Risk Dynamic Index *(FRDI)* from land cover and environmental variables and Normalized Difference Vegetation Index (NDVI) estimates derived from Advanced Very-High-Resolution Radiometer (*AVHRR*-NOAA) sensor data. (Adapted from Hernandez-Leal et al. 2006)

fire occurrence by means of the included land cover and environmental variables within a logistic multivariate regression that incorporates past fire ignition events that took place between 1992 and 1995. The FRDI goes a step beyond FRSI in that it is a dynamic index that takes into account areas' gradual increasing in water stress (i.e., decrease in NDVI values), thus dynamically illustrating the changing probability of ignition across the landscape. The FRDI model, which integrates the multiple fire risk variables of FRSI, and further incorporates the NDVI variables derived from the AVHRR satellite sensor as an indirect indicator of vegetation water stress, improved the identification of the fire risk level on Tenerife Island, Spain.

An FRDI probability map is produced by utilizing NDVI values along with the FRSI model outputs. For the NDVI data, seven NDVI multitemporal composite images were analyzed, covering a period from June 20–30 to August 21–30, 1995. These 10-day composite images were used in order to obtain cloud-free scenes. The AVHRR-NOAA data were acquired from the Global Land 1 km AVHRR Project dataset (AVHRR-NOAA data are available at http://edc2.usgs.gov/1KM/1kmhomepage.php). The AVHRR-NOAA dataset is composed of five-channel, 10-bit, raw AVHRR data, at 1.1-km resolution (at nadir) for every daily afternoon pass over all land and coastal zones using data from NOAA's polar-orbiting TIROS. The data were initially to be collected continuously for 18 consecutive months, from April 1, 1992 through September 30, 1993. However, the acquisition period was extended to September 30, 1996 and is currently available through the USGS AVHRR Global Land 1 km internet portal. Advantages of using data from AVHRR-NOAA in fire risk modeling include the satellite's global coverage, and high temporal resolution. In addition there are two overpasses per day (see Deshayes et al. 1998 and Illera et al. 1996).

By incorporating the NDVI values into the FRSI model, areas of vegetation water stress may be accounted for in that decreases in the NDVI values across adjacent time periods may be used as a surrogate for water stress in areas of previous fire events.

Other studies have also used optical remote sensing for fire risk mapping (see Chuvieco and Congalton 1989; Jaiswal et al. 2002). Remote sensing data can play a vital role in identifying wildland fires, delineating burn area, deriving indices of burned and non-burned vegetation, and recording the frequency at which different vegetation types are affected. Hernandez-Leal et al. (2006) demonstrated that FRDI could have a vital role in the prevention of fires, supporting existing fire propagation models, or developing effective strategies for extinguishing an ongoing fire event through the identification of the dryer, water-stressed areas.

Remotely sensed data can collect timely measurements over larger fire-prone areas. Optical remotely sensed data have been used to provide estimates of fire characteristics that are relevant for ecosystem and fire fuel load parameters; such as species composition, biomass estimates, landscape structure, fire history, fuel moisture content, and fuel availability (Arroyo et al. 2008). With the evolution of newer and higher resolution sensors, improved estimates of vegetation species and cover class may serve to further enhance model performance, as well as wildland fire carbon contributions to global greenhouse gas emissions.

Case Study: Using Remote Sensing to Battle Catastrophic Wildfires
Battling large-scale, catastrophic wildfires has been aided substantially by near real-time satellite data. The Active Fire Mapping Program, a satellite-based fire detection and monitoring program managed by the USDA Forest Service Remote Sensing Applications Center (RSAC) located in Salt Lake City, UT, US, provides near real-time detection and characterization of wildland fire conditions in a geospatial context for the US and Canada.

Data collected daily by the NASA's MODIS satellite are currently the primary remote sensing data source of this program. The Active Fire Mapping Program leverages state-of-the-art technologies to acquire image data directly from orbiting spacecraft in order to minimize the chance of outdated information and to deliver fire geospatial products as rapidly as possible to incident managers, emergency response personnel, and other users of this information.

Satellite image data are continually relayed to the RSAC, where it is integrated and processed with current fire intelligence information and other key geographic strata provided by US and Canadian fire management agencies to produce imagery and data products. The results are a suite of "value-added" geospatial products that provide accurate and current assessments of current fire activity, fire intensity, burned area extent and smoke conditions throughout the US and Canada. Fire mapping and visualization products, fire detection GIS datasets and live data services, multispectral image subsets, and analytical products/summaries are some of the data products supplied by the RSAC. This type of data is crucial to wildland firefighters.

The **Newby Lake Fire** (Figs. 4.25–4.26) near the Okanogan–Wenatchee National Forest in Washington (US) and the Canadian border was ignited by lightning on July 2, 2015. Located in a high elevation area, it was difficult to evaluate and manage. In addition, high winds made the fire spread rapidly. However, within a day, the Incident Management Team was able to provide detailed information to the public on a website: (http://www.inciweb.org/incident/4355/). For some incidents, such as the **Thunder City Fire**, 3-D images are provided to depict elevation (Fig. 4.27).

4.5 Optimizing Sustainable Food and Fiber Production through Remote Sensing

Precision agriculture—information-based management of agricultural production systems—combines satellite remote sensing and GIS to be able to increase food and fiber production to a growing population. The goals of precision agriculture are to (1) optimize the use of resources to reduce farmer costs and increase sustainability of the farm operation, (2) reduce harmful environmental impacts, and (3) improve the quality of the work environment and the social aspects of farming and ranching (Pierce and Nowak 1999).

Fig. 4.25 Newby Lake fire aerial photo. (© www.inciweb.org)

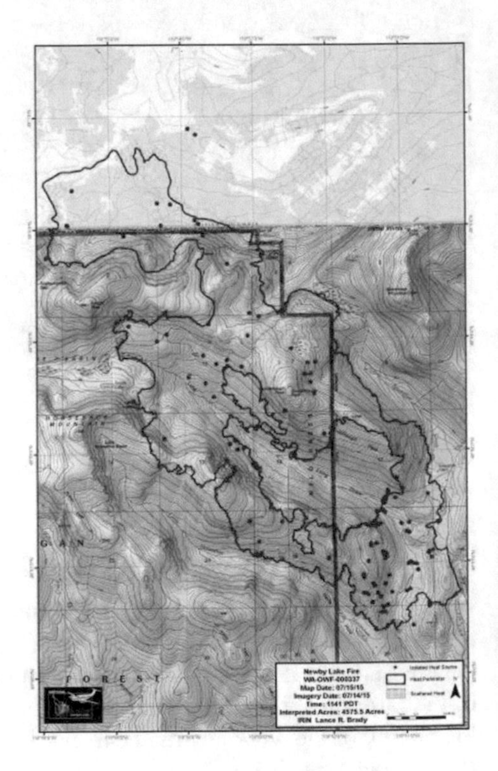

Fig. 4.26 Newby Lake fire showing heat detected by NASA MODIS on July 15, 2015. (© PNW Team 3, www.inciweb.org)

Fig. 4.27 Thunder City Fire 3-D maps of the perimeter on July 23, 2013. The terrain elevation is exaggerated to make it more visible. (©Google Earth/US Forest Service)

Remote sensing has been used in a variety of agricultural applications for well over 20 years. Some applications are fairly simple while others are complex and innovative. Due to the lack of cartographic data, many Argentinean farm managers and agronomists use satellite images, in combination with soil maps, to determine the productive aptitude of soils before buying or renting land (Bongiovanni and Lowenberg-DeBoer 2006). Remotely sensed images offer high-quality data, frequent recurrence cycles, and cost effectiveness.

Entire fields are easily captured in one image, allowing farmers to detect biophysical variability in crops such as water stress, lack of nutrients, diseases, weeds and insect damage. Much of the remotely sensed data used in precision agriculture to detect variability in crop conditions rely on spectral characteristics, predominantly using the red (RED, chlorophyll *a* and *b* absorption) and near-infrared (NIR, internal leaf scattering) wavelength bands (Moran et al. 1997). Various combinations of the two bands—**spectral vegetation indices**—are highly sensitive to vegetation cover and density (Nemani et al. 2006). Successful mapping of LAI—the area of leaves per unit area of ground (m^2/m^2)—using IKONOS imagery was shown by Johnson et al. (2003) to be a useful indicator of plant growth and health, as an equilibrium generally exists between climate–soil conditions and LAI (Johnson et al. 2003; Nemani et al. 2006). Lack of foliage—either fewer or smaller leaves—usually means that plants are stressed by insects or disease and are dying; more vigorous plants will have more foliage and different spectral reflectance.

Spatial data have been used extensively to increase productivity and yield in croplands—enabling farmers to apply the right treatment in the right place at the right time (Pierce and Nowak 1999; Robert 2001; Stafford 2006) by adapting fertilizer distribution to varying soil conditions. More recently, additional practices have evolved to improve the sustainability of the food supply. Adapting production inputs site-specifically within a field and individually for each animal allows for better use of resources to maintain the quality of the environment while reducing costs.

Innovative approaches in applied remote sensing have been successfully implemented to enable product traceability and advance the management of a wide range of agricultural products including viticulture, orchards, livestock and animal fiber production, and turf management.

4.5.1 Improving Wine Harvest and Quality

Aerial views of vineyards provide a better perspective of canopy conditions than relying on ground observations. This, in turn, allows for better management prescriptions. The NDVI, as described in Chap. 3, has been demonstrated as an effective measure of vine health (Johnson et al. 1996; Lamb 2000).

A number of studies using radiometric, high-resolution spectral analysis have demonstrated that grape varieties have detectable systematic differences in reflectance intensity and in the spectral form (Cecchini et al. 2006; Hall et al. 2002, 2003; Lacar et al. 2001; Luz and Fonseca 2003) that matches up with vineyard inventory data.

More recently, Cemin and Ducati (2011) conducted a study to see whether the NDVI could distinguish not just the unique spectral qualities of particular grape varieties, but whether their *terroir* could also be detected. *Terroir*—the relationship between the attributes of a grape and the particular location it is grown—is a determinant of wine quality, identity, and organoleptic descriptors (Cemin and Ducati 2011; van Leeuwen and Seguin 2006). The ability to distinguish a *terroir's* imprint on the spectra of the same grape cultivar is a powerful tool for certifying the geographic origin and authenticity of the product (Blauth and Ducati 2010; Fregoni 2010). The researchers examined plots in Bordeaux and Champagne (France), the Aconcagua and Colchagua valleys in Chile, and the Encruizilhada do Sul (Rio Grande do Sul State, Brazil) using all nine spectral bands from ASTER images to investigate whether there were detectable differences in cabernet sauvignon, pinot noir, merlot, and chardonnay grapes at each location. Cemin and Ducati (2011) confirmed that the four grape varieties did in fact have characteristic spectra that could be distinguished through the NDVI. Additionally, they found that region grouping indicated some spatial influences on spectral behavior reminiscent of *terroir* effects; however, they were using known field data. Remote sensing techniques are sensitive to changes in reflectance; therefore, further investigation would be worthwhile.

Oenoview®, a joint commercial venture between *Institut Coopératif du Vin* (Groupe ICV) and Infoterra, a subsidiary of EADS Astrium has benefited French

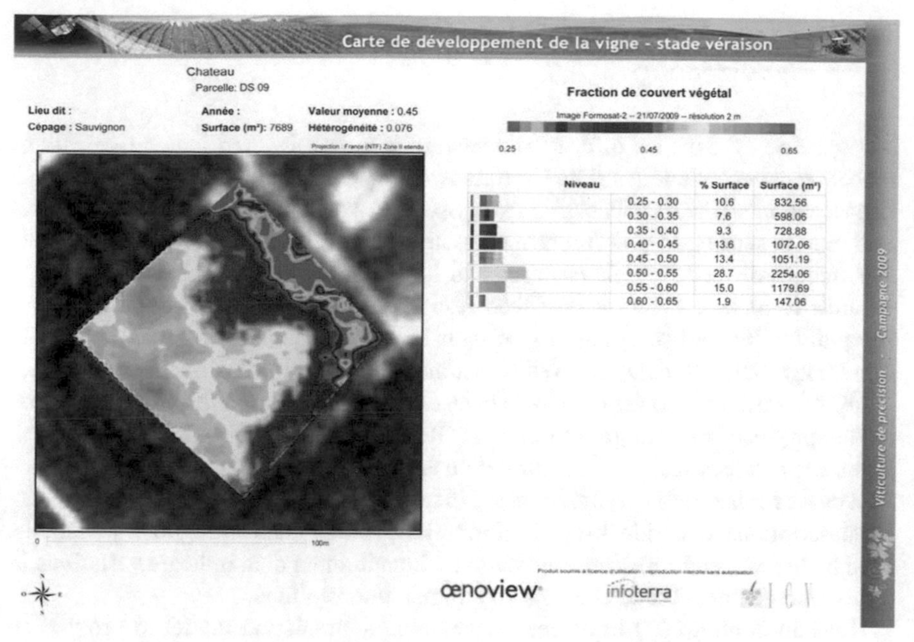

Fig. 4.28 Example of Oenoview® image. (http://www.spotimage.com)

viticulturalists. Using aerial and multispectral images (covering nearly 3000 km^2) taken at the rate of 1000 plots every 8 s, Oenoview has a precision range of approximately 3 km^2 (Samuel 2011). The images are used to map the surface area and variation of leaf canopy across vineyards—a vital statistic to identify subtle nuances in vineyard vigor, soil moisture levels, bunch and grape weight, and the presence of minerals. Wine growers provide information on their fields, vines, and grape varieties. SPOT-5 (Fig. 4.28) and Formosat-2 images are taken in July and August during the period of *véraison* (when grapes begin to ripen, about 3–4 weeks before grape harvest).

Wine grapes require a moderate amount of water stress to produce top quality wine, although at the expense of reduced grape yield (Chaves et al. 2010; Serrano et al. 2012). Berry quality largely depends on achieving an optimum balance between sugars and titratable acidity at harvest. Spectrally, green areas of dense foliage reveals excess soil moisture, while red to purplish areas indicate less moisture, thus pointing to a good vintage. Spatial data are then supplied in a GIS-ready format to be used by the growers to determine when grapes are mature for harvesting and assess variations in grape quality within plots. Rather than acquire plot information empirically over several decades, the data allow growers to separate out lower-grade bunches for *vin de table* and reserve the superior grapes for higher end *grands crus*. Vats containing as little as 20% immature grapes lessen the quality of wine at the end of the process, resulting in lower economic returns (Samuel 2011). The data also allow growers to optimize pruning, fertilizer inputs, and irrigation needs for each vineyard.

4.5.2 Using Remote Sensing to Optimize Grazing and Improve Wool Quality

Australia's vast expanse of crop and pasture land has fostered innovative uses of remote sensing techniques. The Commonwealth Scientific and Industrial Research Organization's (CSIRO) Pastures from Space™ program uses spatial informatics and remote sensing to provide farmers with a suite of tools to accurately estimate the amount of feed in their pastures and how quickly their pastures are growing in order to make efficient use of feed resources. This enables farmers to increase the profitability and sustainability of their lands, which has seen extreme drought conditions. Satellite data are used to estimate pasture biomass or "feed on offer" (FOO™). Pasture biomass data have been combined with climate and soil information to produce "pasture growth rate" (PGR™) estimates at the national, regional, farm, and paddock scale. This information is updated weekly on the Internet, broadcast on Australian radio, and signposted in regional areas. Farmers can obtain weekly subscriptions for paddock-level information to assist them with grazing rotations, feed budgeting, and fertilizer applications. Longitudinal data indicate variations in vegetation due to climatic changes and human interventions.

Whelan et al. (2007) have been developing a simulation model to provide an improved and more precise forecast of wool production across the Australian continent. The goal is to predict wool fiber diameter of the wool clip before it is shorn directly from remotely sensed data by using monthly NDVI values and fiber diameter profiles of sheep at the paddock level. Researchers also hope to predict the effects of modeled future climate change on future wool supplies.

4.6 Exploring and Monitoring Oil, Gas, and Mineral Resources

Ortho-rectified high-resolution satellite images are indispensable in the search for and the development of new oil and gas fields and mineral deposits (Fig. 4.29). Using airborne magnetic or ground gravity surveys to facilitate interpreting subsurface features, geologists and engineers are able to narrow down the search area for performing more detailed 2-D (Fig. 4.30) or 3-D seismic surveys (Fig. 4.31) for exploratory drilling.

Satellite imagery is also used to develop digital elevation models (DEMs) to plan the least environmentally damaging route for pipelines, access roads, and facilities by creating a computer visualization of the site area (Fig. 4.32). The use of satellite remote sensing has greatly reduced the time and cost involved in reconnaissance and the placement of oil wells. Remotely sensed spatial data also allow managers and government officials the ability to monitor wellfields (Fig. 4.33) and obtain information on the extent of oil spills and how best to mitigate and remediate these areas.

Fig. 4.29 California's Mountain Pass rare-earth elements mine in the Mojave desert. The *bright green* areas in the northwest are settling ponds, where mining wastewater is held until the suspended particles sink to the bottom. Rare-earth elements are used in applications such as hybrid cars and wind turbines. (©ALI/EO-1/NASA)

Fig. 4.30 Example of a 2-D seismic survey taken in Ghana, West Africa. (©World Oil Online)

High-resolution providers, such as GeoEye-1, WorldView-2, Worldview-1, QuickBird, IKONOS, SPOT-5, LANDSAT 7+ETM, and ASTER, have been extensively employed in the quest for surface indicators of subsurface oil and gas along with risk assessment and well field monitoring (to identify security and environmental problems).

Fig. 4.31 Example of a 3-D seismic survey. (Courtesy of Russian Oil & Gas Technologies; http://www.rogtecmagazine.com/)

Fig. 4.32 Digital elevation model (DEM) of Tunisian oilfield. IKONOS—0.8 m. (© 2010 GeoEye; http://www.satimagingcorp.com)

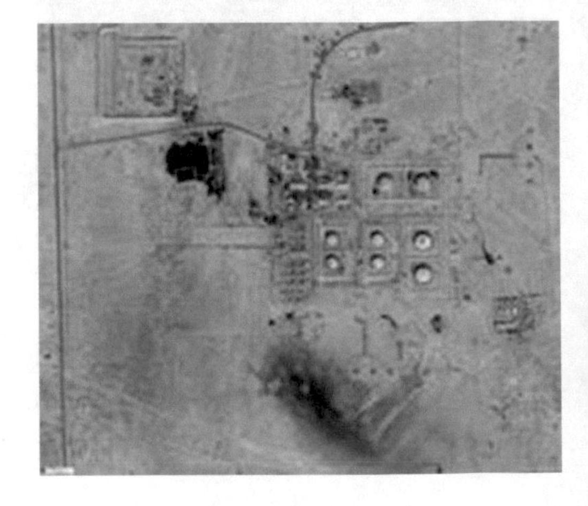

Fig. 4.33 Oil production field, Iraq (IKONOS—0.8 m—© 2010 GeoEye; http://www.satimagingcorp.com/svc/exploration.html)

4.7 Using Remote Sensing for Humanitarian and Peace-Keeping Operations

Tracking the movements of refugees and internally displaced people (IDP), verifying burned and razed villages, documenting the existence of mass graves, and other humanitarian work would be utterly impossible without very-high-resolution satellite imagery. Very high spatial resolution (VHSR) imagery is particularly crucial in assessing the extent of violent conflict, environmental or social justice cases, forced displacement, or other human rights atrocities where field-based assessments are all but impossible in remote, prohibited, dangerous, and/or inaccessible locations. The United Nations High Commissioner for Refugees (UNHCR) and the Internal Displacement Monitoring Center (IDMC) methodology for characterizing and monitoring the nearly 34 million forcibly displaced people worldwide uses GIS, aerial, and satellite imagery (IDMC 2008; UNHCR 2001).

Refugees and IDPs are a very mobile and fluid type of human settlement, living in small makeshift shelters comprised of thatched huts, clay constructions with flat roofs, and tent-like structures covered with bright materials (e.g., UN tent fabrics or plastic sheets). Estimating the total number of dwellings is extremely complex because of the spectral variability and reflectance of the building material. Bright soil backgrounds and scattered scrub vegetation can make overestimation of rooftops a common classification error in arid regions (false alarms) in the automatic recognition of built-up areas. Rooftops can be under- or overestimated in arid regions as they are misclassified as scattered scrub vegetation. As well, the spatial distribution of scattered vegetation and their contrast with the background is very similar to the spatial distribution of the built-up structures in some part of the human settlement, including scattered buildings and refugee camps. In particular, the size of tree canopies and their average distance in some cases are similar to the size of built-up structures and their average distance, and then producing overestimation of built-up areas. However, the ability to determine how many IDPs there are is critical to planning and managing effective humanitarian responses. Remotely sensed data allow for continuous monitoring and corroboration of on-the-ground reports with the aim of decreasing or preventing humanitarian disaster and human rights crimes in politically unstable and chronic conflict areas.

Satellite images have effectively been used in legal proceedings and diplomatic negotiations to provide compelling, visual proof and resolve current and future conflicts over highly contested territories. SPOT images were used to achieve the 1995 Dayton Peace Accord (Johnson 2001). Imagery from India's IRS-1C and US IKONOS satellites was used to monitor human activity and mitigate potential conflicts in Spratly Islands in the South China Sea (Gupta and Bernstein 2001). International humanitarian relief efforts in Thailand, Mexico, Africa, and Europe—particularly Kosovo, Macedonia, and Albania—were made possible by imagery provided by India's IRS-1C, Russia's KVR-1000, and France's SPOT satellites to monitor and mitigate the plight of refugees (Bjorgo 2001).

The well-known Eyes on Darfur project, a collaborative initiative between Amnesty International and the American Association for the Advancement of Science (AAAS), is one case study of an active monitoring project. Images allow the ability to track vulnerable villages and monitor changes that may have occurred (Fig. 4.34).

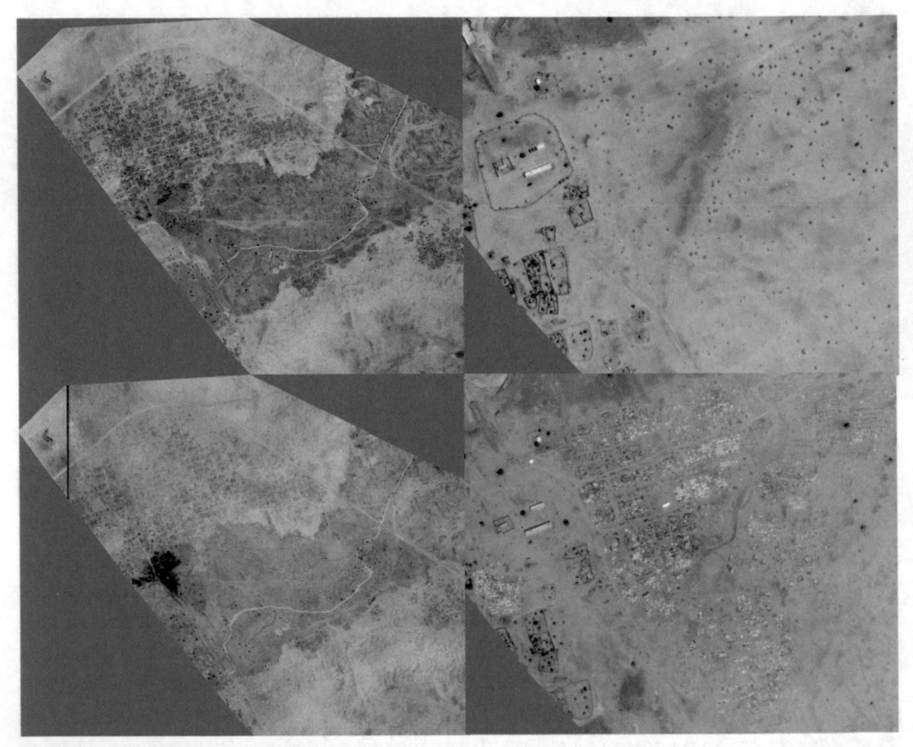

Fig. 4.34 © DigitalGlobe, Inc. Eyes on Darfur. Satellite images of Tawila, Sudan taken on May 3, 2003 (before attack, *top*) and February 23, 2007 after attacks *(bottom)*, showing loss of villages and presence of internally displaced people's (IDP) camps

The potential of geospatial visualization for the international human rights community was highlighted by the September 2004 use of QuickBird satellite imagery by the US Agency for International Development (USAID) and the US Department of State in confirming the extent of ethnic cleansing in Darfur. These agencies were able to view and interpret conditions on the ground in Darfur when eyewitness reporting was difficult to verify. Used in conjunction with classified sources, analyses of the imagery led to confirmation of reports of widespread destruction of villages, livestock, and crops as part of an ethnic cleansing campaign. The remotely sensed imagery also proved effective in diplomatic campaigns and alerting the global public.

4.8 Archaeology and Cultural Heritage

The evolution of satellite image analysis has vastly transformed the field of archaeology, allowing researchers to exploit more fully an enormous wealth of data of the earth's surface and subsurface contained in various types of satellite images along with aerial photography. Archaeologists can examine a broad spectrum of

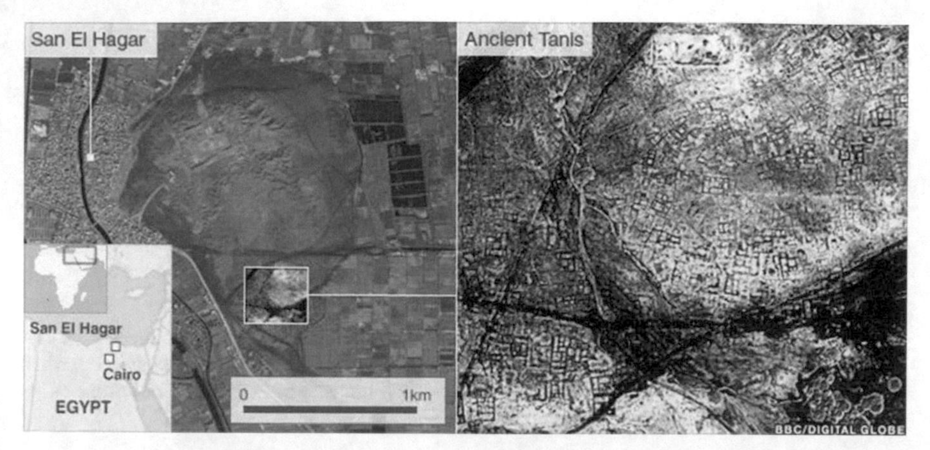

Fig. 4.35 © DigitalGlobe/BBC. The infrared image on the *right* used by the SS-SEPE team reveals the ancient city of Tanis near the modern city of San El Hagar

reflectivity signatures and bands within the remotely sensed data to focus on perspective archaeological sites to determine if there are subsurface structures.

The South Sinai Survey and Excavation Project in Egypt (SS-SEPE) led by Drs. Gregory Mumford and Susan Parcak of University of Alabama, Birmingham (US), used supervised and unsupervised land use/land cover applications of Landsat TM images as well as QuickBird imagery to detect vegetation signatures throughout Egypt. This allowed the researchers to isolate water sources in an arid environment and thus identify existing and potential archaeological sites. Infrared images can differentiate between various subsurface materials. Since ancient Egyptians used mud bricks in building construction, which are substantially denser than the surrounding soils, the shapes of houses, temples, and tombs can be detected most easily using winter images (Pringle 2011). Sites meeting the classification parameters are then "ground-truthed" to verify already identified archaeological sites (i.e., assess the accuracy of the classification) and conduct initial excavations on potential (but unknown) sites. As a consequence of creating this specialized classification, the enormous amount of time and high costs associated with surface detection of vegetation and water signatures in a vast geographical area can be significantly reduced, and the possibility for new discoveries greatly increased (Parcak and Mumford 2010).

This innovative technique rewarded the SS-SEPE team with one of the most significant discoveries in Egypt—in May 2011, 17 previously unknown pyramids along with more than 1000 tombs and 3000 settlements (Figs. 4.35 and 4.36) were revealed (Cronin 2011; Pringle 2011).

The vastness and remoteness of Mongolia, along with the lack of surface remains, has meant that it has been largely unexplored by archaeologists. Starting in 2008, a German research team led by Dr. Jan Bemman of the University of Bonn, used a combination of aerial photographs, satellite images, and surveys. Digital surface models were converted into interactive 3-D models of the area, and superconducting quantum interference devices (SQUIDS) on the ground were used

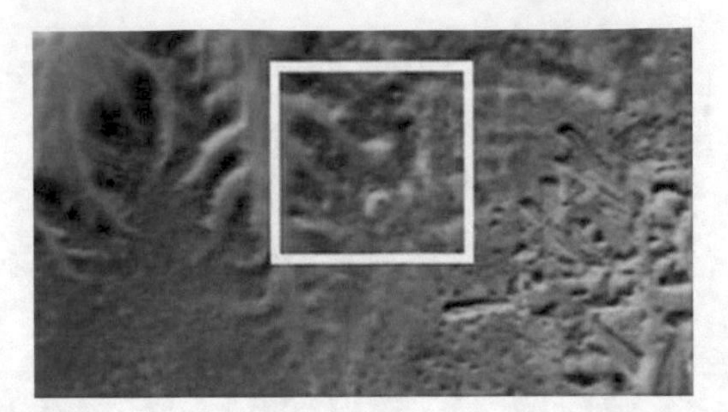

Fig. 4.36 © Sarah Parcak, SS-SEPE. The Landsat Thematic Mapper (TM) image reveals the outlines of a buried pyramid, located in the *center* of the box. Validation of the technology came when the team excavated a 3000-year-old house that the satellite imagery had detected; the outline of the structure was almost identical to that of the remotely sensed image

Fig. 4.37 ©Jan Bemmann. Satellite images reveal the rectangular walled city of Kharbalgas, approximately 30 km^2

to obtain rapid, high-resolution images of the magnetic field. The combination of advanced techniques allowed for detection of compacted construction materials underneath the soil. In 2011, the team detected a massive settlement (Fig. 4.37) in the Orkhon Valley of central Mongolia dating from the eighth and ninth centuries C.E. at the time of the Uighur (Uyghur) Empire (AAAS 2011).

Safeguarding archaeological sites and antiquities has been aided by remotely sensed images. Authorities can monitor images to detect if a site has been looted and contact Interpol to watch out for antiquities that may be sold illegally.

4.9 Summary

This chapter provides just a few of the many terrestrial applications of remote sensing. While not exhaustive, presented here are examples of case studies found in the literature that use data directly acquired from satellite and airborne remote sensing, or data derived from remotely sensed data. It is notable to consider that as remote sensing technologies evolve, resolutions increase, and access improves, the attractiveness of a synoptic tool like remote sensing will also increase. The next chapter will provide examples of how remote sensing is used to better understand atmospheric phenomena.

4.10 Review Questions

1. Terrestrial applications of remote sensing are very diverse and include mapping, change detection, monitoring, modeling, and other observations of land uses stemming from human activities. Why do you think remote sensing has played and continues to play such a vital role in large geographic-area observations?
2. One of the most frequently used applications of remote sensing is the creation of LULC maps that are created from imagery and used to categorize natural and human-made features into representative groups or "classes". Why might resource managers and researchers use LULC maps to gain important information about their region or environment of study?
3. The direct impacts on urban growth or expansion on physical, ecological, and social resources have made research on urban sprawl an issue of increased interest. What advantages might remote sensing provide decision makers at the regional scale to complement traditional census sources?
4. Large animal movement, migration patterns, and habitat utilization patterns are largely tied to the landscape. However, traditional field and aerial-based approaches are often limited to observing only a few animals at a time, within a limited area, and also limited by cost, labor, and the time necessary to complete the study. How might satellite remote sensing be useful in monitoring these types of large animal movement activities?
5. Remote sensing of water and water-related resources is a growing area of study. Previous studies have shown advantages in using spectral remotely sensed data for applications, including surface water supply, reservoir mapping and monitoring, water clarity and quality mapping and modeling, runoff forecast modeling, drainage network mapping, hydrologic research, watershed characterization, and surface water impairment. However, what are some of the challenges in applying spectral-based remote sensing to these types of water-related studies?
6. Deforestation in the Amazon region of South America is driven by many factors, including, social, economic, political, and developmental. Applications of remote sensing to determine patterns and detect large area or regional deforestation status in this region have been numerous. Discuss a few remote sensing techniques that might be useful in determining large area deforestation in this region?
7. The NDVI, derived from multispectral, remotely sensed data, has been used as a traditional tool for forest vegetation health assessments. How does the NDVI work, as it relates to the plant's leaf structure, to give investigators an indication of forest vegetation health?

8. Remote sensing has been shown in several studies to play a vital role in the study of wildfires. Discuss some of the advantages of using remote sensing applications in detecting and understanding the spread of wildfires across the landscape?
9. Name one active and one passive remote sensing data can be used in mapping the Burn Severity in wildland fires.
10. How is allometry used to estimate biomass and carbon? List the four main sources of uncertainty that Chave et al. (2004) identified. Which one is most important?
11. Name three landscape parameters that are used for modeling smoke emissions from wildland areas.
12. Provide an interpretation of Fig. 4.15 explaining the use of remotely sensed data to estimate biomass.
13. The effective tracking of the movements of refugees and IDP presents a huge challenge for the government and peace-keeping organizations in many developing nations. How has remote sensing played a role in tracking and understanding the movements of these groups of people?
14. Speculate on the additional use of remote sensing as it applies to the study of the world's archeology.

References

Ackleson, S. G. and Klemas V. 1987. Remote-sensing of submerged aquatic vegetation in lower Chesapeake Bay: a comparison of Landsat MSS to TM imagery. *Remote Sensing of Environment* 22, 235–248.

American Association for the Advancement of Science (AAAS). 2011. Beneath a barren steppe, a Mongolian surprise. New Focus. *Science* 332 (6028), 416–17.

Armstrong, R. 1993. Remote sensing of submerged vegetation canopies for biomass estimation. *International Journal of Remote Sensing* 14, 621–627.

Arroyo, L.A., C. Pascual, J.A. Manzanera. 2008. Fire models and methods to map fuel types: The role of remote sensing. *Forest Ecology Management*. 256(6):1239–1252. doi:http://dx.doi.org/10.1016/j.foreco.2008.06.048.

Asner, G.P., E.N. Broadbent, P.J.C. Oliveira, M. Keller, D.E. Knapp, and J.N.M. Silva. 2006. Condition and fate of logged forests in the Brazilian Amazon. *Proceedings of the National Academy of Sciences of the United States* 103(34): 12947–12950.

Bahro, B. and K. Barber. 2004. Fireshed assessment: An integrated approach to landscape planning. USDA Forest Service Pacific Southwest Region. RS-TP-017.

Bergen, K.M. and M.C. Dobson. 1999. Integration of remotely sensed radar imagery in modeling and mapping of forest biomass and net primary production. *Ecological Modelling*. 122:257–274.

Barrachina, M., J. Cristobal, and A.F. Tulla. 2015. Estimating above-ground biomass on mountain meadows and pastures through remote sensing. 2015. *International Journal of Applied Earth Observation and Geoinformation*. 38:184–192.

Bjorgo, E. 2001. "Supporting Humanitarian Relief Operations," In Baker, O'Connell, and Williamson, eds., *Commercial Observation Satellites: At the Leading Edge of Global Transparency*. Santa Monica, CA and Bethesda, MD: RAND and the American Society for Photogrammetry and Remote Sensing. pp. 408–422.

Blauth, D.A. and J.R. Ducati. 2010. A web-based system for vineyards management, relating inventory data, vectors and images. *Computers and Electronics in Agriculture*. 71: 182–188.

Bongiovanni, R. and J. Lowenberg-DeBoer. 2006. Argentina. In A.Srinivasan, ed., *Handbook of Precision Agriculture: Principles and Applications*. NY: The Haworth Press, Inc. pp. 615–633. DOI: 10.1300/5627_21

Bowman, D.M.J.S., J.K. Balch, P. Artaxo, W.J. Bond, J.M. Carlson, M.A. Cochrane, C.M. D'Antonio, R.S. DeFries, J.C. D, S.P. Harrison, F.H. Johnston, J.E. Keeley, M.A. Krawchuk, C.A. Kull, J.B. Marston, M.A. Moritz, I.C. Prentice, C.I. Roos, A.C. Scott, T.W. Swetnam, G.R. van der Werf, and S.J. Pyne. 2009. Fire in the Earth System. *Science*. 324(5926):481–484.

British Broadcasting Corporation (BBC). 2008. BBC News: Science/Nature: Images reveal "rapid forest loss." http://news.bbc.co.uk/2/hi/science/nature/7431589.stm

Brivio, P.A., C. Giardino, and E. Zilioli. 2001. Validation of satellite data for quality assurance in lake monitoring applications. *Science of the Total Environment* 268:3–18.

Carpenter, S.R., N.F. Caraco, D.L. Correll, R.W. Howarth, A.N. Sharpley, and V.H. Smith. 1998. Nonpoint pollution of surface waters with phosphorus and nitrogen. *Ecological Applications*. 8(3):559–568.

Cecchini, F., M. Morassut, and E.G. Moruno. 2006 Free amino acid composition of must from 7 Vitis vinifera L. cv. In Latium (Italy). In: van Leeuwen, C., J-P Roby, E. Marguerit, and J. Fanet. (eds) Proceedings of VI International Terroir Congress, ENITA, Bordeaux, pp. 424–428.

Chave, J., R. Condit, S. Aguilar, A. Hernandez, S. Lao, and R. Perez. 2004. Error propagation and scaling for tropical forest biomass estimates. *Philosophical Transactions of the Royal Society of London. Series B: Biological Sciences*. 359(1443):409–420.

Chaves, M.M., O. Zarrouk, R. Francisco, J.M. Costa, T. Santos, A.P. Regalado, M.L. Rodrigues, and C.M. Lopes. 2010. Grapevine under deficit irrigation: hints from physiological and molecular data. *Annals of Botany*. 105, 661–676. DOI: 10.1093/aob/mcq030

Chemin, G. and J.R. Ducati. 2011. Spectral discrimination of grape varieties and a search for *terroir* effects using remote sensing. *Journal of Wine Research*. 22(1): 57–78. DOI: 10.1080/09571264.2011.550762.

Chen, M., Q. Zhuang, D.R. Cook, R. Coulter, M. Pekour, R.L. Scott, J.W. Munger, and K. Bible. 2011. Quantification of terrestrial ecosystem carbon dynamics in the conterminous United States combining a process-based biogeochemical model and MODIS and AmeriFlux data, Biogeosciences, 8, 2665-2688, doi:10.5194/bg-8-2665-2011.

Chuvieco, E., and R.G. Congalton. 1989. Application of remote sensing and geographic information systems to forest fire hazard mapping. *Remote Sensing of Environment*. 29(2):147–159.

Chuvieco, E., J. Salas, and C. Vega. 1996. Remote sensing and GIS for long-term fire risk mapping. In: A review of remote sensing methods for the study of large wildland fires. (Ed. E Chuvieco) pp. 91–108. Megafires Project ENV-CT96-0256. Alcala de Henares, Spain.

Cocks, T., R. Jenssen, A. Stewart, I. Wilson, and T. Shields, 1998. The HyMap airbornehyperspectral sensor: the system, calibration and performance. In: Presented at 1stEARSEL Workshop on Imaging Spectroscopy, Zurich, October 1998.

Cronin, F. 2011. Egyptian pyramids found by infra-red satellite images. BBC News. 24 May 2011. http://www.bbc.co.uk/news/world-13522957. Accessed 25 May 2011.

Curran, L.M. and S.N. Trigg. 2006. Sustainability science from space: quantifying forest disturbance and land-use dynamics in the Amazon. *Proceedings of the National Academy of Sciences of the United States*. 103(34): 12663–12664.

Curran, L.M., S.N. Trigg, A.K. McDonald, D. Astiani, Y.M. Hardiono, P. Siregar, I. Caniago, and E. Kasischke. 2004. Lowland forest loss in protected areas of Indonesian Borneo. *Science* 303: 1000–1003.

Dahl, T.E. 1990. Wetland losses in the United States 1780's to 1980's. U.S. Department of the Interior, Fish and Wildlife Service, Washington, D.C.

Dobson, M. C., F. T. Ulaby, T. LeToan, A. Beaudoin, E. S. Kasischke, and N. Christensen. 1992. Dependence of Radar Backscatter on Coniferous Forest Biomass. *IEEE Transactions on Geoscience and Remote Sensing*. 3(2):412–415.

Dekker, A.G., and S.W.M. Peters. 1993. The use of the Thematic Mapper for the analysis of eutrophic lakes: a case study in the Netherlands. *International Journal of Remote Sensing*. 14:799–821.

Deshayes, M., E. Chuvieco, and D. Cocero. 1998. Evaluation of different NOAA-AVHRR derived indices for fuel moisture content estimation: interest for short-term fire risk assessment. D.X. Viegas (Ed.), Proceedings of the III International Conference on Forest Fire Research—14th Conference on Fire and Forest Meteorology, Coimbra, Portugal, November 1998. ADAI, Coimbra. pp. 1149–1167.

Dusseux, P., L. Hubert-Moy, T. Corpetti, and F. Vertes. 2015. Evaluation of SPOT imagery for the estimation of grassland biomass. *International Journal of Applied Earth Observation and Geoinformation*. 38:72–77.

Englhart, S., V. Keuck, and F. Siegert. 2011. Aboveground biomass retrieval in tropical forests-The potential of combined X-and L-band SAR data use. *Remote sensing of environment*. 115(5): 1260–1271.

Feierabend, J.S. and J.M. Zelazny. 1987. Status report on our nation's nation's wetlands. National Wildlife Federation, Washington, D.C. pp. 22–46.

Feliciano, E., S. Wdowinski, and M. Potts. 2014. Assessing Mangrove Above-Ground Biomass and Structure using Terrestrial Laser Scanning: A Case Study in the Everglades National Park. *Wetlands*:1–14. doi: 10.1007/s13157-014-0558-6.

Franklin, S.E. 2010. Remote Sensing for Biodiversity and Wildlife Management: Synthesis and Applications. McGraw-Hill Education, Inc. London, England. pp. 346. ISBN 0-07-162247-0.

Fregoni, M. 2010. Le aree viticole storiche nel mondo: I loro vitigni, la loro protezione e la tipicita' dei vini in esse ottenuti. In: Tomasi, D. (ed.) Proceedings of VIII International Terroir Congress; CRA, Conegliano, Italy, 3–22 May.

Gong, P. J. Wang, L. Yu, Y. Zhao, Y. Zhao, L. Liang, Z. Niu, X. Huang, H. Fu, S. Liu, C. Li, X. Li, W. Fu, C. Liu, Y. Xu, X. Wang, Q. Cheng, L. Hu, W. Yao, H. Zhang, P. Zhu, Z. Zhao, H. Zhang, Y. Zheng, L. Ji, Y. Zhang, H. Chen, A. Yan, J. Guo, L. Yu, L. Wang, X. Liu, T. Shi, M. Zhu, Y. Chen, G. Yang, P. Tang, B. Xu, C. Giri, N. Clinton, Z. Zhu, J. Chen, and J. Chen. 2013. Finer resolution observation and monitoring of GLC: first mapping results with Landsat TM and ETM + data. *Int. J. Remote Sens.* 34(7): 2607–2654.

Goodale, C.L. 2002. Forest carbon sinks in the Northern Hemisphere. *Ecological Applications*. 12(3):891–899.

Houghton, R.A., F. Hall, and S.J. Goetz. 2009. Importance of biomass in the global carbon cycle. *Journal of Geophysical Research: Biogeosciences*. 114(G2):2005–2012.

Guild, L.S., W.B. Cohen, and J.B. Kauffman. 2004. Detection of deforestation and land conversion in Rondônia, Brazil using change detection techniques. *International Journal of Remote Sensing*. 25(4):731–750.

Gupta, V. and A. Bernstein. 2001. "Keeping an Eye on the Island: Cooperative Remote Monitoring in the South China Sea," In: Baker, O'Connell and Williamson, eds., *Commercial Observation Satellites: At the Leading Edge of Global Transparency*. Santa Monica, CA and Bethesda, MD: RAND and the American Society for Photogrammetry and Remote Sensing. pp. 341–345 and 353–354.

Hall, A., D.W. Lamb, B. Holzapfel, and J. Louis. 2002. Optical remote sensing applications in viticulture-a review. *Australian Journal of Grape and Wine Research* 8:36–47.

Hall, A., J. Louis, and D.W. Lamb. 2003. Characterizing and mapping vineyard canopy using high-spatial-resolution aerial multispectral images. *Computers and Geosciences* 29:813–822.

Hansen, M.C., S.V. Stehman, P.V. Potapov, T.R. Loveland, J.R.G. Townshend, R.S. DeFries, K.W. Pittman, B. Arunarwati, F. Stolle, M.K. Steininger, M. Carroll, and C. DiMiceli. 2008. Humid tropical forest clearing from 2000 to 2005 quantified by using multitemporal and multiresolution remotely sensed data. Proceedings of the National Academy of Sciences of the United States. 105(27): 9439–9444.

Hansen, M.C., S.V. Stehman, and P.V. Potapov. 2010. Quantification of global gross forest cover loss. Proceedings of the National Academy of Sciences of the US. 107(19): 8650–8655.

Hargrove, W.W., J. Spruce, J. Kumar, and F.M. Hoffman. 2012. ForWarn Forest Disturbance Change Detection System Provides a Weekly Snapshot of US Forest Conditions to Aid Forest Managers. American Geophysical Union, Fall Meeting 2012. December 2012.

Hernandez-Leal, P.A., M. Arbelo, and A. Gonzalez-Calvo. 2006. Fire risk assessment using satellite data. *Advances in Space Research*. 37(4):741–746.

Hester, D. B., H. I. Cakir, S. A. C. Nelson, and S. Khorram. 2008. Per-pixel classification of high spatial resolution satellite imagery for urban land cover mapping, *Photogrammetric Engineering & Remote Sensing*. 74(4):463–471.

Hester, D.B., H.I. Cakir, S.A.C. Nelson, S. Khorram, and H.M. Cheshire. 2009, High-resolution land cover change detection based on fuzzy uncertainty analysis and change reasoning. International Journal of Remote Sensing. 31(2): 455–475.

Houghton, R.A. 2005. Aboveground Forest Biomass and the Global Carbon Balance. *Global Change Biology*. 11(6):945–958.

Houghton, R.A., F. Hall, and S.J. Goetz. 2009. Importance of biomass in the global carbon cycle. *Journal of Geophysical Research:Biogeosciences*. 114(G2):1–13.

Hyyppä, J., H. Hyyppä, P. Litkey, X. Yu, H. Haggrén, P. Rönnholm, U. Pyysalo, J. Pitkänen, and M. Maltamo. 2004. Algorithms and methods of airborne laser scanning for forest measurements. *International Archives of Photogrammetry, Remote Sensing and Spatial Information Sciences*. 36(8/W2):82–89.

IDMC. 2008. Guidance on profiling internally displaced people. http://www.internal-displacement.org/8025708F004BE3B1/(httpInfoFiles)/7620BC6F3CED0587C125739B00383CE0/$file/IDP_Profiling_Guidance_2008.pdf

Illera, P., A. Fernandez, and J.A. Delgado. 1996. Temporal evolution of the NDVI as an indicator of forest fire danger. *International Journal of Remote Sensing*. 17:1093–1105.

Ioannis, M. and M. Braun. 2014. Land Use and Land Cover Mapping in Europe. Practices and Trends. Springer Publishers. ISSN 1567-3200/ISBN 978-94-007-7968-6/ISBN 978-94-007-7969-3 (eBook). DOI 10.1007/978-94-007-7969-3.

Jaiswala, R.K., S. Mukherjeeb, K.D. Rajuc, and R. Saxenad. 2002. Forest fire risk zone mapping from satellite imagery and GIS. *International Journal of Applied Earth Observation and Geo information*. 4(1):1–10.

Jensen, J.R., S. Narumanlani, O. Weatherbee, and H.E. Mackey, Jr. 1993. Measurement of seasonal and yearly cattail and waterlily changes using multidate SPOT panchromatic data. *Photogrammetric Engineering and Remote Sensing*. 52:31–36.

Jensen, J.R. 2000. Remote sensing of the environment an earth resource perspective. Prentis-Hall, Inc. Upper Saddle River, NJ. pp. 544.

Johnson, L., B. Lobitz, R. Armstrong, R. Baldy, E. Weber, J. DeBenedictis, and D. Bosch. 1996. Airborne imaging aids vineyard canopy evaluation. *California Agriculture*. 15(4):14–18.

Johnson, L., D. Roczen, S. Youkhana, R. Nemani, and D. Bosch. 2003. Mapping vineyard leaf area with multispectral satellite imagery. *Computers and Electronics in Agriculture*. 38: 33–44.

Johnson, R.G. 2001. "Supporting the Dayton Peace Talks," in Baker, O'Connell and Williamson, eds., *Commercial Observation Satellites: At the Leading Edge of Global Transparency*. Santa Monica, CA and Bethesda, MD: RAND and the American Society for Photogrammetry and Remote Sensing. pp. 299, 302.

Keane, R.E., R. Burgan, and J. van Wagtendonk. 2001. Mapping wildland fuels for fire management across multiple scales: Integrating remote sensing, GIS, and biophysical modeling. *International Journal of Wildland Fire*. 10(4):301–319.

Kerr, J.T. and M. Ostrovsky. 2003. From space to species: ecological applications for remote sensing. *Trends in Ecology and Evolution*. 18(6):299–305.

Keywood, M., M. Kanakidou, A. Stohl, F. Dentener, G. Grassi, C.P. Meyer, K. Torseth, D. Edwards, A.M. Thompson, U. Lohmann, and J. Burrows. 2013. Fire in the Air: Biomass Burning Impacts in a Changing Climate. *Critical Reviews in Environmental Science and Technology*. 43(1): 40–83.

Khorram, S., 1982. A Remote Sensing-Aided Procedure for Site-Specific Estimation of Net Radiation Over Large Areas. *Journal of Applied Photographic Engineering, SPSE*. 89(1): 31–36.

Khorram, S. and H.M. Cheshire. 1985. Remote sensing of water quality in the Neuse River Estuary, North Carolina. *Photogrammetric Engineering and Remote Sensing*. 51:329–341.

Khorram, S., J. F. Knight, and H. I. Cakir. 2004. Thematic Accuracy Assessment of Regional Scale Land-Cover Data, In: Remote Sensing and GIS Accuracy, eds. Ross Lunetta and John Lyon, Chapter 7, pp. 107–124.

Khorram, S., F. Koch, C.F. van der Wiele, and S.A.C. Nelson. 2012. Remote Sensing. Springer. NY. ISBN 978-1-4614-3102-2. doi. 10.1007/978-1-4614-3103-9.

Khorram, S., C.F van der Wiele, and H. Yuan. 2011. Development of a Modified Neural Network Based Land Cover Classification System Using Automated Data Selector and Multiresolution Remotely-Sensed Data. *Geocarto International.* 26(6): 435–457.

Khorram, S., G. Biging, M. Potts, P. Gong, J. Ediriwickrama, and Y.L. Chen. 2014. Development of New Geospatial Tools for Wildland Risk Reduction. Final Report, NASA Headquarters, 26 p, ESPM, University of California, Berkeley.

Kloiber, S.M., T.H. Anderle, P.L. Brezonik, L. Olmanson, M.E. Bauer, and D.A. Brown. 2000. Trophic state assessment of lakes in the Twin Cities (Minnesota, US) region by satellite imagery. *Archive Hydrobiologie Special Issues Advances in Limnology*. 55:137–151.

Koh, L.P., J. Miettinen, S.C. Liew, and K. Ghazoul. 2011. Remotely sensed evidence of tropical peatland conversion to oil palm. Proceedings of the National Academy of Sciences of the US. 108(12): 5127–5132.

Kuemmerle, T., O. Chaskovsky, J. Knorn, V.C. Radeloff, I. Kruhlov, W.S. Keeton, and P. Hostert. 2009. Forest cover change and illegal logging in the Ukrainian Carpathians in the transition period from 1988 to 2007. *Remote Sensing of Environment*. 113: 1194–1207.

Kurvonen, L. J. Pulliainen, and M. Hallikainen. 1999. Retrieval of Biomass in Boreal Forests from Multitempotal ERS-1 and JERS-1 SAR Images. *IEEE Transactions on Geoscience and Remote Sensing*. 37(1):198–205.

Lacar, F.M., M.M. Lewis, and I.T. Griersom. 2001. Use of hyperspectral reflectance for discrimination between grape varieties. Proceedings of Geoscience IEEE 2001 Geoscience and Remote Sensing Symposium: Scanning the Present and Resolving the Future, Sydney, IEEE. 6, pp. 2878–2880.

Lamb, D.W. 2000. The use of qualitative airborne multispectral imaging for managing agricultural crops-a case study of southeastern Australia. *Australian Journal of Experimental Agriculture*. 40:725–738.

Langmann, B., B. Duncan, C. Textor, J. Trentmann, and G.R. van der Werf. 2009. Vegetation fire emissions and their impact on air pollution and climate. *Atmospheric Environment* 43(1):107–116.

Laporte, N.T., J.A. Stabach, R. Grosch, T.S. Lin, and S.J. Goetz. 2007. Expansion of industrial logging in Central Africa. *Science*. 316:1451.

Lathrop, R.G. and T.M. Lillesand. 1986. Utility of Thematic Mapper data to assess water quality. *Photogrammetric Engineering and Remote Sensing*. 52, 671–680.

Latifi, H., F.E. Fassnacht, F. Hartig, C. Berger, J. Hernandez, P. Corvalan, and B. Koch. 2015. Stratified aboveground forest biomass estimation by remote sensing data. *International Journal of Applied Earth Observation and Geoinformation*. 38:229–241.

Latifi, H., F. Fassnacht, and B. Koch. 2012. Forest structure modelling with combined airborne hyperspectral and LiDAR data. *Remote Sensing of Environment*. 121:10–25.

Lehmann, A. and J.B. Lachavanne. 1997. Geographic information systems and remote sensing in aquatic botany. *Aquatic Botany*: 58, 195–207.

LeToan, T. A. Beaudoin, J. Riom, D. Guyon. 1992. Relating forest biomass to SAR data. *IEEE Transactions on Geoscience and Remote Sensing*. 30(2):403–411.

Lillesand, T.M., W.L. Johnson, R.L. Deuell, O.M. Linstrom, and D.E. Meisner. 1983. Use of Landsat data to predict the trophic state of Minnesota lakes. *Photogrammetric Engineering and Remote Sensing*. 49:219–229.

Linkie, M., R.J. Smith, and N. Leader-Williams. 2004. Mapping and predicting deforestation patterns in the lowlands of Sumatra. *Biodiversity and Conservation*. 13:1809–1818.

Loehr, R.C. 1974. Characteristics and comparative magnitude of nonpoint sources. *Journal of Water Pollution Control Fed*. 46:1849–1872.

Lucas, L.A. 2007. Combinatorial screening of the effect of temperature on the microstructure and mobility of a high performance polythiophene semiconductor. *Applied physics letters*. 90(1):012112.

Lucas, R.M., A.C. Lee, and P.J. Bunting. 2008. Retrieving forest biomass through integration of CASI and LiDAR data. *International Journal of Remote Sensing*. 29(5):1553–1577.
Luckman, A. 1997. A study of the relationship between radar backscatter and regenerating tropical forest biomass for spaceborne SAR instruments. *Remote Sensing of Environment*. 60(1):1–13.
Ross, S., Lunetta, J.G., Lyon, B. Guindon, and C.D. Elvidge. 1998. North American Landscape Characterization Dataset Development and Data Fusion Issues. *Photogrammetric Engineering & Remote Sensing*. 64(8):821–829.
Luz, N.B. and D.L. Fonseca. 2003. Viabilidade na identificação de variedades de videiras pelas diferenças observadas na resposta espectral das folhas. In: Proceedings of the III Colóquio Brasileiro de Ciências Geodésicas, Curitiba, Brazil, Universidade Federal do Paraná, pp. 1–12.
Madden, C.J. and R.D. DeLaune. 1987. Chemistry and nutrient dynamics. the ecology of the barataria basin, louisiana: an estuarine profile. W.H. Conner and J.W. Day eds. Fish and Wildlife Service, Biological Report. 85(7.13):18–30.
Mitchard, E.T.A. 2009. Using satellite radar backscatter to predict above-ground woody biomass: A consistent relationship across four different African landscapes. *Geophysical Research Letters*. 36:23.
Menzel, W.P. 2006. Remote Sensing Applications with Meteorological Satellites. NOAA Satellite and Information Service. University of Wisconsin, Madison, WI. pp. 307.
Misbari, S. and M. Hashim. 2014. Total aboveground biomass (TAGB) estimation using IFSAR: speckle noise effect on TAGB in tropical forest. *IOP Conference Series: Earth and Environmental Science*. 18:1–6.
Moore, G.K. 1980. Satellite remote sensing of water turbidity. *Hydrobiological Sciences*. 25:407–421.
Moran, E.F. 1993. Deforestation and land use in the Brazilian Amazon. *Human Ecology*. 21(1):1–21.
Moran, M.S., Y. Inoue, and E.M. Barnes. 1997. Opportunities and limitations for image-based remote sensing in precision crop management. *Remote Sensing of Environment*. 6: 319–346.
Munyati, C. 2000. Wetland change detection on the Kafue Flats, Zambia, by classification of a multitemporal remote sensing image dataset. *International Journal of Remote Sensing*. 21(9):1787–1806.
Narumalani, S., J.R. Jensen, J.D. Althausen, S.G. Burkhalter, and H.E. Mackey Jr. 1997. Aquatic macrophyte modeling using GIS and logistic multiple regression. *Photogrammetric Engineering and Remote Sensing*. 63:41–49.
National Aeronautics and Space Administration (NASA). 1998. NASA's Earth Observing System—EOS AM-1. Greenbelt: NASA Goddard Space Flight Center. pp. 43.
National Aeronautics and Space Administration (NASA). 2011. World of Change: Amazon Deforestation: Feature Articles. NASA Earth Observatory. http://earthobservatory.nasa.gov/Features/WorldOfChange/deforestation.php
Nelson, S.A.C., K.S. Cheruvelil, and P.A. Soranno, 2006. Remote sensing of freshwater macrophytes using Landsat TM and the influence of water clarity. *Aquatic Botany*. 85: 289–298.
Nelson, S.A.C., P.A. Soranno, K.S. Cheruvelil, S.A. Batzli and D.L. Skole. 2003. Regional assessment of lake water clarity using satellite remote sensing 2003. *Journal of Limnology*. 62 (Suppl. 1):27–32.
Nelson, S.A.C. 2010. Review: Remote Sensing for Biodiversity and Wildlife Management by Steven E. Franklin, McGraw-Hill Publishers, London, England, 2010, 346 pp. *Photogrammetric Engineering & Remote Sensing*. 76(12): 1310–1312.
Nemani, R.R., L.F. Johnson, and M.A. White. 2006. Application of remote sensing and ecosystem modeling in vineyard management. In A. Srinivasan, ed. Handbook of Precision Agriculture: Principles and Applications. NY: The Haworth Press, Inc. pp. 413–429. DOI: 10.1300/5627_15
Olivier, J.G., G. Janssens-Maenhout, J.A. Peters. 2012. Trends in global CO2 emissions 2012 report. PBL Netherlands Environmental Assessment Agency. Hague, Netherlands. pp. 40
Pandey, U. 2010. Potential of Envisat ASAR data for woody biomass assessment. Tropical Ecology. 51(1):117.

Papathanassiou, K., S.R. Cloude. 2003. The effect of temporal decorrelation on the inversion of forest parameters from Pol-InSAR data. International Geoscience and Remote Sensing Symposium, pp. III: 1429–1431.

Parcak, S. and G. Mumford. 2010. South Sinai Survey and Excavation Projects in Egypt. SEPE: Satellite Imaging. http://www.deltasinai.com/image-00.htm. Accessed 2 September 2011.

Pedlowski, M.A., E.A.T. Matricardi, D. Skole, S.R. Cameron, W. Chomentowski, C. Fernandes, and A. Lisboa. 2005. Conservation units: a new deforestation frontier in the Amazonian state of Rondônia, Brazil. *Environmental Conservation*. 32(2): 149–155.

Penuelas, J., J.A. Gamon, K.L. Griffen, and C.B. Field. 1993. Assessing community type, plant biomass, pigment composition, and photosyntheic efficiency of aquatic vegetation on spectral reflectance. *Remote Sensing of Environment*. 46:110–118.

Petersen, G.W., J.M. Hamlett, G.M. Baumer, D.A. Miller, R.L. Day, and J.M. Russo. 1991. Evaluation of agricultural nonpoint pollution potential in Pennsylvania using a geographical information system. Final Report ME89279. Environmental Resources Research Institute, Pennsylvania State University, PA. pp 60.

Pierce, F.J. and P. Nowak. 1999. *Advances in Agronomy*. 67:1.

Pringle, H. 2011. Satellite imagery uncovers up to 17 lost Egyptian pyramids. 27 May 2011. AAAS Science Now. http://news.sciencemag.org/sciencenow/2011/05/satellite-imagery-uncovers-up-to.html?etoc&elq=c8aa9b3355d44873bf50dac382fafb07 Accessed 1 June 2011.

Rahman, H. 2001. Influence of atmospheric correction on the estimation of biophysical parameters of crop canopy using satellite remote sensing. *International Journal of Remote Sensing*. 22(7):1245–1268.

Ricketts, P.J. 1992. Current approaches in geographic, information systems for coastal management. *Marine Pollution Bulletin*. 25(1–4):82–87.

Robert, P.C. 2001. *Plant Soil*. 247, 143.

Samuel, H. 2011. Satellites help French winemakers pick a perfect harvest. The Telegraph (UK). http://www.telegraph.co.uk/news/worldnews/europe/france/8673485/Satellites-help-French-winemakers-pick-a-perfect-harvest.html. Accessed 1 August 2011.

Santoro, M., A. Shvidenko, I. McCallum, J. Askne, and C. Schmullius. 2007. Properties of ERS-1/2 coherence in the Siberian boreal forest and implications for stem volume retrieval. *Remote Sensing of Environment*. 106(2):154–172.

Sasser, C.E., M.D. Dozier, J.G. Gosselink, and J.M. Hill. 1986. Spatial and temporal changes in Louisiana's Barataria Basin marshes, 1945–1980. *Environmental Management*. 10(5):671–680.

Saugier, B., J. Roy, and H.A. Mooney. 2001. Estimations of global terrestrial productivity: converging toward a single number. Terrestrial global productivity. pp. 543–557.

Serrano, L., C. Gonázlez-Flor, and G. Gorchs. 2012. Assessment of grape yield and composition using the reflectance based water index in Mediterranean rainfed vineyards. *Remote Sensing of Environment*. 118:249–258. DOI: 10.1016/j.rse.2011.11.021

Shearman, P.L., J.A., B. Mackey, J.E. Bryan, and B. Lokes. 2009. Forest conversion and degradation in Papua New Guinea 1972–2002. *Biotropica*. 41(3): 379–390.

Song, C., C.E. Woodcock, K.C. Seto, M.P. Lenney, and S.A. Macomber. 2001. Classification and change attention using Landsat TM data: when and how to correct atmospheric effects?*Remote Sensing of Environment*. 75:230–244.

Spruce, J., W.W. Hargrove, J. Gasser, J. Smoot, and P. Kuper. 2014. MODIS NDVI Change Detection Techniques and Products Used in the Near Real Time Forwarn System for Detecting, Monitoring, and Analyzing Regional Forest Disturbances. American Geophysical Union, Fall Meeting 2014. December 2014.

Spruce, J., W.W. Hargrove, J. Gasser, and S.P. Norman. 2013. Monitoring Regional Forest Disturbances across the US with Near Real Time MODIS NDVI Products included in the ForWarn Forest Threat Early Warning System. American Geophysical Union, Fall Meeting 2013. December 2013.

Stafford, J.V. 2006. The role of technology in the emergence and current status of precision agriculture. In: A. Srinivasan, ed., Handbook of Precision Agriculture. NY: The Haworth Press, Inc. pp. 19–56. DOI: 10.1300/5627_02

Stephens, B.B. 2007. Weak northern and strong tropical land carbon uptake from vertical profiles of atmospheric CO2. *Science*. 316(5832):1732–1735.

Stratton, R.D. 2006. Guidance on spatial wildland fire analysis: models, tools, and techniques. Gen. Tech. Rep. RMRS-GTR-183. Fort Collins, CO: U.S. Department of Agriculture, Forest Service, Rocky Mountain Research Station. pp. 15. http://www.fs.fed.us/rm/pubs/rmrs_gtr183.pdf

Stroppiana, D., P.A. Brivio, J. Gragoire, C. Liousse, B. Guillaume, C. Granier, A. Mieville, M. Chin, and G. Patron. 2010. Comparison of global inventories of CO emissions from biomass burning derived from remotely sensed data. *Atmospheric Chemistry and Physics*. 10(24):12173–12189.

Tim, U.S. and R. Jolly. 1994. Evaluating agricultural nonpoint-source pollution using integrated geographic information systems and hydrologic/water quality model. *Journal of Environmental Quality*. 23:25–35.

Townsend, P.A., R.H. Gardner, T.R. Lookingbill, and C.C. Kingdon. 2012. Remote Sensing and Landscape Pattern Protocol for Land-Cover Monitoring of Parks, Natural Resource Report NPS/NCRN/NRR—2012/575.

Turner, W., S. Spector, N. Gardiner, M. Fladeland, E. Sterling, and M. Steininger. 2003. Remote sensing for biodiversity science and conservation. *Trends in Ecology and Evolution*. 18(6):306–314.

UNHCR. 2001. Handbook for Emergencies, 3rd Edition. http://www.unicef.org/emerg/files/UNHCR_handbook.pdf.

van Leeuwen, C. and G. Seguin. 2006. The concept of *terroir* in viticulture. *Journal of Wine Research*. 17(1): 1–10.

Verbyla, D.L. 1995. Satellite Remote Sensing of Natural Resources. NY: CRC Lewis. pp. 198.

Verlinden, A. and R. Masogo. 1997. Satellite remote sensing of habitat suitability for ungulates and ostrich in the Kalahari of Botswana. *Journal of Arid Environments*. 35:563–574.

Watt P.J. and D.N.M. Donoghue. 2005. Measuring forest structure with terrestrial laser scanning. *International Journal of Remote Sensing*. 26(7):1437–1446.

Wear, D.N., T.G. Turner, and R.J. Naiman. 1998. Land cover along an urban-rural gradient: implications for water quality. *Ecological Applications*. 8(3):619–630.

Weise, D.R. and C.S. Wright. 2014. Wildland fire emissions, carbon and climate: Characterizing wildland fuels. *Forest Ecology Management*. 317:26–40.

Whelan, B.M., J.A. Taylor, and A.B. McBratney. 2007. Potential Management Classes: Is there value in their delineation? In Proceedings of the 2007 Australasian Precision Agriculture Symposium (15-21). Australian Centre for Precision Agriculture, The University of Sydney, NSW.

Wickham, J., C. Homer, J. Vogelman, A. NcKerrow, R. Mueller, N. Herold, and J. Coulston. 2014. The Multi-Resolution Land Characteristics (MRLC) Consortium—20 Years of Development and Integration of US National Land Cover Data. *Remote Sensing*. *6*(8):7424–7441.

Wildland Fire Executive Council (WFEC) 2009. Quadrennial Fire Review (QFR) Final Report 2009. USDA Forest Service. http://www.forestsandrangelands.gov/strategy/building.shtml

Additional Reading

Remote Sensing Applications Center—USDA Forest Service, Active Fire Mapping Program http://activefiremaps.fs.fed.us/

Hernandez-Leal, P.A., A. Gonzalez-Calvo, M. Arbelo, A. Barreto and L. Arvelo-Valencia, Synergy of GIS and Remote Sensing data in forest fire danger modelling. Grupo de Observación de la Tierra y la Atmósfera (GOTA), Departamento de Física, Universidad de La Laguna, La Laguna 38200, Canary Islands (SPAIN). http://www.earsel.org/SIG/FF/6th-workshop/forest-fires-2007/58-141_Fire%20Risk%20Estimation/Synergy%20of%20GIS%20and%20Remote%20Sensing%20data%20in%20forest%20fire%20danger%20modelling.pdf

Cocks, T., R. Jenssen, A. Stewart, I. Wilson, and T. Shields. 1998. The HyMap airbornehyperspectral sensor: the system, calibration and performance. In: Presented at 1stEARSEL Workshop on Imaging Spectroscopy, Zurich, October 1998. http://www.hyvista.com/wp_11/wp-content/uploads/2011/02/EARSEL98_HyMap.pdf

Chuvieco, E., I. Aguado, and A.P. Dimitrakopoulos. 2004. Conversion of fuel moisture content values to ignition potential for integrated fire danger assessment. Can. J. For. Res. 34:2284–2293. http://www.geogra.uah.es/images/Documentos/emilio/PDF/chuvieco2004d.pdf

Encyclopedia of Earth: Monitoring Forest Fire Danger with Remote Sensing. http://www.eoearth.org/view/article/154673/

Maeda, E.E., A.R. Formaggio, Y.E. Shimabukuro, G.F.B. Arcoverde, A. Lima. 2009. Forest fire risk mapping in the Brazilian Amazon using MODIS images and artificial neural networks. Anais XIV Simposio Brasileiro de Sensoriamento Remoto, Natal Brasil. INPE, pp. 1425–1432. http://marte.sid.inpe.br/col/dpi.inpe.br/sbsr@80/2008/11.17.09.57/doc/1425-1432.pdf

Ashraf, M.A., J. Maah, and I. Yusoff. 2011. Introduction to Remote Sensing of Biomass, Biomass and Remote Sensing of Biomass, Dr. Islam Atazadeh (Ed.), ISBN: 978-953-307-490-0, InTech, DOI: 10.5772/16462. Available from: http://www.intechopen.com/books/biomass-and-remote-sensing-of-biomass/introduction-to-remote-sensing-of-biomass

Relevant Websites

Vegetation Health Web: http://www.orbit.nesdis.noaa.gov/smcd/emb/vci

Sea Turtle Tracking: www.seaturtle.org/tracking/

World Working Group on Birds of Prey and Owls: www.raptors-international.de

Invasive Grass behind Largest Wildfires-land use/land cover classification showed that cheatgrass correlated with wildfire incidents: http://news.sciencemag.org/sciencenow/2012/12/science-shot-invasive-grass-behin.html?ref=em

Painted lady butterfly migration thru radar: http://news.sciencemag.org/sciencenow/2012/10/sci-enceshot-butterfly-migration.html?ref=hp

Multi-generational long-distance migration of insects: studying the painted lady butterfly in the Western Palaearctic: http://onlinelibrary.wiley.com/doi/10.1111/j.1600-0587.2012.07738.x/abstract

Matching multiscale remote sensing data to interdisciplinary conservation needs: the case of chimpanzees in western Tanzania: http://www.isprs.org/proceedings/XXXIV/part1/paper/00059.pdf

Biologging technologies: new models for conservation: http://www.int-res.com/abstracts/esr/v10/

Science Today: Biologging: http://www.calacademy.org/sciencetoday/biologging/

Tracking Progress: Success and failure of biologging in protecting the global ocean. AAAS Annual Meeting, 16–20 February 2012: http://aaas.confex.com/aaas/2012/webprogram/Session4538.html

Summary on the special issue on biologging: http://www.eurekalert.org/pub_releases/2010-03/su-esr030310.php

Biologging sockeye salmon in British Columbia: VIDEO: http://membercentral.aaas.org/multimedia/videos/biologging-sockeye-salmon-british-columbia

Brunello under fire-Tuscan wine a different grape variety than label. http://www.winespectator.com/webfeature/show/id/Brunello-Under-Fire_4075

Widespread devastation found in 2010 Amazon Megadrought: http://news.sciencemag.org/climate/2012/12/widespread-devastation-found-2010-amazon-megadrought

Chapter 5
Atmospheric Applications of Remote Sensing

The discovery that the stratospheric ozone layer can be eroded by human activities along with the existence of a larger-than-expected hole over Antarctica (Fig. 5.1) more than 30 years ago propelled the use of remote sensing of Earth's atmospheric structure for more than weather forecasting; it became central to observation research and a tool for the development of environmental policy such as the Montreal Protocol on Substances that Deplete the Ozone Layer.

These concurrent discoveries spurred the application of satellite remote sensing to atmospheric and climate modeling as a way of better understanding our atmosphere and grappling with the impacts on both human and natural ecosystems.

Monitoring gases, air pollutants, radiation, water vapor, and other data using remotely sensed imagery has led to a better understanding of the dynamic physical and chemical processes in our atmosphere—both regionally and globally. LiDAR has been used to monitor trace gases from the ground level up to the stratosphere with a high-range resolution. Radar has been used for weather forecasting and tracking trends in other weather-related phenomena. Forecasting air quality and predicting climatic changes are now a societal imperative as airborne pollutants harm human and natural systems.

This chapter illustrates some of the many ways remote sensing has been used to advance our knowledge of atmospheric phenomena and airborne vectors; bring about critical advances in public health, safety, and welfare; and generate early warning systems.

S. Khorram et al., *Principles of Applied Remote Sensing*,
DOI 10.1007/978-3-319-22560-9_5

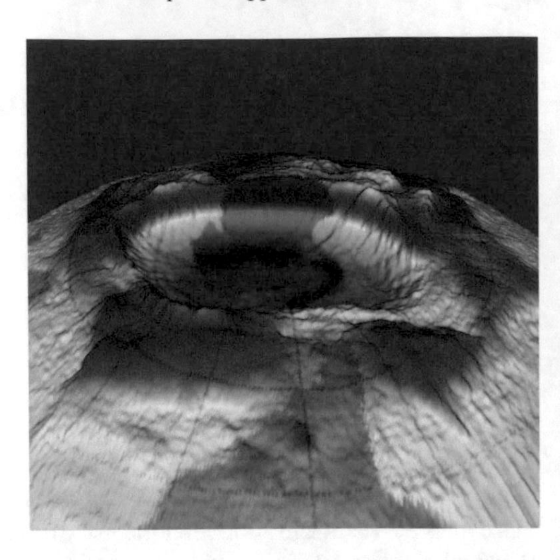

Fig. 5.1 Using a mapping spectrometer, NASA's Goddard Space Flight Center rendered this 3-D image of the hole in the ozone layer above Antarctica. Scientists discovered the hole in the 1980s; the phenomenon reappears during austral spring. (©NASA Goddard)

5.1 Weather Forecasting and Extreme Weather Events

The local weather forecasts that you rely on each day are the most commonplace atmospheric applications of remotely sensed data. In the US, the Geostationary Operational Environmental Satellites (GOES) have 4-km spatial resolution and 30-min temporal resolution. The images viewed on television in the US are from GOES satellites; METEOSAT is used throughout Europe and METEOR in Russia.

The basic mechanisms of how the ocean and atmosphere interact to form weather have been known for a long time. What is rather poorly understood are the *microphysics*—the relatively small interactions between temperatures and movements of water droplets and ice particles within a band of clouds—that can rapidly gain strength within a larger system, transforming an ordinary storm into an extreme weather event[1].

Meteorological scientists depend on both satellite and airborne sensors to better understand the phase changes of water—from ocean vapor to ice crystals—within the clouds of a storm system.

Figures 5.2, 5.3 and 5.4 depict Hurricane Frances along the eastern US. The true color composite was collected by SeaWIFS satellite and the classified map of the same hurricane was collected by GOES-12.

[1] Extreme weather includes unusual, severe, or unseasonal weather; weather at the extremes of the historical distribution—the range that has been seen in the past. Extreme events are typically based on a location's recorded weather history and defined as lying in the most unusual 10% (http://www.ncdc.noaa.gov/climate-information/extreme-events).

Fig. 5.2 True color composite of Hurricane Frances taken by SeaWIFS satellite on August 31, 2004. Tropical Storm Gaston is visible at the top of the image as it moves out over the Atlantic, south of Long Island (NY, US). (© NASA)

5.1.1 Measuring Precipitation from Space

Hydrometeorological events—either too much or too little precipitation—have severe consequences such as flooding, deadly landslides, and devastating droughts. Understanding the variability of precipitation events is an essential component in developing improved models to predict extreme weather scenarios. Space-borne sensors provide the tools for obtaining quantitative precipitation measurements

Fig. 5.3 Hurricane Frances depicted using GOES-12 visible image. (© NASA)

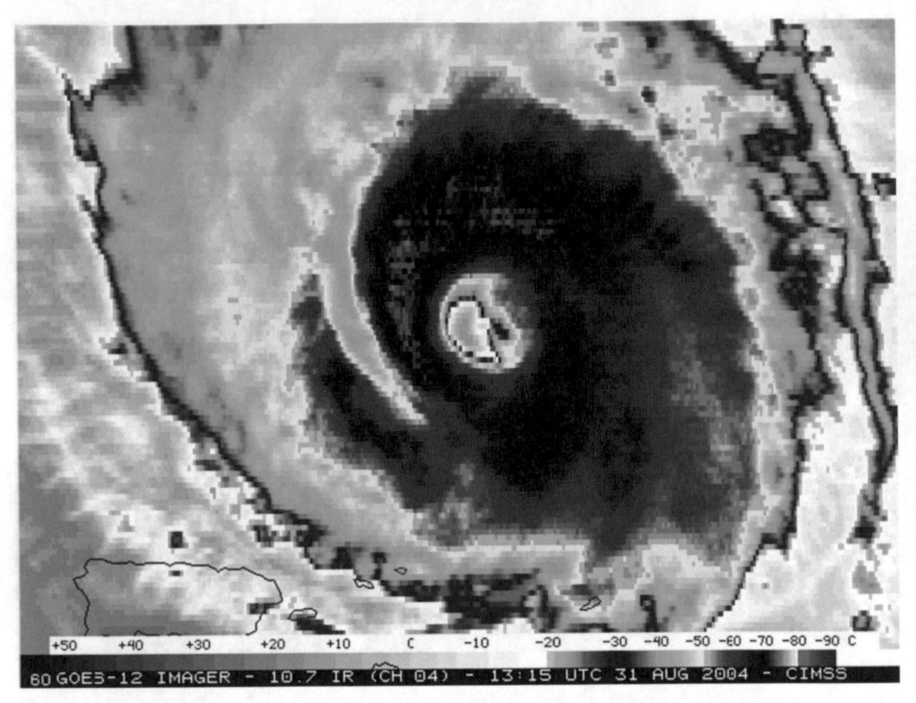

Fig. 5.4 Hurricane Frances depicted using GOES-12 Colorized IR image. *Red* shades indicate the severity and the wind velocity, with *dark red* being the highest wind velocity. (© NASA)

globally on sub-daily to decadal time scales (Kirschbaum and Hou 2011). Remotely sensed precipitation data products are particularly useful in locales with a lack of rain measurement instruments or radar networks, where extreme meteorological events regularly cause significant agricultural losses.

In 1997, the Tropical Rainfall Measuring Mission (TRMM) was launched as a means of confirming the validity of tropical rainfall estimates. Subsequently, the Global Precipitation Measurement (GPM) Mission improves upon the TRMM by using a number of satellites along with an advanced radar–radiometer system on a GPM "core satellite" to deliver near real-time estimates of rain and snow every 2–4 h globally. GPM was designed to advance scientific understanding of the Earth's water and energy cycle to improve predictions of weather and to track trends.

5.2 Public Health

Health providers, health effects researchers, and public health officials rely on environmental data to study and understand the geographic, environmental, and meteorological differences in disease and the effects of airborne pollutants. Geospatial technologies support enhanced understanding of how pollutants and vectors travel within the atmosphere. Health-related monitoring (Fig. 5.5) focuses on measuring

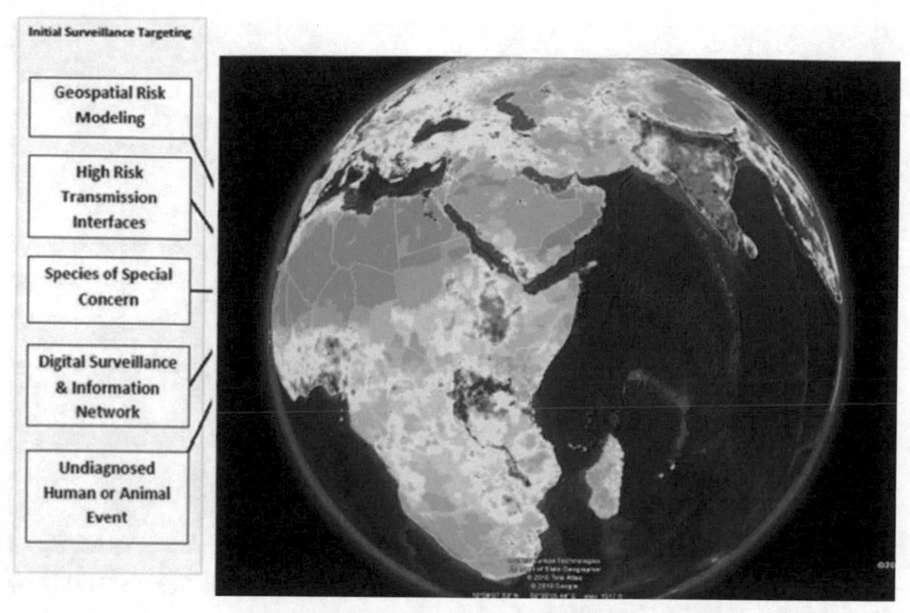

Fig. 5.5 Geospatial data is used to monitor environmental health impacts globally

airborne pollutants to better understand human and ecosystem health effects, and develop strategies to lessen the effects through better collaborative partnerships, environmental policies, and land planning measures.

5.2.1 Measuring Air Pollution to Understand Human and Ecosystem Health Impacts

Regulatory initiatives have improved air quality in North America and Europe; however, developing countries and rapidly growing megacities are still grappling with massive air pollution challenges. Many complex scientific questions remain, calling for innovative research. It has become increasingly clear that exposure to pollutants harms human and natural ecosystem health. The US Environmental Protection Agency (EPA) has set National Ambient Air Quality Standards (NAAQS) for six major pollutants (criteria pollutants): ozone, particulate matter <2.5 μm ($PM_{2.5}$), carbon monoxide (CO), sulfur dioxide (SO_2), nitrogen oxides (NOx), and lead. Poor air quality results in a variety of adverse respiratory and cardiovascular effects, cancer, and reduced lifespans. Air pollution also has been linked to adverse effects on ecosystems through wet (e.g., fog and precipitation) and dry (e.g., wind) deposition processes. Sulfur can lead to acidification of lakes and streams and damage plants and soils. Nitrogen in the atmosphere has been found to harm fish and other aquatic life. Elevated concentrations of ground-level ozone are also extremely harmful to human and ecosystem health.

Fig. 5.6 Hazy skies over eastern China on January 6, 2012 extending south from Beijing to near the Yellow Sea. (©MODIS/Aqua/NASA)

Remotely sensed data are integral to harmonizing weather, air quality, and climate models with aquatic and terrestrial ecosystem models to better understand the atmospheric–biogeochemical drivers of ecosystem exposure and the resulting effects. The Dutch–Finnish Ozone Monitoring Instrument (OMI), a part of NASA's Aura satellite, is a UV–visible spectrometer observing reflected sunlight in the 270–500 nm range that can measure criteria pollutants such as ozone, NO_2, SO_2, and aerosols. Aerosol optical depth (AOD), retrieved from MODIS Terra and Aqua satellites, measures light extinction by aerosol scattering and absorption in the atmospheric column. AOD values are highly correlated with $PM_{2.5}$ concentrations (Al-Saadi et al. 2005; Liu et al. 2004; van Donkelaar et al. 2006; van Donkelaar et al. 2010) (Figs. 5.6 and 5.7).

Case Study: Measuring China's Air Pollution

A team of researchers from Yale University and Battelle Memorial Institute mapped ground-level $PM_{2.5}$ concentrations in China using total column AOD from MODIS and MISR (Multiangle Imaging Spectroradiometer) satellites and coincident vertical profiles from the GEOS-Chem global chemical transport model. AOD measurements and ground-based $PM_{2.5}$ measurements were related using algorithms and models. Satellite-based air quality measurements can help to fill in spatial and information gaps where ground-based (in situ) monitoring stations are not available or practical. Satellite observations provide consistent, repeated data recording that allow for comparison over time and locations.

The data indicate that all but four provinces have average annual exposures above levels recommended by the World Health Organization (WHO).

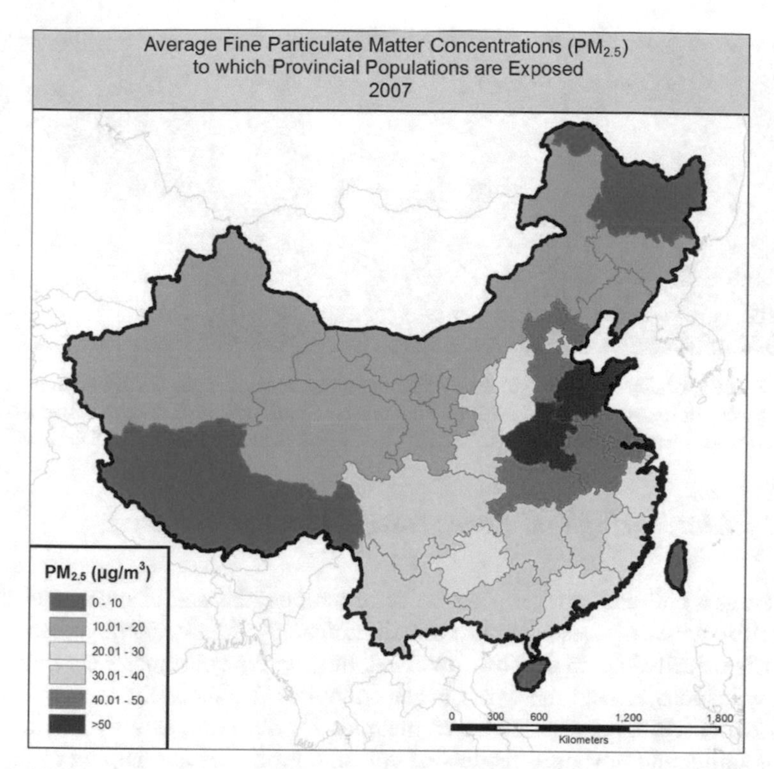

Fig. 5.7 2007 Annual average population-weighted $PM_{2.5}$ concentrations in China. (©Angel Hsu)

5.3 Appraising and Predicting Episodic Events

Episodic events are geologic and weather events that occur on an irregular, but periodic basis. Predicting the intensity of these often-catastrophic events is tricky—too many false alarms and people ignore warnings, while delayed information can result in avoidable tragedy. Episodic events include volcanic eruptions, earthquakes, dust storms, wildfires, and other such dramatic phenomena.

Satellite remote sensing has been used for both early warning of human and natural disasters and for post-disaster assessments. Images gathered during and in the immediate aftermath of an earthquake, flood, fire, or any other natural or human disaster have dramatically extended the ability of international organizations, government agencies, and information nongovernmental organizations (NGOs) to quickly gather, analyze, and distribute reliable. Images illustrate the extent of damage and locations that are impacted, providing a vital tool for emergency responders and aid agencies to properly size up the situation and develop appropriate responses. This information guides efforts on the ground such as pinpointing still-passable roads and bridges and finding safe locations for mobile medical units or refugee shelters.

Fig. 5.8 **a** Popocatépetl volcano in the Puebla region of Mexico emits a thick column of water vapor, gases, and ash on April 21, 2012. (© Hector Aiza Ramirex/Corbis Response Team, NASA). **b** A True-color image from MODIS-Terra on 20 April 2012. (© Jeff Schmaltz/MODIS Land Rapid)

5.3.1 Monitoring and Forecasting Volcanic Activity

Numerous small earthquakes, surface deformations, thermal changes, and the release of magmatic gases (Fig. 5.8) are all indicators of an eminent volcanic eruption. Historically, due to the costs involved, only active volcanoes were monitored; most volcanoes around the world are not monitored effectively, if at all (Sparks et al. 2012). Utilizing satellite-based methods for detecting activity has increased substantially, and advanced models of volcanic processes are helping to interpret monitoring data. The challenges include first distinguishing between magmatic, hydrothermal, and even atmospheric errors, and then figuring out which processes will culminate in eruption (Sparks et al. 2012).

Interferometric synthetic aperture radar (InSAR) uses the phase component of radar images to determine the position of the Earth's surface. *Digital elevation models* (DEMs)—crucial in predicting pyroclastic flows and lahars—with centimeter-scale accuracy are produced from simultaneously recorded images from different radars. Deformation is measured using time-separated images. Satellite data provide a global perspective, mapping tectonic strain across continents and allowing the exploration of volcanoes in remote, underfunded, or inaccessible locations. Volcanoes long thought to be dormant or inactive have been revealed as showing signs of activity (Fig. 5.9). Space-based observations enable limited financial and staff resources to be used more efficiently for locations that require detailed monitoring.

Thermal and radiant flux (power) changes—vital information taken from fumaroles and hot springs by intrepid scientists—have been measured using infrared hyperspectral data (Cipar et al. 2012). Using visible through short-wave infrared (Vis-NIR-SWIR) optical spectra data recorded in the 900–2500 nm wavelength range by the ARTEMIS sensor (flown on the TacSat-3 spacecraft), Cipar and his colleagues estimated the temperature and heat flux of an active lava lake within the Halema'uma'u crater of Kilauea caldera (Hawaii, US). Elevated radiance in the NIR–SWIR wavelength regions recorded a portion of the blackbody radiation function from small, hot areas of the lava lake, which were inverted to determine

Fig. 5.9 This InSAR image shows a pulse of uplift during 2004 and 2006 at Mount Longonot, Kenya, a volcano previously believed to be dormant. The image, from the ESA satellite Envisat, is draped over a digital elevation model from the Shuttle Radar Topography Mission. Each complete color cycle *(fringe)* represents 2.8 cm of displacement toward the satellite (Biggs et al. 2009) The distance between craters is ~35 km

the temperature and power output of the crater (Cipar et al. 2012). Corrections in the NIR–SWIR were needed to account for absorption and scattering by atmospheric constituents, primarily water vapor.

Monitoring gases—particularly SO_2—contained within volcanic ash clouds (Fig. 5.10 and Fig. 5.11) through remote sensing has yielded valuable information for air quality and public health, furthered the science in volcanology, enabled characterization of aviation hazards, and validated the global circulation model (GCM) in climate science (Kreuger et al. 2009; Rix et al. 2009). Sulfate in air pollution is well known for producing acid rain and fog, which in turn acidifies vegetation and water bodies, and causes adverse health effects. Hyperspectral instruments are able to easily detect SO_2 and differentiate between volcanic sources (which typically are launched into the stratosphere), and plumes from sulfate ore smelters, power plants, and industry (which are found at lower altitudes of the boundary layer and troposphere; Figs. 5.10 and 5.11).

Satellite data from thermal infrared instruments have been used to assess volcanic ash clouds and dispersal for several decades, but in order to make reasonably reliable forecasts of ash concentrations, an estimate of the thickness of the volcanic cloud is necessary. Volcanic ash is potentially dangerous for aircraft because, among other problems, the micron-size silicate particles melt below the operating temperatures of the engines, collect on turbine components, and block air flow to the point that engines may potentially shut down (International Civil Aviation Organization (ICAO) 2001; Kreuger et al. 2009). Information is insufficient about the exact tipping point of atmospheric ash concentrations that would cause engine shutdown; making judgments about aviation risk is further complicated by the differing characteristics of jet engines, the amount of time spent flying through ash clouds, and the characteristics of the ash cloud itself in terms of composition, particle size, spatial distribution, dispersion, particle removal rates and patterns, etc. (Prata and Prata 2012). Consequently, aviation authorities typically use a very cautious approach

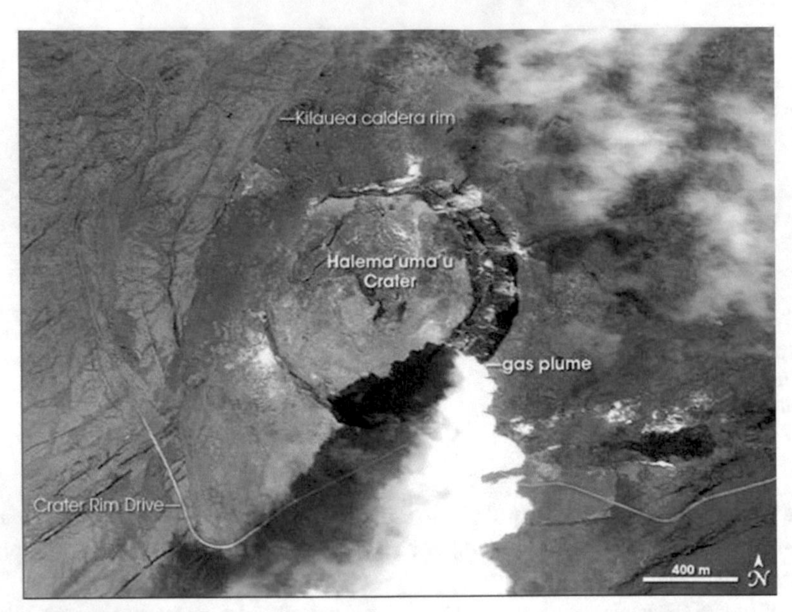

Fig. 5.10 On April 2, 2008, the U.S. Geological Survey's Hawaiian Volcano Observatory cautioned residents downwind of Kilauea Volcano that the Halema'uma'u vent in the summit crater was emitting hazardous amounts of sulfur dioxide (SO_2). The levels of the gas were high enough that even people without preexisting respiratory problems could be affected, and the National Park Service closed *Crater Rim Drive*. This image of the summit of Kilauea volcano shows a white plume streaming south–southwest from the active vent on March 29, 2008. Throughout the eruption, which began on March 19, emissions alternated between *gray–brown* plumes (dominated by ash) and *white* plumes (which contained rock dust and ash, gas, including SO_2, and steam). (© Formosat-2, Taiwan National Space Organization)

and close airspace, creating massive travel disruptions. Using multispectral satellite measurements from the Spin Enhanced Visible and Infrared Imager (SEVIRI), Prata and Prata (2012) performed a retrospective study of the April–May 2010 Eyjafjallajökull volcano (Iceland) eruption to develop ash mass loading forecasts in order to improve ash/aviation risk assessments.

5.3.2 Using Remote Sensing for Early Warning of Dust Storms

Dust storms (Fig. 5.12) are common episodic events in dryland areas—deserts and semi-arid rangelands—around the world. In addition to severely reducing visibility and drastically impacting air quality, dust storms have deleterious effects on human health, particularly children and elderly populations, and people with respiratory or pulmonary illnesses.

Dust storms reduce visibility (Figs. 5.13 and 5.14) and irritate eyes, nasal passages, and lungs; they also contain an enormous variety of harmful pollutants.

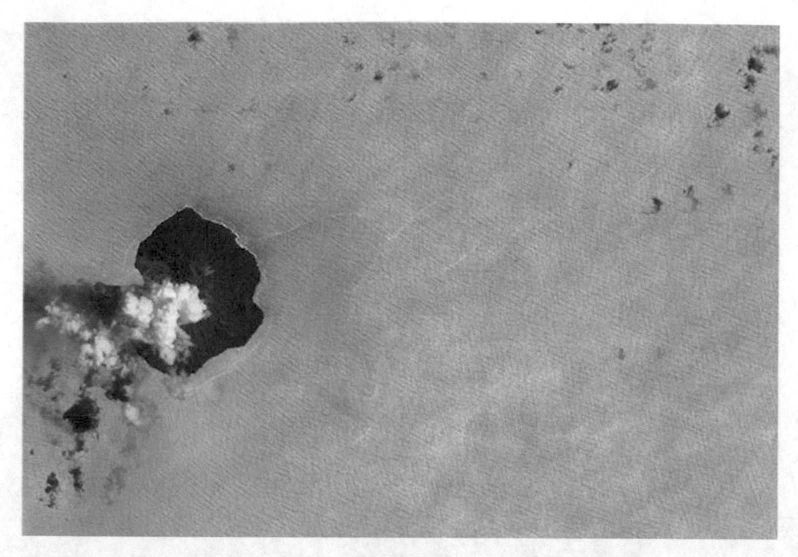

Fig. 5.11 Tinakula is a small, volcanic island in the South Pacific, located about 1400 miles northeast of Brisbane, Australia. This natural-color satellite image shows a plume of volcanic gas, possibly mixed with a bit of ash, rising above the island's summit. Over the past decade, satellites have detected intermittent "thermal anomalies" on the island that suggest eruptions have taken place, but eyewitness observations are infrequent. Around the island, the reflection of sunlight on the ocean—sunglint—gives the surface a milky appearance that makes the wave patterns readily visible. (© ALI/E-01/NASA)

Plant pollens, particulate matter, minerals such as phosphorous, pesticide residues, industrial pollutants (e.g., heavy metals and toxic chemicals), radioactive particles, as well as carcinogenic hydrocarbon compounds and living microorganisms—bacteria, fungi, and viruses—transmit diseases thousands of miles across continents to humans and livestock based on wind patterns and steering currents (Centeno 2011; Sprigg et al. 2009).

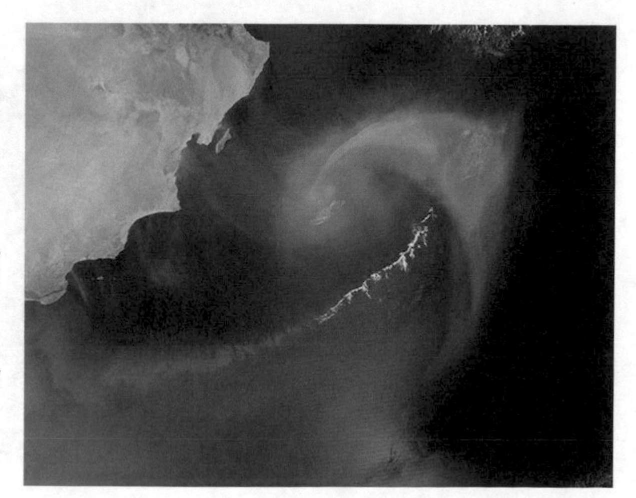

Fig. 5.12 Dust blowing over the Arabian Sea originating from a sand sea area known as the Empty Quarter or Rub' al Khali. The Empty Quarter—which contains roughly half as much sand as the entire Sahara desert—covers parts of Saudi Arabia, Yemen, Oman, and the United Arab Emirates and makes the Arabian Peninsula one of the world's most prolific dust-producing region. (©MODIS/Aqua/NASA)

Fig. 5.13 A dust storm strikes Phoenix, Arizona (US) on July 6, 2011. (© Nick Oza, The Arizona Republic/AP)

Annual volumes of dust transported from African deserts have been estimated at over 1 billion tons (Centeno 2011). Dust-borne particles from the African Sahara and Sahel deserts provide a much-needed source of minerals for nutrient-poor Amazon rainforest soils (Bristow et al. 2010; Koren et al. 2006; Swap et al. 1992) and phytoplankton in the Atlantic Ocean (Jickells et al. 2005; Mahowald et al. 2005), but have been found to contribute to coral reef bleaching (Shinn et al. 2001) and degraded air quality in South America, the southern US, and the Caribbean (Gyan et al. 2005; Prospero 1999; Prospero et al. 2001). Korea and Japan experience

Fig. 5.14 A dust storm moves through Phoenix, Arizona (US) on July 6, 2011. An enormous wall of dust turned the sky brown, creating dangerous driving conditions. (© David Kadlubowski, The Arizona Republic/AP)

storms of yellow dust from China and dust clouds from the Gobi, and Taklimakan deserts in Asia reach the northwestern US (Centeno 2011; Struck 2008).

Seasonal outbreaks of meningitis and silicosis lung disease in Africa and Kazakhstan have been attributed to dust storms. Thus, if there is a pandemic, such as avian flu, West Nile virus, or other disease, it can very quickly move into other areas (Struck 2008). Prolonged droughts and climatic changes can increase the occurrence of dust storms. Scientists had originally thought that fungi, bacteria, and viruses contained within dust clouds would die on exposure to ultraviolet radiation and cold temperatures; however, viable microbes have been detected in dust tracked as far as 6500 km (Prospero et al. 2005; Struck 2008).

This sobering discovery prompted the UN's World Meteorological Organization (WMO) to develop a satellite-based modeling system to track dust storms globally and alert those in its path. In addition, researchers Stanley Morain of University of New Mexico (Albuquerque, NM, US) and William Sprigg of University of Arizona (Tucson, AZ, US) collaborated with NASA and the WMO to develop better forecasting capabilities from satellite data. Since 2004, NASA's Public Health Applications in Remote Sensing (PHAiRS) project has worked to create a forecasting model that takes into account wind speed and direction, near-surface temperature, surface topography, surface roughness, and the proportion of land to water using data from NASA's Terra and Aqua satellites. The resulting model has been able to predict the timing of two out of three dust storms in Phoenix, AZ (US).

The ultimate goal is to produce a reporting system that public health officials can use to warn the public approximately 48 h in advance and anticipate dust-related pandemics (Sprigg et al. 2009).

5.4 Global Climate Change

Remote sensing plays a prominent and ongoing role in the monitoring of phenomena associated with global climate change. For instance, in keeping with the methodology and findings of the IPPC Fourth Assessment Report, NASA's Global Climate Change Web site (http://climate.nasa.gov/index.cfm) provides summary data about five key quantitative indicators, all of which are currently tracked by satellite-based sensors: (1) the extent of arctic sea ice, (2) the extent of land ice, (3) sea level rise, (4) the atmospheric carbon dioxide concentration, and (5) global temperature. Below, we provide examples where specific spaceborne sensors have been applied to provide information about these five interrelated topic areas.

Arctic Sea Ice In the Arctic Ocean, sea ice reaches its minimum annual extent each September. In September 2007, Arctic sea ice fell to its lowest extent ever, which was 24% lower than the previous record minimum (September 2005) and 37% below the historical average (Comiso et al. 2008; Giles et al. 2008). Giles et al. (2008) used data from the RA-2 radar altimeter aboard the European Space Agency's Envisat satellite to look at sea ice thickness anomalies in the Arctic region. They

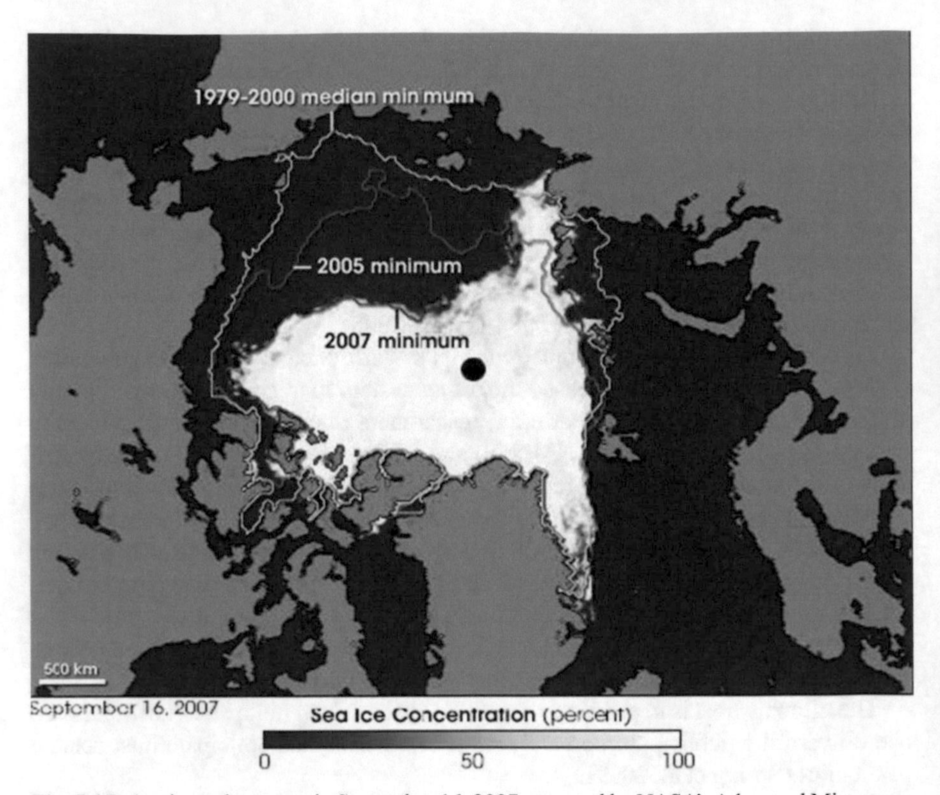

Fig. 5.15 Arctic sea ice extent in September 16, 2007 as record by NASA's Advanced Microwave Scanning Radiometer—Earth Observing System (AMSR-E) sensor mounted on the Aqua satellite. Blue indicates open water, white indicates high sea ice concentration, and turquoise indicates loosely packed sea ice. Previous minimum sea ice extents are shown for comparison. The black circle represents a lack of data as the North Pole is beyond the maximum latitude observed by the satellite. (© NASA and U.S. National Snow and Ice Data Center (http://earthobservatory.nasa.gov/IOTD/view.php?id=8126))

found that average ice thickness during winter 2007–2008 was 0.26 m less than the average winter-season thickness for the previous 6 years (winter 2002–2003 to winter 2006–2007). The decline in ice thickness was particularly pronounced in the Western Arctic, where the winter 2007–2008 thickness was 0.49 m below the average of the previous 6 years.

Figure 5.15 depicts an image of the Arctic region captured on September 16, 2007 by the Advanced Microwave Scanning Radiometer for EOS (AMSR-E), a sensor on NASA's Aqua satellite. The image contains three contour lines depicting annual minimum sea ice extents as reported by the U.S. National Snow and Ice Data Center. The red line is the September 2007 minimum (i.e., the record-low extent), while the green line indicates the September 2005 minimum and the yellow line indicates the median minimum extent between 1979 and 2000.

Land Ice In the context of climate change, the phrase "land ice" especially refers to the major ice sheets of Antarctica and Greenland. Temperatures on the Antarctic

Fig. 5.16 Envisat Advanced Synthetic Aperture Radar (ASAR) image of the Larsen B ice shelf, acquired 22 March 2007. Previous historical extents, determined using synthetic aperture radar (SAR) imagery from the European Space Agency's ERS satellite. (© European Space Agency (http://esa-multimedia.esa.int/images/EarthObservation/Envisat/Larsen-2007.jpg))

Peninsula have risen more than 0.1 °C per decade for the past 50 years (Steig et al. 2009), while temperatures have risen sharply in Greenland since 1990 (Hanna et al. 2008). As a result of these rising temperatures, the Antarctic and Greenland ice sheets have been losing mass at an accelerating rate. Based on data from NASA's GRACE (Gravity Recovery and Climate Experiment) satellite mission, Velicogna (2009) estimated that the mass loss of the Greenland ice sheet increased from 137 Gt/year in 2002–2003 to 286 Gt/year in 2007–2009. Similarly, in Antarctica, the mass loss increased from 104 Gt/year in 2002–2006 to 246 Gt/year in 2006–2009.

On the Antarctic Peninsula, accelerating ice loss led to two major collapses of the Larsen Ice Shelf in the last two decades. First, synthetic aperture radar (SAR) images from the European Space Agency's ERS-1 satellite documented the breaking away of a 4200 km^2 section of the Larsen-A ice shelf in January 1995, which disintegrated almost completely within a few days (Rott et al. 1996). Similarly, ERS SAR images documented the rapid breakup of 2300 km^2 of the Larsen-B ice shelf during 1 week in March 2002, representing a large portion of the decrease in the ice shelf's total area from approximately 11,500 km^2 in 1995 to less than 3000 km^2 in 2003 (Rack and Rott 2004). Figure 5.16 is an image of the Larsen-B ice shelf, ac-

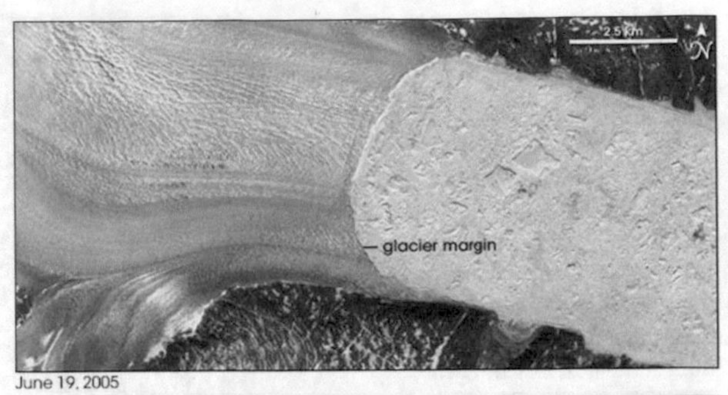

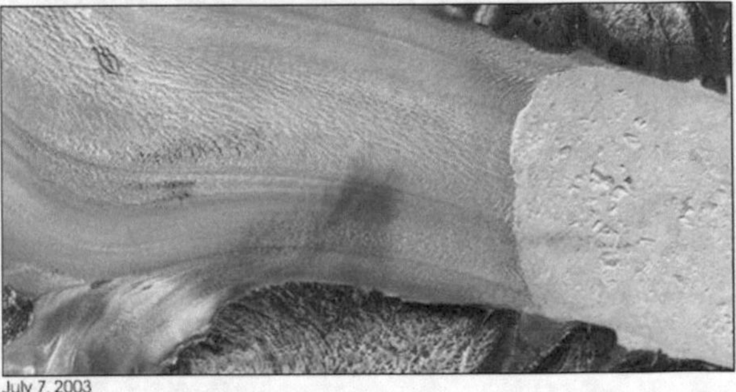

Fig. 5.17 Time series of Advanced Spaceborne Thermal Emission and Reflection Radiometer (ASTER) images of the Helheim glacier, Greenland, acquired in June 2005 *(top)*, July 2003 *(middle)*, and May 2001 *(bottom)*. http://earthobservatory.nasa.gov/IOTD/view.php?id=6207. (©Jesse Allen, NASA Earth Observatory)

quired in 2007 by the Advanced Synthetic Aperture Radar (ASAR) sensor onboard the European Space Agency's Envisat satellite. Previous ice extents at a series of dates between 1992 and 2002 were derived from ERS SAR or Envisat ASAR imagery. As the image suggests, the Larsen-B shelf has continued to decline steadily after its 2002 collapse.

While rising temperatures have increased surface melting of the Greenland Ice Sheet in the last couple of decades, Howat et al. (2005) argued that much of the ice sheet's rapid thinning (up to 10 m/year) may be due to recent acceleration of outlet glaciers (i.e., glaciers that flow through narrow fjords and out to sea). The researchers used RADARSAT and Advanced Spaceborne Thermal Emission and Reflection Radiometer (ASTER) imagery to look at ice flow velocities on Greenland's Helheim Glacier. They found that the outlet glacier's margin retreated landward more than 7.5 km between 2000 and 2005, as illustrated by Fig. 5.17. Because of

Fig. 5.18 Global map of sea level trends during the 1992–2011 time period, constructed from satellite altimetry data. (© NOAA)

this retreat, the mass of ice that restricted the glacier's speed was released, allowing the glacier to accelerate. Increased ice melt due to rising temperatures may lead to a feedback loop: thinning ice near the glacier's margin leads to increased calving (i.e., creation of icebergs) and further glacial retreat, releasing more ice and allowing the glacier to accelerate, which causes even more thinning (http://earthobservatory.nasa.gov/IOTD/view.php?id=6207).

Sea Level Rise Sea level rise is caused by the thermal expansion of sea water due to warming temperatures, as well as melting of nonpolar glaciers, ice caps, and the polar ice sheets (IPCC 2007a; Vaughan 2005; also see previous discussion). In the twentieth century, the global average sea level rose approximately 1.7 mm/year (Church and White 2006); moreover, sea level rise accelerated significantly (by 0.013 ± 0.006 mm/year) during this period. In the last two decades, a series of satellite altimetry missions have facilitated global mapping of ocean surface topography, including sea surface height: TOPEX/Poseidon (launched in 1992), Jason-1 (launched in 2001), and Jason-2 (launched in 2008). The Jason-2 satellite is operated by the Ocean Surface Topography Mission (OSTM), which is a joint effort of NASA, the National Oceanic and Atmospheric Administration (NOAA), the European Organisation for the Exploitation of Meteorological Satellites (EUMETSAT), and France's Centre National d'Etudes Spatiales (CNES).

Figure 5.18 shows a map of sea level trends for the period between 1992 and 2011, constructed from data acquired by the TOPEX/Poseidon, Jason-1, and Jason-2 satellites. The large amount of spatial variation in sea levels reflects some influence of nonuniform changes in temperature and salinity, as well as changes

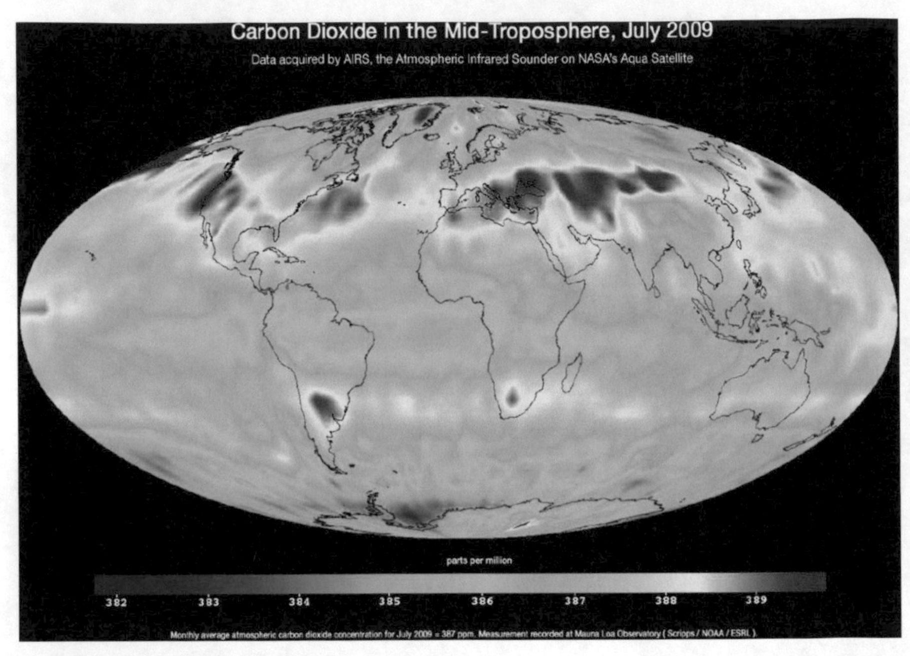

Fig. 5.19 Map of mid-troposphere CO_2 concentrations, constructed using data acquired by the Atmospheric Infrared Sounder (AIRS) sensor during July 2009. Image courtesy of NASA/Jet Propulsion Laboratory

in ocean circulation (IPCC 2007b). Overall, the global sea level has risen approximately 3.1 mm/year since 1993 (IPCC 2007a).

Atmospheric Carbon Dioxide Concentration Carbon dioxide (CO_2) is the most prominent of the "greenhouse gases" implicated in the warming of the Earth's surface. Global CO_2 emissions come from natural (e.g., volcano eruptions) and anthropogenic sources; by far, the largest source of anthropogenic emissions is fossil fuel use (IPCC 2007a). Other greenhouse gases with anthropogenic sources, such as methane (CH_4) and nitrous oxide (N_2O), have also been linked to increases in global temperature during the industrial era, and particularly in the last 50 years (IPCC 2007a).

A principal tool in current greenhouse gas monitoring and research is the Atmospheric Infrared Sounder (AIRS) instrument, which is mounted on NASA's Aqua satellite. The AIRS sensor was designed to measure the levels of atmospheric gases that influence global climate, including CO_2, carbon monoxide, methane, and ozone; it also provides profiles of atmospheric and surface (i.e., sea- and land-surface) temperature, relative humidity and water vapor, and various cloud properties (Aumann et al. 2003). Figure 5.19 is an image of global CO_2 concentrations, created with data acquired by the AIRS sensor during July 2009. The observed concentration patterns result from the transportation of CO_2 around the Earth by the general circulation of the atmosphere. The northern hemisphere mid-latitude jet stream serves as a north-

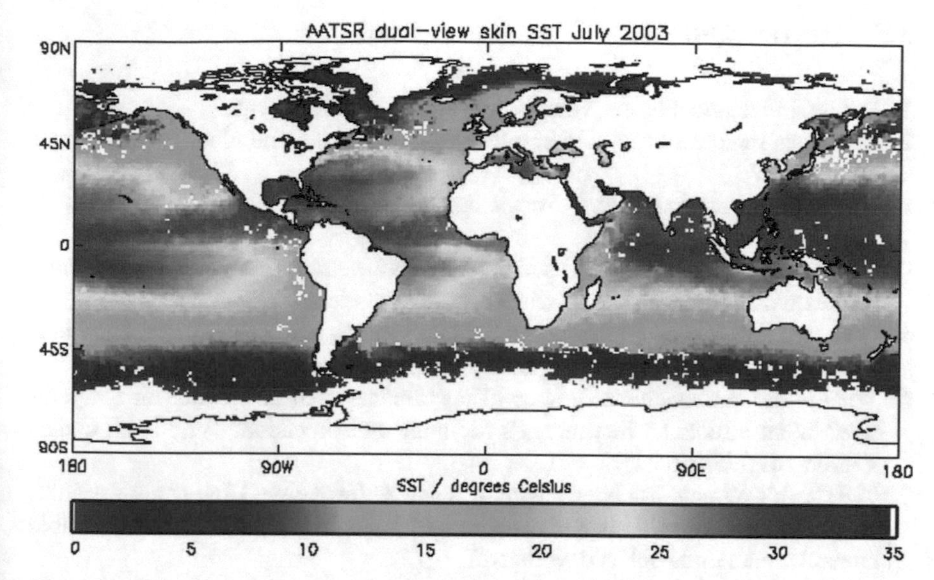

Fig. 5.20 Global map of sea surface temperature (SST) derived from data collected by the Advanced Along-Track Scanning Radiometer (AATSR) during July 2003. Image courtesy of the European Space Agency (data prepared by Anne O'Carrol, Meteorological Office, UK)

ern limit of enhanced CO_2. A belt of enhanced CO_2 circles the globe in the southern hemisphere, in this case corresponding to the flow of the southern hemisphere mid-latitude jet stream. This belt is fed by biogenesis activity in South America (i.e., the release of CO_2 through the respiration and decomposition of vegetation), forest fires in South America and Central Africa, and clusters of gasification plants in South Africa and power generation plants in southeastern Australia (http://photojournal.jpl.nasa.gov/catalog/PIA12339).

Global Surface Temperature One of the most reliable parameters for tracking climate change is sea surface temperature (SST). In short, heat is a major determinant of global climate, and the oceans serve a major role as massive heat reservoirs (http://science.nasa.gov/earth-science/oceanography/physical-ocean/temperature/). Sensors that provide long-term SST data suitable for climate change detection include the Advanced Very High Resolution Radiometers (AVHRR) onboard NOAA's polar-orbiting satellites, as well as the Along-Track Scanning Radiometer (ATSR) series of instruments onboard the European Space Agency's ERS-1, ERS-2, and Envisat satellites (Good et al. 2007). To cite one application example, Good et al. (2007) used AVHRR data to estimate global trends in sea surface temperature for a 20-year period (1985–2004). They found that daytime and nighttime temperatures increased 0.18 (±0.04) and 0.17 (±0.05) °C per decade, respectively. Figure 5.20 shows an example of a global SST map derived using data, from July 2003, that were collected by the Advanced Along-Track Scanning Radiometer (AATSR) sensor onboard Envisat.

5.5 Review Questions

1. Define, in technical terms, what is an extreme weather event.
2. How does remote sensing aid in either the precision or the accuracy of weather forecasting?
3. Define the term, "pollution" from a human and global (i.e., plant and animal) perspective.
4. How is remotely sensed data used to harmonize weather, air quality, and climate models with aquatic and terrestrial ecosystem models?
5. Give two specific examples where remote sensing can be used to predict episodic events. What instruments or techniques are used?
6. What is the mathematic tolerance of instruments used to monitor air temperature? What affects an instrument's accuracy and precision? Why is an understanding of this critical?
7. The Earth's climate has been changing since its formation. How many variables should be monitored? How many degrees of freedom will there be in a workable climate model once one is developed?

References

Al-Saadi, J., J. Szykman, R.B. Pierce, C. Kittaka, D. Neil, D.A. Chu, L. Remer, L. Gumley, E. Prins, L. Weinstock, C. MacDonald, R. Wayland, F. Dimmick, and J. Fishman. 2005. Improving national air quality forecasts with satellite aerosol observation. *Bulletin of the American Meteorological Society* 86, 1249–1261. doi:10.1175/BAMS-86-9-1249

Aumann, H.H., M.T. Chahine, C. Gautier, M.D. Goldberg, E. Kalnay, L.M. McMillin, H. Revercomb, P.W. Rosenkranz, W. L. Smith, D.H. Staelin, L.L. Strow, and J. Susskind. 2003. AIRS/AMSU/HSB on the Aqua Mission: design, science objectives, data products, and processing systems. *IEEE Transactions on Geoscience and Remote Sensing* 41(2): 253–264.

Biggs, J., E.Y. Anthony and C.J. Ebinger. 2009. Multiple inflation and deflation events at Kenyan volcanoes, East African Rift. *Geology* 37(11): 979–982.

Bristow, C.S., K.A. Hudson-Edwards, and A. Chappell. 2010. Fertilizing the Amazon and equatorial Atlantic with West African dust. *Geophysical Research Letters* 37: L14807. doi:10.1029/2010GL043486

Centeno, J.A. 2011. Chemical and pathology studies of particulate matter—a medical geology perspective. In: Budge, A.M. and S.A. Morain, eds., International Society for Photogrammetry and Remote Sensing. Advances in Geospatial Technologies for Health. Proceedings. XXXVIII-8/C23. 12–13 September 2011, Santa Fe, NM, US. pp. 22.

Church, J. A. and N.J. White. 2006. A 20th century acceleration in global sea-level rise, *Geophysical Research Letters* 33: L01602.

Cipar, J.J., G.P. Anderson, and T.W. Cooley. 2012. Halema'uma'u Crater, Hawaii, using a space-based hyperspectral imager. IEEE *Journal of Selected Topics in Applied Earth Observations and Remote Sensing* 1–8. doi: 10.1109/JSTARS.2012.2184086

Comiso, J.C., C.L. Parkinson, R. Gersten, and L. Stock. 2008. Accelerated decline in the Arctic sea ice cover, *Geophysical Research Letters* 35: L01703.

Giles, K.A., S.W. Laxon, and A.L. Ridout. 2008. Circumpolar thinning of Arctic sea ice following the 2007 record ice extent minimum. *Geophysical Research Letters* 35: L22502.

Good, S.A., G.K. Corlett, J.J. Remedios, E.J. Noyes, and D.T. Llewellyn-Jones. 2007. The global trend in sea surface temperature from 20 years of Advanced Very High Resolution Radiometer data. *Journal of Climate* 20: 1255–1264.

Gyan, K., W. Henry, S. Lacaille, A. Laloo, C. Lamsee-Ebanks, S. McKay, R.M. Antoine and M. A. Monteil. 2005. African dust clouds are associated with increased paediatric asthma accident and emergency admissions on the Caribbean island of Trinidad. *Int. J. Biometeorol.* 49: 371–376. doi:10.1007/s00484-005-0257-3

Hanna, E., P. Huybrechts, K. Steffen, J. Cappelen, R. Huff, C. Shuman, T. Irvine-Fynn, S. Wise, and M. Griffiths. 2008. Increased runoff melt from the Greenland Ice Sheet: a response to global warming. *Journal of Climate* 21: 331–341.

Howat, I.M., I. Joughin, S. Tulaczyk, and S. Gogineni. 2005. Rapid retreat and acceleration of Helheim Glacier, east Greenland, *Geophysical Research Letters* 32: L22502.

Intergovernmental Panel on Climate Change (IPCC). 2007a. Climate Change 2007: Synthesis Report. Contribution of Working Groups I, II and III to the Fourth Assessment Report of the Intergovernmental Panel on Climate Change [Core Writing Team, Pachauri, R.K and Reisinger, A. (eds.)]. IPCC, Geneva, Switzerland.

Intergovernmental Panel on Climate Change (IPCC). 2007b. Climate Change 2007: The Physical Science Basis. Contribution of Working Group I to the Fourth Assessment Report of the Intergovernmental Panel on Climate Change [Solomon, S., D. Qin, M. Manning, Z. Chen, M. Marquis, K.B. Averyt, M. Tignor and H.L. Miller (Eds.)]. Cambridge University Press, Cambridge, United Kingdom and New York, NY, US.

International Civil Aviation Union (ICAO). 2001. Manual on volcanic ash, radioactive material and toxic chemical clouds, ICAO Doc. 9766-AN/954, Montreal, Quebec, Canada.

Jickells, T.D., Z.S. An, K. K. Andersen, A.R. Baker, G. Bergametti, N. Brooks, J.J. Cao, P.W. Boyd, R.A. Duce, K.A. Hunter, H. Kawahata, N. Kubilay, J. laRoche, P.S. Liss, N. Mahowald, J.M. Prospero, A.J. Ridgwell, I. Tegen, and R. Torres. 2005. Global iron connections between desert dust, ocean biogeochemistry, and climate. *Science* 308: 67–71, doi:10.1126/science.1105959

Kirschbaum, D. and A. Hou. 2011. Advancing extreme weather monitoring from space: from TRMM to GPM. Earthzine: fostering earth observation and global awareness. Online available: http://earthzine.org/2011/04/18/advancing-extreme-weather-monitoring-from-space-from-trmm-to-gpm/

Koren, I., Y.J. Kaufman, R. Washington, M.C. Todd, Y. Rudich, J. Vanderlei Martins, and D. Rosenfeld. 2006. The Bodélé depression: a single spot in the Sahara that provides most of the mineral dust to the Amazon forest. *Environ. Res. Lett.*, 1(1), 014005, doi:10.1088/1748–9326/1/1/014005

Kreuger, A.J., N.A. Krotkov, K. Yang, S. Carn, G. Vincente, and W. Schroeder. Applications of satellite-based sulfur dioxide monitoring. 2009. IEEE *Journal of Selected Topics in Applied Earth Observations and Remote Sensing* 2: 293–298.

Liu, Y., R.J. Park, D.J. Jacob, L. Qinbin, V. Kilaru, and J.A. Sarnat. 2004. Mapping annual mean ground-level $PM_{2.5}$ concentrations using Multiangle Imaging Spectroradiometer aerosol optical thickness over contiguous United States. *J. Geophys. Res.* 109: D22206. doi:10.1029/2004JD005025

Mahowald, N.M., A.R. Baker, G. Bergametti, N. Brooks, R.A. Duce, T.D. Jickells, N. Kubilay, J.M. Prospero, and I. Tegen. 2005. Atmospheric global dust cycle and iron inputs to the ocean. *Global Biogeochem. Cycles* 19: GB4025. doi:10.1029/2004GB002402

NOAA. National Centers for Environmental Information. *Extreme Events*. http://www.ncdc.noaa.gov/climate-information/extreme-events. Accessed 29 May 2015.

Prata, A.J. and A.T. Prata. 2012. Eyjafjallajökull volcanic ash concentrations determined using spin enhanced visible and infrared imager measurements. *Journal of Geophysical Research-Atmospheres* 117: 1–24. doi:10.1029/2011JD016800

Prospero, J.M. 1999. Long-term measurements of the transport of African mineral dust to the southeastern United States: Implications for regional air quality. *J. Geophys. Res.* 104: 15,917–15,927. doi:10.1029/1999JD900072

Prospero, J.M., I. Olmez, and M. Ames. 2001. Al and Fe in PM 2.5 and PM 10 suspended particles in south-central Florida: The impact of the long range transport of African mineral dust. *Water Air Soil Pollut.* 125: 291–317. doi:10.1023/A:1005277214288

Prospero, J.M., E. Blades, G. Mathison, and R. Naidu. 2005. Interhemispheric transport of viable fungi and bacteria from Africa to the Caribbean with soil dust. *Aerobiologia* 21: 1–19. doi:10.1007/s10453-004-5872-7

Rack, W. and H. Rott. 2004. Pattern of retreat and disintegration of the Larsen B ice shelf, Antarctic Peninsula. *Annals of Glaciology* 39: 505–510.

Rix, M., P. Valks, N. Hao, J. van Geffen, C. Clerbaux, L. Clarisse, P-F. Coheur, D.G. Loyola, T. Erbertseder, W. Zimmer, and S. Emmadi. 2009. Satellite monitoring of volcanic sulfur dioxide emissions for early warning of volcanic hazards. IEEE *Journal of Selected Topics in Applied Earth Observations and Remote Sensing* 2: 196–206.

Rott, H., P. Skvarca, and T. Nagler. 1996. Rapid collapse of northern Larsen ice shelf, Antarctica. *Science* 171: 788–792.

Shinn, E.A., G.W. Smith, J.M. Prospero, P. Betzer, M.L. Hayes, V. Garrison, and R.T. Barber. 2000. African dust and the demise of Caribbean coral reefs. *Geophysical Research Letters* 27: 3029–3032.

Sparks, R.S.J., J. Biggs, and J.W. Neuberg. 2012. Monitoring Volcanoes. *Science* 335: 1310–11.

Sprigg, W., S. Morain, G. Pejanovic, A. Budge, W. Hudspeth, and B. Barbaris. 2009. Public-health applications in remote sensing. SPIE. http://spie.org/documents/Newsroom/Imported/1488/1488_5434_0_2009-02-09.pdf. Accessed 30 October 2011.

Steig, E.J., D.P. Schneider, S.D. Rutherford, M.E. Mann, J.C. Comiso, and D.T. Shindel. 2009. Warming of the Antarctic ice-sheet surface since the 1957 International Geophysical Year. *Nature* 457: 459–462.

Struck, D. 2008. Dust storms overseas carry contaminants to U.S. Washington, D.C.: *Washington Post*. 6 February 2008.

Swap, R., M. Garstang, S. Greco, R. Talbot, and P. Kållberg. 1992. Saharan dust in the Amazon Basin. *Tellus Series B, Chemical and Physical Meteorology* 44: 133–149. doi:10.1034/j.1600-0889.1992.t01-1-00005.x

van Donkelaar, A., R.V. Martin, and R.J. Park. 2006. Estimating ground-level $PM_{2.5}$ using aerosol optical thickness determined from satellite remote sensing. *J. Geophys. Res.* 111: D21201, doi: 10.1289/ehp.0901623.

van Donkelaar, A., R.V. Martin, M. Brauer, R. Kahn, R. Levy, C. Verduzco, and P.J. Villeneuve. 2010. Global estimates of ambient fine particulate matter concentrations from satellite-based aerosol optical depth: development and application. *Environmental Health Perspectives* 118(6): 847–855, doi:10.1029/2005JD006996

Vaughan, D.G. 2005. How does the Antarctic Sheet affect sea level rise? *Science* 308: 1877–1878.

Velicogna, I. 2009. Increasing rates of ice mass loss from the Greenland and Antarctic ice sheets revealed by GRACE. *Geophysical Research Letters* 36: L19503.

Additional Reading

Centeno, J.A. 2008. Natural disasters and their long-term impacts on the health of communities. *J Environ Monit.* 10: 266.

Centeno, J.A. 2008. Impacts of the natural environment on human health. *Interciencia*; 33(3)169–171.

Cook, A.G., P. Weinstein, and J.A. Centeno. 2005. Health effects of dust—Role of trace elements and compounds. *Biol Trace Element Research*.103: 1–15.

Helz, R.L. USGS Fact Sheet: Monitoring Ground Deformation from Space.

Lu, Z., T. Masterlark, C. Wicks, Jr., W. Thatcher, D. Dzurisin, and J. Power. 2002. Interferometric Synthethic Aperture Radar Studies of Alaska Volcanoes. International Society for Photogrammetry and Remote Sensing. 10–15 November 2002, Denver, CO, US. Proceedings. Volume XXXIV, Part 1. pp 9.

Selinus, O., R.B. Finkelman, and J.A. Centeno. 2007. The Medical Geology Revolution. *Geosciences*. 5:108–109.

Relevant Websites

Weather Europe: http://www.sat24.com/

South African Weather: http://www.weatherphotos.co.za/index.html

Volcanoes: US Geological Survey, VHP InSAR Research Group: http://volcanoes.usgs.gov/activity/methods/insar/index.php

Remote sensing for natural disasters—Science Development Network: http://www.scidev.net/en/features/remote-sensing-for-natural-disasters-facts-and-figures.html

Chapter 6
Observing Coastal and Ocean Ecosystems

6.1 Introduction

Marine environments contain substantial biological diversity, deliver vital ecosystem services[1], supply valuable natural resources, and are a core component of our weather and climate system. However, the ocean environment is complex and ever-changing. Examining how our oceans, atmosphere, and landmasses interact would be virtually impossible without the use of a wide variety of sensors and platforms. Satellite observation sensors work in concert with in situ sensors (e.g., buoys and high-frequency radars), research vessels and ships of opportunity, aircraft, gliders (unmanned underwater robots), autonomous undersea vehicles (AUVs), drifters, animal telemetry, and tripod LiDAR to provide cohesive information regarding deep ocean, coastal, and shelf areas in order to understand the complexity, function, and structure of these systems (Chapron et al. 2008).

There is a large variety of sensors that provide remotely sensed data for coastal and marine applications such as:

- Color scanners: used for ocean color mapping (chlorophyll and suspended sediments, diffused attenuation coefficients, etc.)
- Multispectral data: used for water quality studies (chlorophyll *a*, suspended solids, and turbidity)
- Infrared radiometer data: used for sea surface temperature (SST) and currents mapping
- Synthetic aperture radar: used for surface waves, swells, internal waves, oil slicks, etc.
- Hyperspectral data: used for water quality and pollution detection studies

[1] Ecosystem services as defined by Costanza et al. (1997) and accepted by the United Nations are the benefits people obtain—directly or indirectly—from the habitats, biological, and system properties and processes of ecosystems. Seventeen discrete benefits have been identified: atmospheric gas regulation, soil formation, climate regulation, water supply, nutrient cycling, waste assimilation, pollination, disturbance regulation, biological control, refugia, food production, raw materials, genetic resources, erosion control/sediment retention, regulation of hydrological flows, recreation, and cultural benefits.

S. Khorram et al., *Principles of Applied Remote Sensing*,
DOI 10.1007/978-3-319-22560-9_6

- Altimeter data: used for sea surface topography, currents, and surface roughness
- Scatterometer data: used for amplitude of short surface waves (surface wind velocity and roughness)
- Microwave radiometer data: used for microwave brightness temperature (salinity, surface temperature, and water vapor).

Various airborne and satellite sensor products, derived from different wavelengths of the electromagnetic spectrum, include (but are not limited to) the following:

- Ocean color (Figs. 6.3 and 6.4)
- Phytoplankton (Figs. 6.5 and 6.6)
- Chlorophyll *a* and other parameters (Fig. 6.9)
- Total suspended sediment (TSS) and turbidity (Figs. 6.8 and 6.10)
- Surface temperature (Figs. 6.11 and 6.12)
- Ocean currents (Fig. 6.12)
- Submerged aquatic vegetation (SAV) (Fig. 6.13)
- Bathymetric data (Fig. 6.14, 6.15, and 6.16)
- Oil slicks and their spatial distributions (Fig. 6.17)
- Coral reefs (Figs. 6.18 and 6.19)

There are other water quality parameters in lakes and inland waters that can be added to this list including: turbidity, suspended solids (organic and inorganic), salinity, and nutrient load. Coastal and ocean waters can be further defined as Case 1 or Case 2. Case 1 waters are characterized by a ratio in which the concentration of phytoplankton (chlorophyll *a*) is higher compared with other dissolved inorganic particles in the water column. In Case 2 waters, the dissolved inorganic particles are higher compared with the level of the chlorophyll *a* concentration in the ratio. Thus, optical properties in Case 1 waters are determined primarily by phytoplankton and related colored dissolved organic matter (CDOM), and optical properties in Case 2 waters are largely influenced by constituents other than phytoplankton concentration. The relative contributions of phytoplankton, CDOM, and suspended inorganic matter to the spectral properties of Case 1 and Case 2 waters are illustrated in Fig. 6.1.

The research community has developed models for estimating the biogeochemical products in open oceans, globally and on a routine basis from satellite data. However, the complexity of coastal water, the atmosphere above the surface of the water, and the variability in time and space require more complex algorithms for handling these nonlinear multivariate bio-optical systems. For these reasons, a single algorithm to be transportable in time and space on an operational basis requires further development. As illustrated through Fig. 6.2 and Table 6.1, the spectral characteristics of coastal and marine water components indicate that no single wavelength range (band) can account for various coastal and sea surface conditions. Therefore, a more complex modeling approach is needed to address them. Surface and subsurface information on ocean chemistry, physics, and biology is only practically possible through global-scale ocean observations.

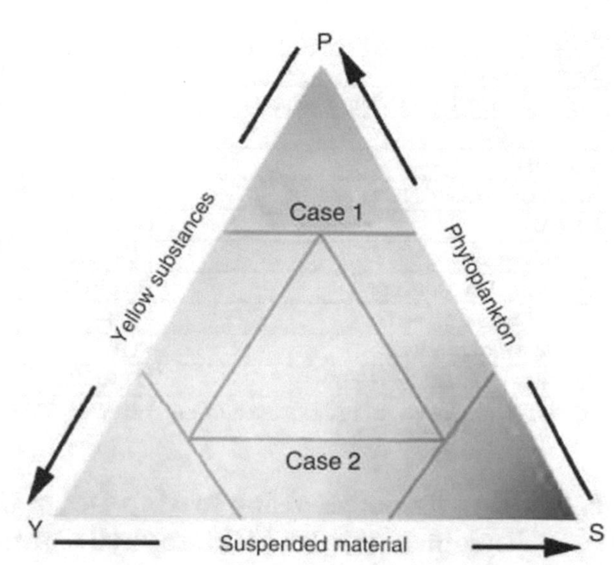

Fig. 6.1 The triangular diagram illustrating the relative contribution of phytoplankton, dissolved organic matter *(yellow substances)* and *suspended material* (non-algal particles), and the division between Case 1 and Case 2 waters. © IOCCG 2000; 2000 QND 2006, after Hoepffner and Zibordi 2009

Figure 6.2 illustrates that the maximum reflectance of water bodies shifts to higher wavelengths in coastal waters with high concentration of non-algal particles and CDOMs.

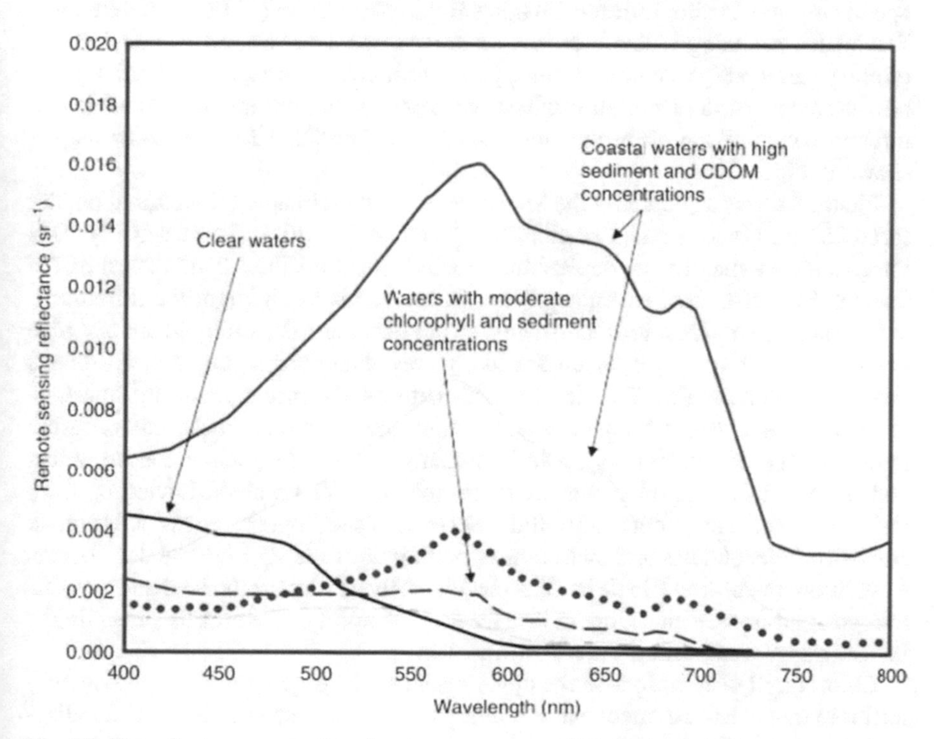

Fig. 6.2 The reflectance properties of various water types. (Adapted from IOC)

Table 6.1 An example of spectral range for coastal waters applications

Product name	Critical spectral range (nm)
Chlorophyll and other pigment concentration	443–445 (max. abs.), 550–560 (min. abs.)
	490, 548, 640 (for other pigments)
Red tides	510
Sediment concentration	530–650, NIR
CDOM	410–420
Seabed reflectance	490–580
Chlorophyll fluorescence	680–685
Aerosol optical properties	700–NIR
Atmospheric corrections	TIR
Sea surface temperature (SST)	NIR; TIR

CDOM colored dissolved organic matter, *NIR* near infrared, *TIR* thermal infrared

6.2 Using Remote Sensing to Map Ocean Color, Phytoplankton, and Chlorophyll Concentration

Airborne and satellite remotely sensed data are commonly used for mapping ocean color. The Ocean Biology Processing Group (OBPG) at NASA's Goddard Space Flight Center conducted a satellite ocean color measurements in June 2015 of the Bering and Chukchi seas (Alaska/Siberia). The OBPG ocean color map was developed using the Visible Infrared Imaging Radiometer Suite (VIIRS) satellite data. The VIIRS scanning radiometer is a sensor on board the Suomi National Polar-orbiting Partnership weather satellite. This sensor collects imagery and radiometric measurements of the land, atmosphere, cryosphere, and oceans in the visible and infrared bands of the electromagnetic spectrum. The OBPG ocean color map is shown in Fig. 6.3.

Figure 6.4 was captured by the Moderate Resolution Imaging Spectroradiometer (MODIS) on NASA's Aqua satellite on December 21, 2010 (Painter 2010). This image shows a massive phytoplankton bloom located off the Atlantic coast of Patagonia. Scientists used seven separate spectral bands to highlight the differences in the plankton communities across this swath of ocean. The milky green and blue phytoplankton blooms displayed are likely coccolithophores, single-celled plants that form calcite scales. Calcite is a carbonate mineral often found in limestone chalk. Blooms of coccolithophores and diatoms are common in these waters during the southern hemisphere's spring and summer where warmer, saltier coastal waters and currents from the subtropics meet the colder, fresher waters flowing up from the south. Where these currents collide—known to oceanographers as a shelf-break front—turbulent eddies and swirls form, pulling nutrients up from the deep ocean. In addition, the nearby Rio de la Plata (La Plata River) runs off the land and deposits nitrogen and iron-laden sediment into the sea just north of the area as shown in the image (http://oceancolor.gsfc.nasa.gov/cms/).

Chlorophyll distributions in the upper ocean are a unique proxy for phytoplankton biomass—a basic component in primary production—as well as a useful indica-

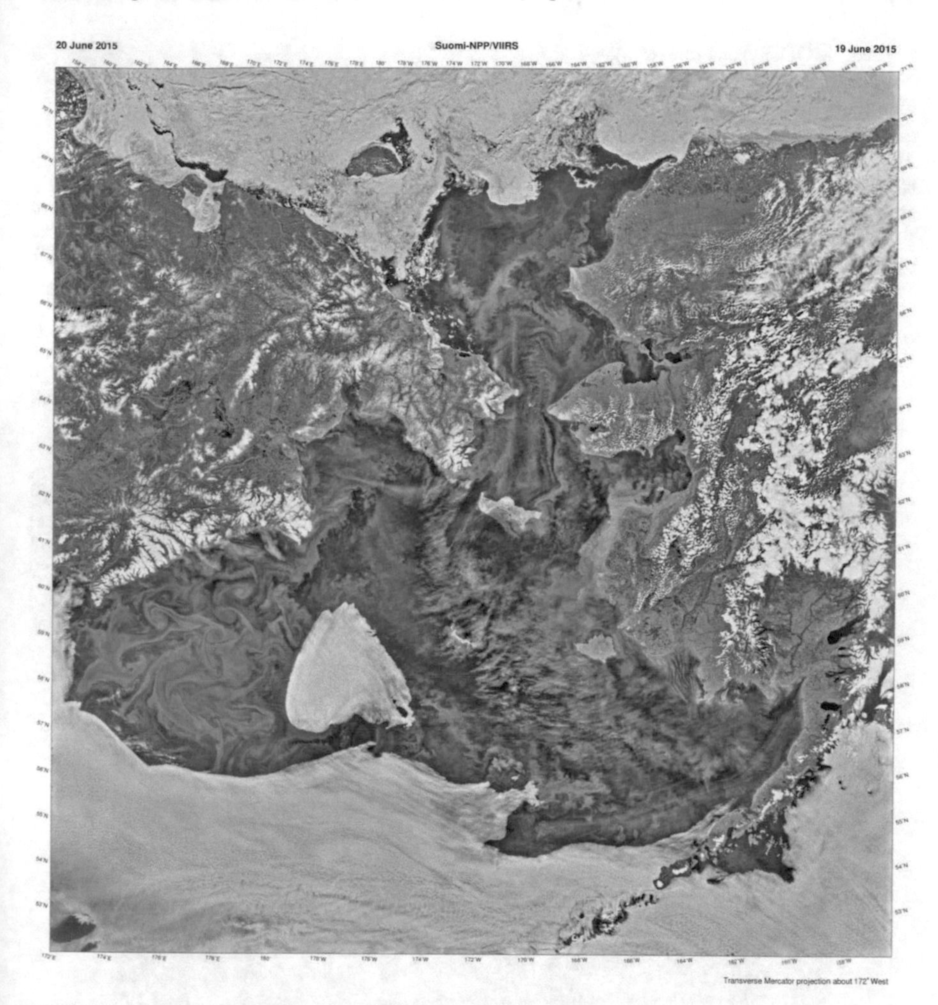

Fig. 6.3 The VIIRS satellite data collected this view on June 19/20, 2015 showing ocean color in the Bering and Chukchi seas (Alaska/Siberia). (Image source: http://oceancolor.gsfc.nasa.gov/cms/)

tor for ocean upwelling. Phytoplankton (Fig. 6.5) serve a crucial biogeochemical function, converting CO_2 to particulate organic matter and releasing oxygen. The rate of marine primary production is determined by temperature, light (strongly influenced by surface turbulent mixing depths), and limiting nutrients—particularly nitrogen, phosphorus, iron, and silicon for some plankton (Doney 2010).

Satellite sensors developed to quantify global ocean chlorophyll levels (IOCCG 2006) and produce biological data include sensors such as the SeaWIFS (Sea-Viewing Wide Field-of-View Sensor). The SeaWIFS sensor was launched on August 1, 1997, onboard GeoEye, Inc.'s OrbView-2/SeaStar satellite. The SeaWIFS, although designed to operate for a period of only 5 years, began collecting data on September 18, 1997 and was retired from operation data on December 11, 2010.

Fig. 6.4 MODIS Aqua image of the massive phytoplankton bloom off the Atlantic coast of Patagonia on December 21, 2010 (NASA 2010). (© NASA image created by Norman Kuring, Ocean Color Web. Caption by Michael Carlowicz)

Fig. 6.5 Phytoplankton bloom swirls a figure-eight in the south Atlantic ocean, about 600 km east of the Falkland Islands, on December 2, 2011. Different types and quantities of phytoplankton exhibit different colors. Once a bloom begins, ocean color sensors can identify its chlorophyll pigment, and therefore its species and toxicity. (© Envisat/ESA)

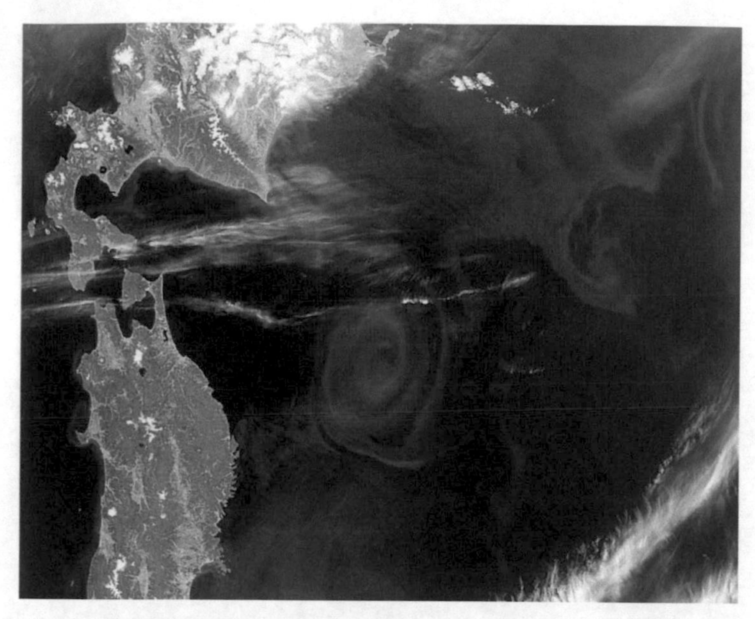

Fig. 6.6 Circulation patterns—A very well-defined spiral eddy is visible off the east coast of Japan. (© SeaWIFS Project, NASA/Goddard Space Flight Center/ORBIMAGE)

The sensor collected data at a 1.1 km local area coverage (LAC) in the following optical bands: 1 (402–422 nm), 2 (433–453 nm), 3 (480–500 nm), 4 (500–520 nm), 5 (545–565 nm), 6 (660–680 nm), 7 (745–785 nm), and 8 (845–885 nm). Figure 6.6 shows SeaWIFS data used to prominently identify a well-defined spiral circulation pattern (eddy) visible off the east coast of Japan.

Optical remote sensing in coastal waters typically provides quantitative estimates of phytoplankton biomass (indexed by chlorophyll concentrations), total suspended matter, and dissolved organic substances, with accuracies as good as the same types remote sensing-derived estimate within open ocean (approximately 30–40 % in biomass) when using regional dedicated algorithms. These components, in turn, influence the light attenuation coefficient in the water which is also directly retrieved from the satellite (Fig. 6.7), providing information on water transparency. The latter represents an important quality parameter to assess the ecological status of lakes and coastal waters. Remotely sensed chlorophyll concentration and light attenuation are also used to estimate water-column primary production, which represents a critical indicator to assess coastal eutrophication, and to analyze the carbon budget directly associated with the continental shelf pump health and resilience.

As shown in Fig. 6.8, from an image captured by the NASA MODIS Aqua satellite, both the Yangtze River and Hangzhou Bay, to the south, are filled with thick brown sediment. The sediment extends both to the north and the south along China's coast, and is joined by sediment from smaller watercourses along the coast. As the sediment flows into the East China Sea, it begins to sink and the reflectance properties change; as less light is reflected, the color becomes greenish, then blue. The city of Shanghai is likely a major contributor to these changes in the Yangtze River.

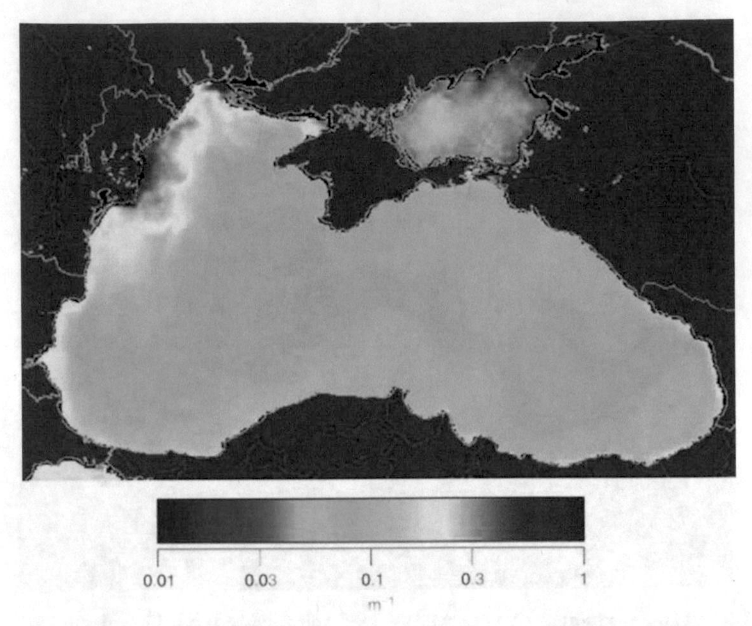

Fig. 6.7 The light attenuation coefficient for the Black Sea derived from SeaWIFS monthly composite (January 2003). The coefficient is calculated using a semi-analytical algorithm to retrieve the inherent optical properties (IOPs) of the water constituents. (Source: Joint Research Centre of the European Commission © Joint Research Centre 2006)

Fig. 6.8 Suspended sediments and the changes in turbidity in the Yangtze River and Hangzhou Bay as mapped by NASA MODIS Aqua satellite. (© MODIS/Aqua/NASA)

6.3 Remote Sensing of Eutrophication and Ocean Hypoxia

Nutrients can be present in two major forms: inorganic (or mineral) and organic (both living and detrital). Nitrogen and phosphorus are the most significant nutrients, and their main species include dissolved (nitrate, nitrite, ammonium, organic N, phosphate, organic P) and particulate (organic N, organic P) components. Silica is also a fundamental nutrient in coastal systems because diatoms, a predominant phytoplankton group in estuaries and coastal systems, require silica in their structural composition. Complex models frequently address all of these properties, accounting for the explicit parameterization of several element cycles, thus tracing each nutrient and carbon source in the system. Figure 6.9 illustrates some of these results, with an example of spatial maps for chlorophyll, silicate acid, and semi-labile organic matter in a mesotidal estuary. It should be noted that complex models are usually based on the assumption of variable internal nutrient composition, allowing for the tracing of the evolution of C:N ratios in time for phytoplankton, and providing relevant clues on the influence of nutrient deficiency.

Urban development and agricultural operations produce high concentrations of nutrients. These nutrients impact coastal watersheds, causing upwellings and algal blooms in many estuaries and coastal waters. Algal blooms are harmful in that

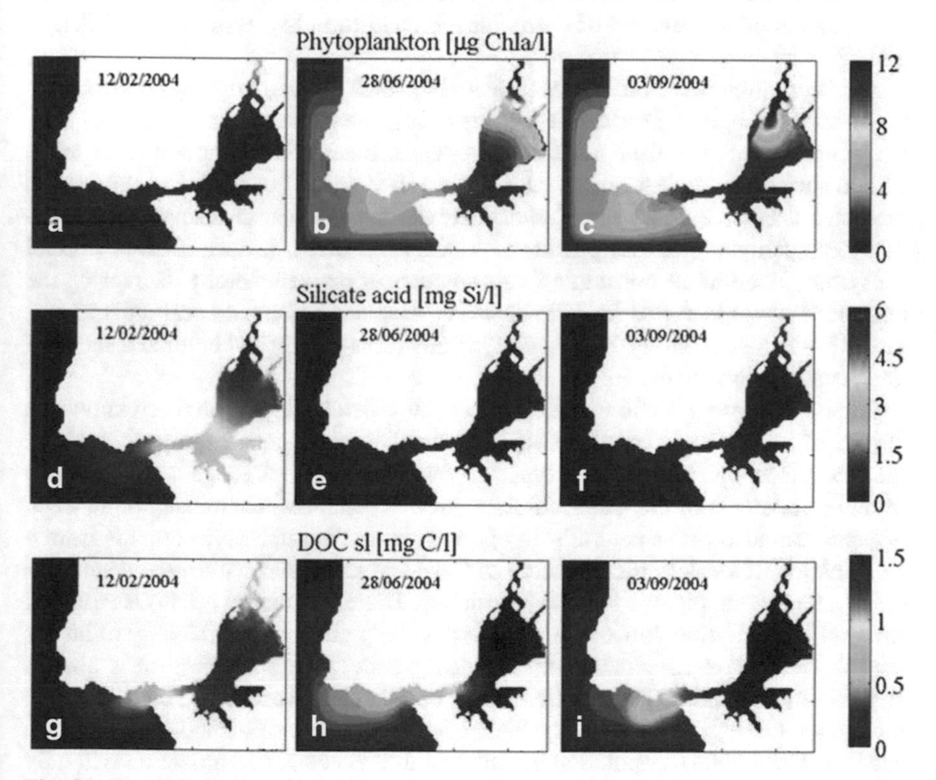

Fig. 6.9 Spatial maps for chlorophyll, silicate acid and semi-labile organic matter in a mesotidal estuarine system. (adopted from Kenov et al. 2014)

Fig. 6.10 After a nearly ice-free winter, Lake Erie (US) was filled with suspended sediment and algae. *Muddy, tan-colored* water along the shoreline reveals sediment from rivers. *Milky green, light blue,* and *white* shades may also be sediment-rich waters as the lake bottom is rich in quartz sand and calcium carbonate (chalk) from limestone. (© Jeff Schmaltz, MODIS/Terra/NASA)

they cause eutrophic conditions, depleting oxygen levels needed by organic life and limiting aquatic plant growth by reducing water transparency. Most algal blooms can be observed from satellites, due to their distinct color, location, or repetitive seasonal appearance. Concentrations of chlorophyll *a* (chl-*a*) and TSS can be used as indicators of the severity of eutrophication and turbidity, respectively (Klemas 2008).

Eutrophication is the biological process by which aquatic primary production is augmented through an increase in the rate of organic matter and nutrients (e.g., nitrogen and phosphorus from fertilizers, sewage effluent, and other pollutants) delivered to surface water (e.g., river, lake, estuary, and coastal waters). Eutrophication promotes excessive algae and plankton growth and can cause a severe reduction in water quality. Algae and plankton blooms (Fig. 6.10), in turn, disrupt normal ecosystem functioning, consuming vast amounts of oxygen thereby decreasing the amount of oxygen needed for fish, shellfish, seagrass beds, and coral reefs to survive (Howarth et al. 2011). The water typically turns cloudy and colored a shade of green, yellow, brown, or red.

Excessive algae growth in coastal waters may also lead to a condition known as "hypoxia", as surface layer algae begin to die, then sink, then decompose, robbing the subsurface water layers of oxygen. Hypoxia may also arise as a consequence of other pollutants in the water column, such as sediment, chemicals, or sewage. Hypoxia can also occur naturally in open-ocean and coastal environments from a combination of weak vertical mixing of fresh and salt water within the water column and/or strong organic matter degradation. Dissolved oxygen (DO) is essential for aerobic respiration; low oxygen levels adversely affect the physiology of higher animals, leading to mass mortality or "dead zones" where marine life is absent. Hypoxia occurs when DO falls below 2 ml of O_2/liter. Worldwide, there are now more than 400 coastal hypoxic systems covering an area of $>245,000$ km^2 (Diaz and Rosenberg 2008). Population growth and further coastal urbanization will only exacerbate coastal hypoxia without careful land and ocean management.

One future area of satellite remote sensing ocean color research is to improve the classification of phytoplankton pigment concentrations, which would then serve to improve our understanding of algal biodiversity and taxonomic composition. This improved understanding would also provide a better understanding of the impacts of pollution and excess nutrients on marine ecosystems (Pan et al. 2010). For example, some pigments are unique biomarkers for particular phytoplankton in coastal waters (e.g., alloxanthin in cryptophytes, peridinin in some dinoflagellates, and zeaxanthin from cyanobacteria). Thus, further improved remote sensing capabilities in the identification and classification of these phytoplankton species may provide vital information on the seasonal and regional distribution of these species (Pan et al. 2010).

Newer and enhanced sensors will further increase our capacity to assess optical water quality characteristics (e.g., chlorophyll *a*) that leads to coastal eutrophication and hypoxia episodes. However, there is still much research to be done with currently available sensor data (e.g., the Landsat 8 sensor, as well as the upcoming launch of the ESA Sentinel-2 satellite sensor) and the expanding availability of hyperspectral data. In addition, an increased understanding of mesoscale ocean features (e.g., fronts, filaments, eddies, Lagrangian coherent structures, and river plumes) may also provide associative data that will be helpful in developing remote sensing-based models of the occurrence, distribution, and duration of ocean and coastal eutrophication and hypoxia episodes. Mesoscale ocean features have been shown to be associated with enhanced productivity and fish aggregation, span spatiotemporal scales that range from one to thousands of kilometers and from hours to weeks. Finally, hydrodynamic-biogeochemical models developed from the remotely sensed optical properties of seawater are capable of providing tools that reveal the present state of the ecosystem in three dimensions, as well as forecasting future conditions in relation to nutrient inputs. Understanding the relationship of the coastal and ocean systems to these nutrient inputs may also aid in assessment and identification of coastal eutrophication and areas of hypoxia (Banks et al. 2012).

6.4 Using Remote Sensing to Map the Sea Surface Temperature and Circulation Patterns

Satellite instruments, other than optical sensors, can provide additional information of importance to coastal water management. Microwave sensors collect data either passively or actively in that part of the electromagnetic spectrum ranging from 3 mm to 1 m wavelength. These instruments provide significant advantages in viewing the Earth under all weather conditions–even through clouds and day and night capabilities. However, their spatial resolution, which depends on the physical dimension of the antenna, is often not adequate for coastal applications. Sophisticated synthetic aperture radars (SARs) are an exception to that rule and can retrieve surface features at extremely fine resolution, of the order of meters.

Numerical ocean circulation models are widely used for such predictions. Although satellites cannot provide direct measurements of ocean surface currents, a

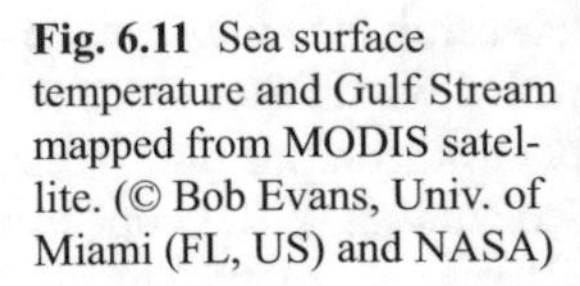

Fig. 6.11 Sea surface temperature and Gulf Stream mapped from MODIS satellite. (© Bob Evans, Univ. of Miami (FL, US) and NASA)

combination of several sensors such as scatterometers, altimeters, and radiometers may be used to infer these currents. Such a product, known as Ocean Surface Currents Analyses Real-time (OSCAR), is currently being freely distributed to the user community (Santoki et al. 2012).

Ocean color measurements have been combined with satellite data on SST and other physical properties of the ocean to explain the processes that regulate primary production in the sea. Examples of MODIS satellite and Japanese ADEOS/OCTS satellite applications for SST mapping are shown in Figs. 6.11 and 6.12, respectively.

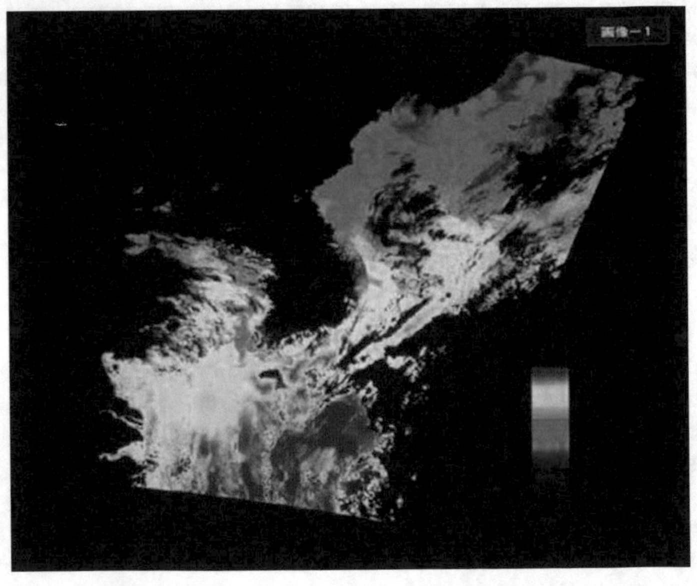

Fig. 6.12 Sea surface temperature mapping from ADEOS/OCTS over the Sea of Japan

6.5 Spatial Analysis of Submersed Aquatic Vegetation

An additional area of growing popularity in marine and freshwater remote sensing studies is the identification, detection, and analysis of spatial distributions of submersed aquatic vegetation (SAV). Aquatic vascular plant populations are found in various freshwater bodies, coastal and ocean environments worldwide. Emergent and submersed aquatic vegetation in these environments provides critical habitat for aquatic species, also a trophic and food source, enhanced water quality, reduced turbidity and sedimentation, bank and shoreline stabilization, and increased oxygen within the water column. However, monitoring, management, and assessment of the environment where these plants are found are greatly hindered by limitations of size and access within the aquatic environment. Understanding the dynamics of aquatic plant populations has been recognized as much more important in more recent years due to global declines in native populations concomitant with invasions of nonnative species through global human movement (Madsen and Wersal 2012; Waycott et al. 2009).

The monitoring, management, and assessment of emergent aquatic vegetation may present less of a challenge, in that as the name implies, these plants have part of their growing cycle above the surface of the water. However, SAV typically remain completely submerged below the water surface. The disappearance of native SAV species or the introduction and dispersal of invasive SAV species can cause severe ecological and economic impacts (Charles and Dukes 2007; Langeland 1996; Pu et al. 2010). Ecological impacts are centered on the fact that the entire food web dynamics can be significantly altered by the addition or removal of a single keystone or exotic species (Hovel and Lipcius 2002; Irlandi et al. 1995). Invasive submersed species, in particular, displace native plants and can shift balanced, heterogeneous ecosystems to monocultures with severely altered food web dynamics (Richardson et al. 2012). Invasive SAV can also promote lethal impacts affecting higher trophic levels. For example, an epiphytic cyanobacterium commonly found on invasive *Hydrilla verticillata* has been shown to produce a neurotoxin(s) that causes avian vacuolar myelinopathy (AVM). AVM is an often-lethal neurological disease of waterfowl and their predators in the southeastern United States, from coots *(Fulica americana)* to bald eagles *(Haliaeetus leucocephalus;* Wilde et al. 2005; Williams et al. 2007). Furthermore, large quantities of invasive SAV biomass are prime habitat for mosquitoes, which can carry a number of diseases (Nichols and Shaw 1986). Major economic loss has been sustained from invasive SAV whose growth inhibits flood control, hydropower generation, irrigation, navigation, and recreation in infested water bodies (Pimentel et al. 2005). Related loss of native SAV species can result in the collapse of fish populations (Wilson et al. 2013). Many have quantified the direct economic losses associated with both native losses and invasive introductions as ranging from tens of thousands to millions of US dollars (e.g., Anderson 1993; Larkum et al. 2006; Pimentel et al. 2005). Indirect economic losses, more difficult to define and sometimes of major importance, only add to this deficit (Pimentel et al. 2005). Thus, the methods by which to detect, monitor, and assess SAV,

Fig. 6.13 The fowling capabilities of hydrilla *(Hydrilla verticillata)* in Lake Okeechobee, Florida *(left)* and in the Albemarle-Pamlico Estuary System, North Carolina *(right)*

including evaluation of the efficacy of management actions, have become increasingly important (Blossey 2004).

Large-scale SAV surveys are rarely performed due to logistical difficulties and the high costs incurred when obtaining survey data over large areas. In past years, remote sensing coupled with modeling and interpolation techniques has shown the potential to be an important tool to obtain survey information on SAV over large areas (Nelson et al. 2006; Valley et al. 2005).

Remote sensing has the potential to be an important tool to obtain survey information on SAV measurements across large geographic areas. In a study conducted in multiple lakes within the state of Michigan, Nelson et al. (2006) found significant relationships between SAV plant cover types and SAV abundance measures using Landsat 5 TM values in logistic regression models.

Data provided through remote sensing applications have a tremendous potential for also detecting the status and distribution of nuisance aquatic vegetation (NAV) over large geographic areas. NAV such as Eurasian watermilfoil *(Myriophyllum spicatum)* now impact many freshwater systems of the US. Other species, such as hydrilla *(Hydrilla verticillata),* although primarily a freshwater invasive aquatic plant, have been shown to tolerate tidal- and brackish-water environments. This tolerance for low salinity environments may threaten the spread of this plant into near - shore coastal environments. These NAV species tend to be highly aggressive aquatic plants that outcompete native aquatic plants, altering habitats for birds and fish, as well as leading to situations of overgrown plant materials that congest waterways, thus limiting travel and recreation (Fig. 6.13). Current field-based surveillance methods for nuisance and invasive submerged plants are labor-intensive, time-consuming, and costly. However, remote sensing approaches provide certain advantages over traditional methods, allowing for repeated image coverage over large areas and synoptic analysis techniques for detection, mapping, and monitoring of submerged aquatic invasive species.

Although the combination of satellite remote sensing and statistical modeling techniques for the application of SAV distribution and abundance monitoring has shown promise (Nelson et al. 2006), the technology still must overcome the difficulties inherent in interpreting reflectance values of water (Lehmann and Lachavanne 1997; Penuelas et al. 1993). For example, clear water provides little atmospheric reflectance, with the majority of the shortwave radiation scattered in the atmosphere and longer wavelengths absorbed within the few millimeters of the water surface. Adding to this challenge, currently, the majority of the optical satellite sensors are

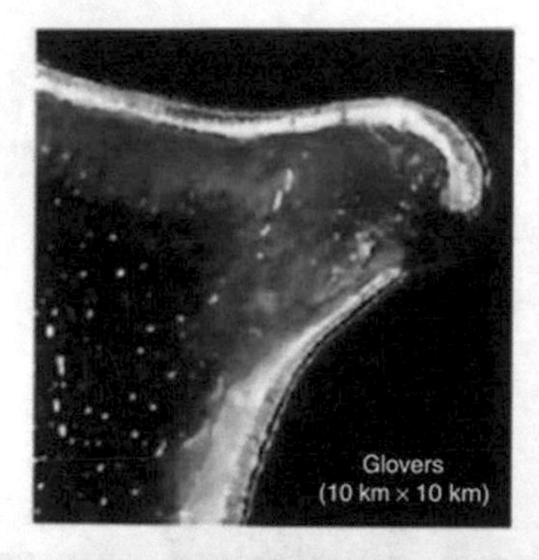

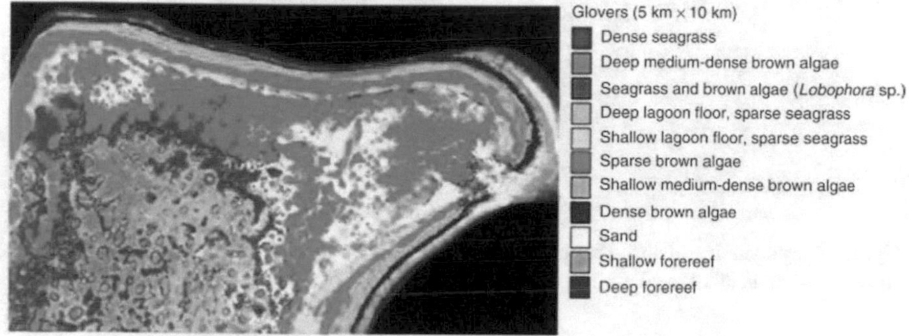

Fig. 6.14 Classification of tropical coral reef environment (Belize, Caribbean Sea). *Top*: RGB color composite based on the *red, green,* and *blue* bands of IKONOS sensor. *Bottom*: Eleven-class scheme classification of the benthic features. Reproduced from Andréfouët et al. (2003). Multi-site evaluation of IKONOS data for classification of tropical coral reef environments. Remote Sensing of Environment 88: 128–143, with permission from Elsevier

still primarily designed for detecting land features. However, despite the potential limitations of using existing sensors, newer sensors with improved spatial, radiometric, and spectral resolutions may, in the near future, prove to be applicable for aquatic studies.

6.6 Remote Sensing of Coastal Bathymetry

In shallow coastal waters, remote sensing is being used to monitor benthic structure and classify bottom types. At present, satellites most commonly exploited for that purpose are equipped with radiometers, such as the SPOT-HRV and Landsat TM series with typically 10–30 m spatial resolution, or even IKONOS with spatial resolution better than 10 m. Even though these sensors have a limited set of wavelengths (three to four broad spectral bands in the visible), the spectral contrast between various bottom structures (e.g., corals and non-coral objects) can be used to construct classification algorithms based on differential reflectance (Wilson 2011) (Fig. 6.14).

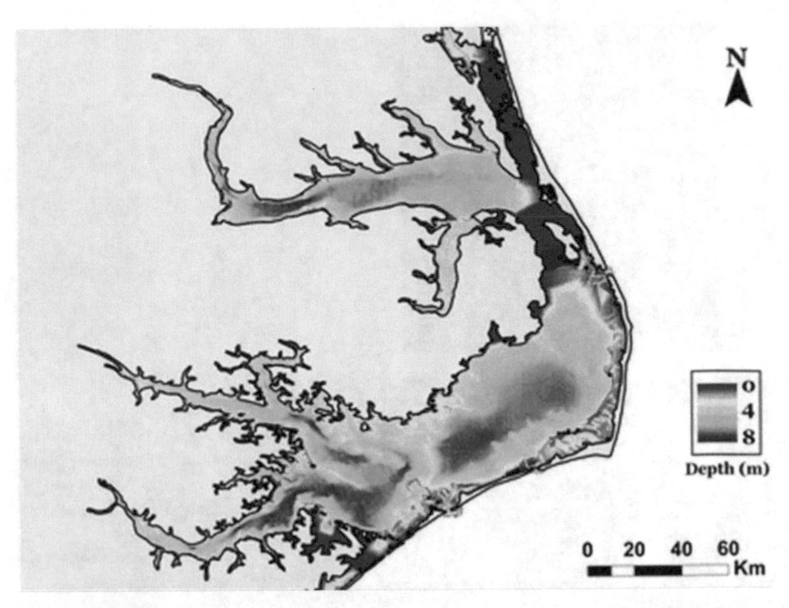

Fig. 6.15 Bathymetry of the Albemarle-Pamlico Estuarine System derived from Landsat data (after Miller et al. 2014)

LiDAR data integrated with global positioning systems (GPS) can be used in obtaining accurate topographic and bathymetric maps, including shoreline positions. LiDAR surveys can produce a 10 cm vertical accuracy at spatial densities greater than one elevation measurement per square meter. This is important for various coastal research applications of LiDAR, such as mapping change along barrier island beaches and other sandy coasts. The ability of LiDAR to rapidly survey long, narrow strips of terrain is very valuable in this application, as beaches are elongate, highly dynamic sedimentary environments that undergo seasonal and long-term erosion or accretion, and are also impacted by severe storms (Stockdon et al. 2002).

An example of bathymetry of the Albemarle–Pamlico Estuarine System derived from LiDAR is shown in Fig. 6.15. The shallow estuaries and broad shoals characteristic of both the Albamarle and Pamlico sounds (after Miller et al. 2014) are illustrated in this figure.

Fig 6.16 depicts another example of bathymetry data, in this case derived from Landsat 8 imagery (Lira and Taborda 2014).

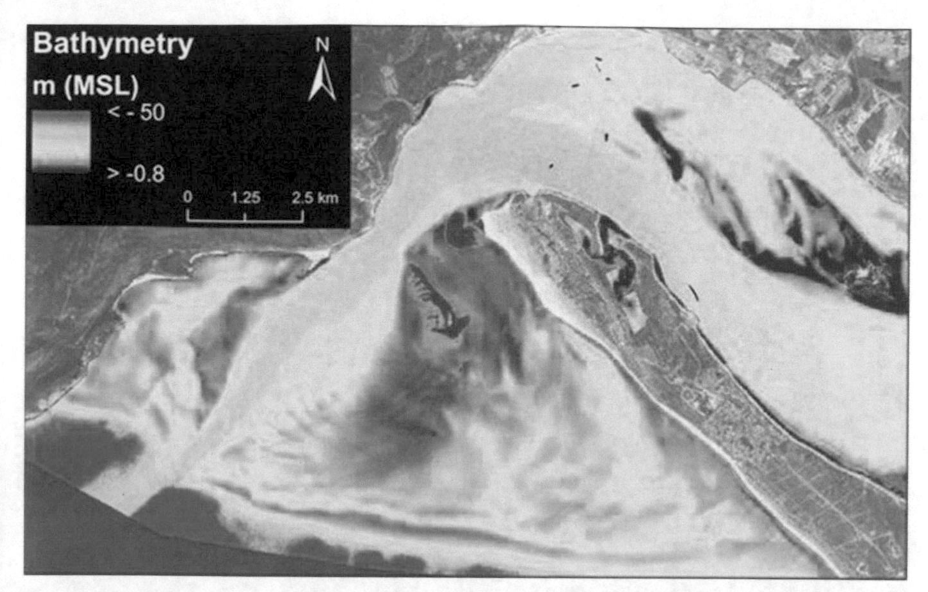

Fig. 6.16 Bathymetric derived data from August 9, 2013 Landsat 8 image of Sado ebb delta, West Portuguese coast (http://earthexplorer.usgs.gov/)

6.7 Remote Sensing of Coral Reefs

Coral reefs are extremely vulnerable to changes in ocean temperature, pollution, and algal levels. Tropical and subtropical corals exist in a symbiotic relationship with zooxanthellae that live within their tissues, provide the coral with oxygen and nutrients, and give the coral its color (DiMassa et al. 2011). When stressed, corals respond by expelling their symbionts, losing their pigmentation in the process. This results in a white or bleached appearance, which occurs most frequently with warm water events (Berkelmans 2002). Warm water events occur when strong solar insolation is coupled with stratification in the water column. Water temperature stratification occurs when there are calm winds, an absence of waves, and low currents (Skirving et al. 2006), creating warm surface temperatures in the shallow waters where tropical corals are found. Prolonged periods of bleaching (i.e., warm temperatures and lack of the algae's nutrients) lead to coral mortality.

One of the threats to coral reefs and coastal waters are oil spills from various oil rigs and oil transportation devices. Oil slicks and their spatial distributions can be readily detected and mapped from remotely sensed data. The massive 2010 oil spill in the Gulf of Mexico, as imaged by the Scan SAR by InfoTerra, is shown in Fig. 6.17.

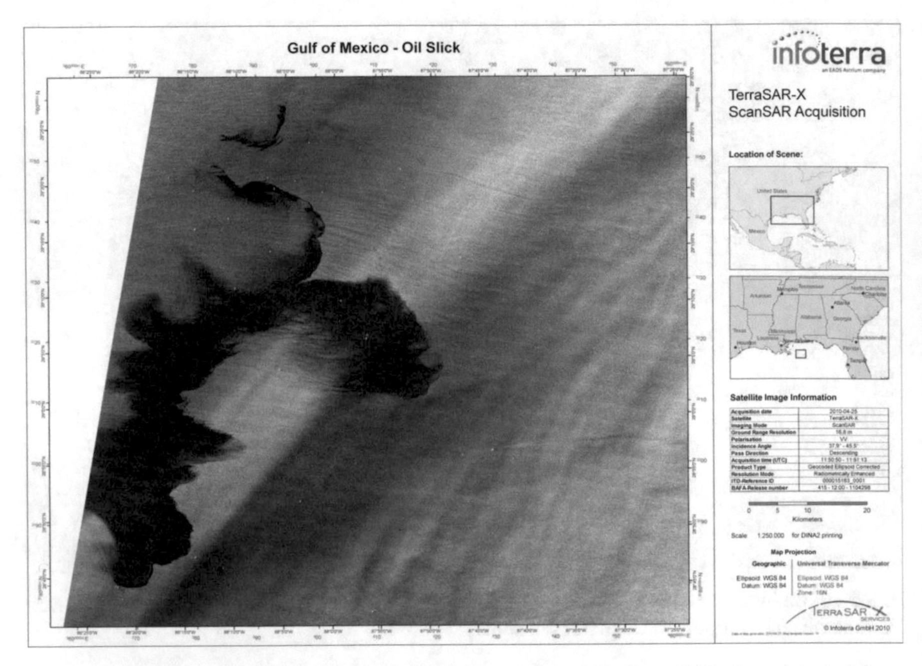

Fig. 6.17 The 2010 oil spill in the Gulf of Mexico, acquired by Scan SAR data. Adopted from InfoTerra

According to the Global Coral Reef Monitoring Network, approximately 19% of the world's coral reefs are dead with another 15% expected to die by 2025. Reefs are important as both fish habitats and buffers against high waves (Pala 2011).

A first step in determining conditions ripe for coral bleaching is to evaluate the magnitude of the current needed to produce tidal mixing. DiMassa et al. (2011) examined surface current speed measured by twin high-frequency radar systems monitoring the Capricorn/Bunker groups of island and reefs in the Great Barrier Reef of Australia and compared it with Simpson–Hunter (1974) threshold values determined throughout the radar coverage area along with local bathymetry measurements. Areas in which the surface currents exceeded the Simpson–Hunter threshold values were determined to be well mixed, even in the absence of wind or waves, while regions in which surface currents did not exceed the threshold values were assumed to have some degree of stratification. The extent of mixing or stratification was used as an indicator for coral bleaching. DiMassa et al. (2011) produced maps which indicate relative susceptibility to coral bleaching based on tidal currents and the generation of tidal mixing fronts.

Remotely sensed data can be used for modeling the resilience of coral reefs. IKONOS and Quickbird satellite data were used in a spatial predictive modeling context by Knudby et al. (2014) to map four of the resilience indicators to predict the stress-tolerant coral taxa, coral diversity, herbivore biomass, and coral recruitment (quantified as the density of juvenile corals), as well as herbivore functional group richness and the live cover of corals and coralline algae, both of which may also influence reef resilience.

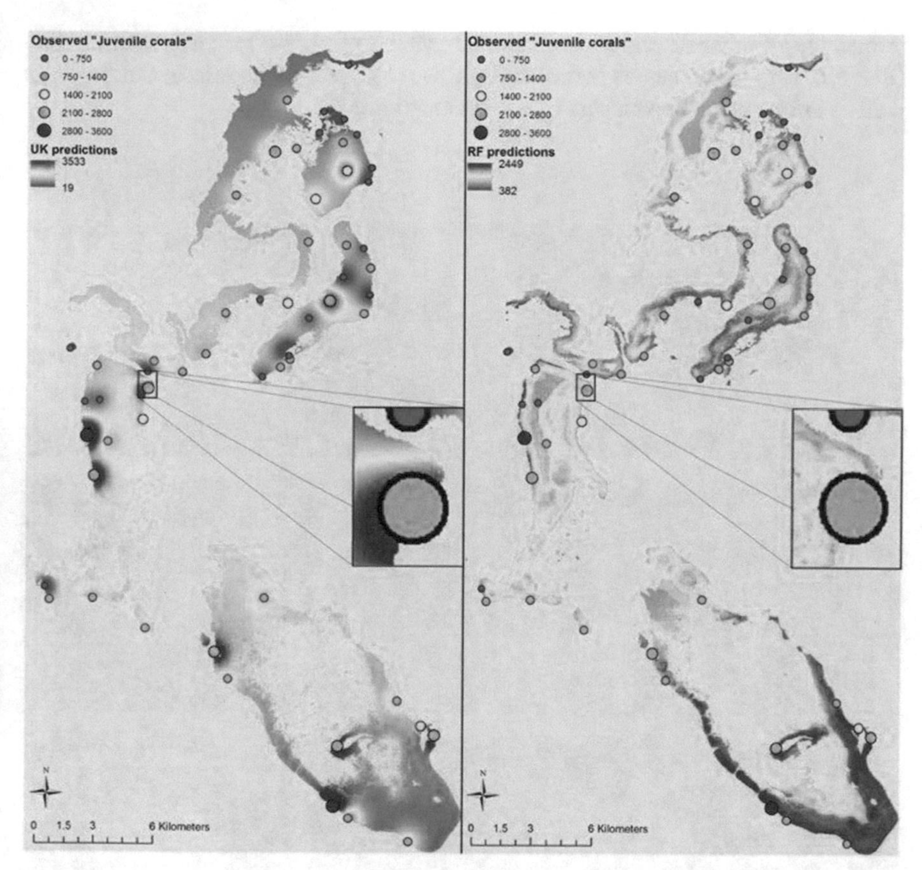

Fig. 6.18 Predictions of the spatial distribution of juvenile coral density across the Kubulau traditional fisheries management area, Fiji (Knudby et al. 2014). Points indicate field observations while the colored background illustrates the model predictions Universal kriging (UK) predictions are shown on the *left*, random forest (RF) predictions are shown on the *right* (after Knudby et al. 2014)

The results from Knudby et al. (2014) suggest that, based on the data available, the spatial distribution of stress-tolerant coral taxa and herbivore biomass could be reasonably well predicted using the random forest model, while the density of juvenile corals could only be poorly predicted using the universal kriging model and coral diversity was essentially unpredictable. An example output, the predicted spatial distribution of juvenile coral density from each of the two models, is provided in Fig. 6.18.

A global assessment of locations most at risk to coral bleaching was performed by Australian scientists, led by Joseph Maina of Macquarie University in Sydney. The researchers used nearly 30 years of satellite and tidal data to examine temperature, wind speed, tidal dynamics, ultraviolet radiation levels, sediment, and nutrient levels, and combined that information with stress-reducing factors to identify which reefs would be likely to survive additional stressors and which ones would be more vulnerable to adverse regional conditions (Maina et al. 2011). The resulting map (Fig. 6.19) assists managers in the strategic use of conservation efforts and funds, as well as informing policymakers of changing conditions.

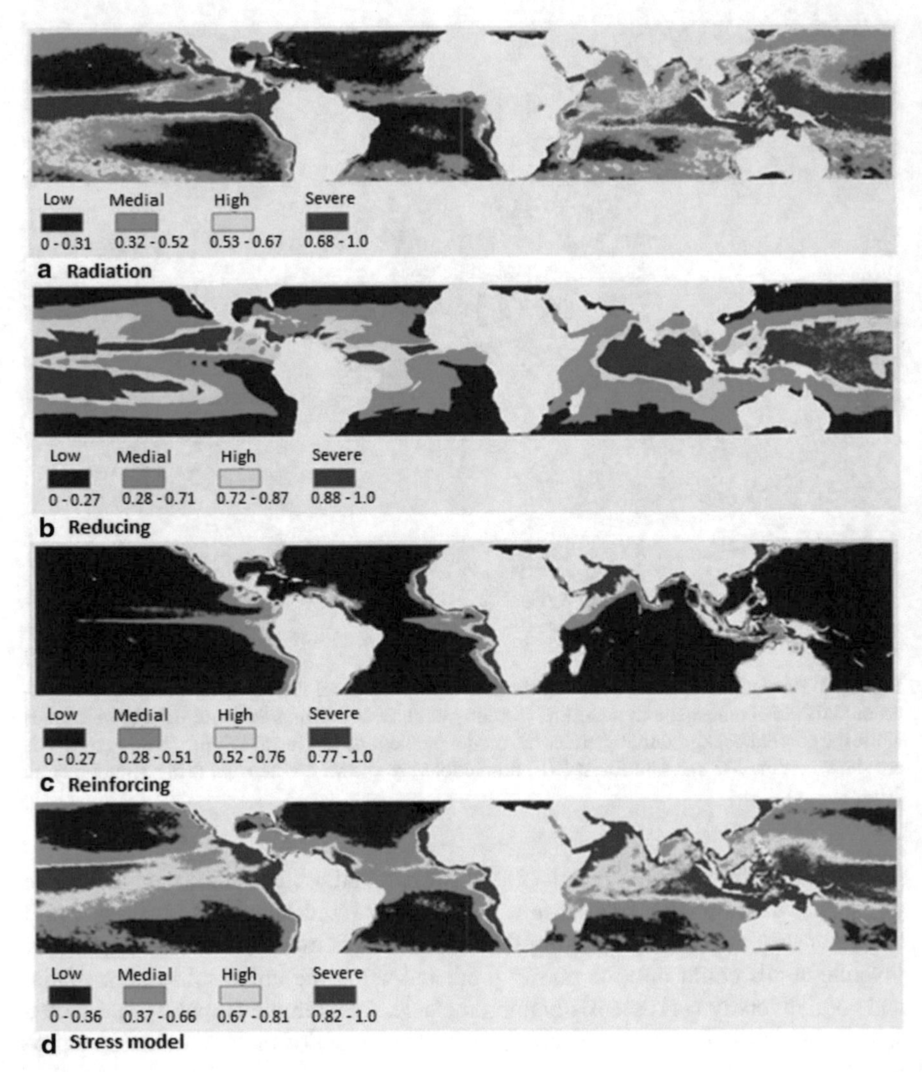

Fig. 6.19 Composite layers for radiation, reducing, reinforcing stress categories, and the overall coral reef stress model. © Maina et al. 2011

Mapping of coral reefs requires high-resolution (1–4 m) imagery (Mishra et al. 2006; Mumby and Edwards 2002). Coral reef ecosystems usually exist in clear water and can be classified to show different forms of coral reef development (i.e., living coral, dead coral, coral rubble, algal cover), bottom structure, and seagrass densities. Higher resolution multispectral imagers (e.g., IKONOS) have been used to map coral reefs; however, hyperspectral imagers should improve the results significantly (Maeder et al. 2002).

6.8 Achieving Sustainable Fisheries and Aquaculture Management

Overfishing of the world's open and coastal regions has been a huge issue ecologically, as well as politically. Countries economically-dependent on ocean harvesting of various fin and shellfish species and their by-products are typically regulated by their own sovereign and national laws. However, if these laws do not provide regulations that prevent overfishing, the impact of this activity can have a devastating effect on the global population of fish species that may be far reaching beyond the jurisdictional boundaries of the country (or countries) involved. The ramifications of these activities may serve to reduce or eliminate populations of targeted harvest species and bycatch. In addition, overharvesting may also serve to pollute fishing grounds from increased boat activity, as well as destroy critical habitat and spawning grounds, and greatly weaken the ability to sustainable native stocks.

Sustainable fisheries techniques largely function by developing management plans that map out critical fisheries habitats and nursery grounds, as well as limit the number of species caught in a particular area to prevent overfishing. Traditional approaches to the use of remote sensing in fisheries relied on aerial photographic imagery that was transferred to fishing vessels at sea with the data being converted into charts showing locations of large shoals of fish (Simpson 1992). Fishing fleets rely on GPS technologies to locate catches that are within reach and fuel costs. Simpson (1992) and Chassot et al. (2011) point out the use of satellite sensors, such as the Advanced Very High Resolution Radiometer (AVHRR), Nimbus-7 coastal zone color scanner (CZCS), SeaWIFS, MODIS, and MERIS have been used to augment stock assessment tools by mapping the distribution of phytoplankton pigment (i.e., chlorophyll *a*) in the ocean. The presence of phytoplankton has been shown to be correlated to the shoaling of many pelagic and non-pelagic fish species, as these species may feed on the phytoplankton or the zooplankton and associated smaller fish that feed on the plankton (Simpson 1992). In addition for improving stock assessments, satellite-derived mesoscale oceanographic features may also be used to improve the understanding of species habitat distributions in larger areas. Finally, other work focuses on food webs, temperature, and salinity gradients for a better understanding of population movements (Hooker et al. 2000).

Satellite remote sensing for sustainable fisheries management, although underutilized, provides important information for assessing sustainable fisheries stocks.

While stock assessment models must rely on the combination of additional data (e.g., chlorophyll, sea surface temperate, salinity, mesoscale ocean structures), many of these model variables can be derived from currently available sensors. In the consideration of climate change, the collection of data is imperative at a comprehensive scale necessary to provide global management decisions for sustainable fisheries, commercial harvest, and stock utilization (Beaugrand et al. 2002). Newer satellite sensors, promising repetitive, synoptic, high-resolution imagery, will serve to improve these global models that are necessary for understanding not only current fish species spatiotemporal distribution, composition, and population dynamics but also how these dynamics will be altered in light of global warming events.

6.9 Ocean Observation Networks

The vastness of our oceans requires a global framework in collaborative data collection. From a remote sensing perspective, most parameters, other than physical, are poorly sensed. For example, remote sensing data are rare to nonexistent for long time series that include data from the deep oceans, under-ice measurements, and geochemical, biological, and ecosystem parameters (Church 2010; Daly 2011). There is, however, an extensive use of proxies and many sensors are not optimized for autonomous systems (Chapron et al. 2008). A major proportion of the current ocean observation system—comprising different observing units and networks—was built and driven by climate observing requirements. The data products help inform climate research and aide in decisions about climate change. However, there are many more motivations than climate and weather for conducting ocean observations such as monitoring biodiversity, regional seas and regional fisheries management organizations, global marine resource agreements and assessments, and the development of ecosystem-based approaches for management of the ocean environment.

Several of the major ocean observation networks are described here. No single asset or suite of sensors addresses all spatial and temporal scales of ecosystem variability; it is necessary to integrate an appropriate mix of fixed and mobile platforms, remote sensing, process studies, models, and data management to advance a better understanding of the effects of climatic phenomena and human activity on ecosystem variability.

6.9.1 Global Ocean Observing System (GOOS)

The Global Ocean Observing System (GOOS) is a voluntary collaborative network of observations, where the components of the system are funded by national sources. As of 2015, global participation in the observing networks in GOOS varies—from a high of 77 countries contributing to tide gauge observations of sea level to about 10 countries involved in the resource-heavy global network of repeat hydrographic lines, measuring every feasible variable from the platform of a research ship. The GOOS system includes in situ and satellite observations along with operational observing networks and sustained research-funded observing networks and platforms, and is designed to serve many different objectives and users. For the essential (ocean) climate variables that can be measured on the surface of the ocean from satellites, GOOS examines the adequacy of commitments to missions in order to maintain a continuous and high-quality climate record.

6.9.2 Australia's Integrated Marine Observing System (IMOS)

As a marine country with the third largest ocean territory on Earth, Australia's weather patterns are highly sensitive to ocean influences. Australia also realizes an enormous economic benefit from its ocean territory. In response to these opportunities and challenges, Australia's Integrated Marine Observing System (IMOS) was created to observe and manage marine resources based on a national strategic approach addressing five major themes: multi-decadal ocean change, weather variability and weather extremes, major boundary currents and inter-basin flows, continental shelf processes, and ecosystem responses. The data are open-access and free of charge. IMOS collaborates with GOOS and other ocean observing networks.

6.9.3 European Marine Observation and Data Network (EMODnet)

The European Marine Observation and Data Network (EMODnet) is the fulfillment of an objective of the European Union's maritime policy to integrate existing, but fragmented and often inaccessible marine data (e.g., SeaDataNet, MyOcean, EuroGOOS, EuroARGO, EuroSITES, and other national and regional ocean observing systems) into interoperable, continuous, and publicly available data streams. EMODnet contains several portals including bathymetry, biology, chemistry, geology, hydrography, physics, and physical parameters.

6.9.4 US Integrated Ocean Observing System (IOOS®)

The US Integrated Ocean Observing System (IOOS®) is a national–regional partnership created to ensure the sustained observation of US coastal areas, oceans, and the Great Lakes, and to develop near real time and retrospective information products from those observations to assist people in their lives and livelihoods. The primary focus is to provide timely information through national and regional collaboration.

The Pacific Islands Ocean Observing System (PacIOOS) is one of 11 regional associations in the IOOS. The Pacific Island region spans 2 days and 8 time zones; it takes 11 h for PacIOOS to traverse from one end to the other. PacIOOS addresses familiar question such as where are the fish today? Which beaches are safe to visit today? Can I bring my vessel into the harbor safely? Is my home going to be inundated? Thematic areas include marine operations, ecosystem information, coastal hazards, water quality, numeric modeling, outreach and education, and data management and visualization. PacIOOS reports to the public and local groups, through its governing council, and directly relate the value of ocean observation to them at a very local and personal level, while still serving data to inform research and science at the global scale. Direct economic benefit, environmental preservation, and public health/safety are primary drivers. Science and research are secondary drivers of regional observing systems. PacIOOS is managed as nonprofit and governed by public stakeholders, providing connectivity to local people and a conduit for their voice.

Physical models include high-resolution wind models, wave models, and ocean circulation models down to 60 m. PacIOOS runs three suites of numerical models. Atmospheric models are run for the HI Island chain and then for each individual island at a higher resolution. Wave models are run basin-wide, and then nested down to the Hawaii regional and island scales. The ocean model is run at the Hawaii regional scale and then nested at high resolution to individual coastlines. It may be possible to expand this modeling capacity in both spatial area and increased resolution in the coming years.

6.10 Review Questions

1. Specifically what instruments are used to map ocean currents? How are they used? What is their fidelity?
2. Describe a commercial activity (practical application) that benefits from ocean color measurements.
3. How is nitrogen and phosphorus in water systems measured?
4. What is the most important (remotely sensed) variable in accessing coral reef health? Why?
5. Define the term "sustainable fisheries."
6. What (remotely sensed) parameters would a local fishery depend on? Who should fund the gathering of this data and how should it be distributed to the fisherman?
7. Compare and contrast two Ocean Observatory Systems.
8. Compare and contrast Case 1 and Case 2 waters.
9. What is eutrophication?
10. Name one type of remotely sensed data that is typically used for detecting and mapping oil spills in coastal marine waters.
11. How is the rate of marine primary production determined and what typical remote sensed data can be used for mapping and modeling this rate?
12. What is the major limitation in using remote sensing to map submersed aquatic vegetation?

13. Name and explain the use of three types of remotely sensed data used in observing coastal and ocean ecosystems.
14. What type of remote sensing data can be used for bathymetry mapping in coastal areas?

References

Anderson, L.W.J. 1993. Aquatic weed problems and management in North America: Aquatic weed problems and management in the western United States and Canada. In: AH Pieterse and KJ Murphy (eds.), *Aquatic Weeds*. Oxford University Press, Oxford. pp. 371–391.

Andréfouët, S., P. Kramer, P., and D. Torres-Pulliza. 2003, Multi-site evaluation of IKONOS data for classification of tropical coral reef environments, *Remote Sensing of Environment*, 88 (2003), pp. 128–143

Banks, A.C., P. Prunet, J. Chimot, P. Pina, J. Donnadille, E. Jeansou, M. Lux, G. Petihakis, G. Korres, G. Triantafyllou, C. Fontana, C. Estournel, C. Ulses, and L. Fernandez. 2012. A satellite ocean color observation operator system for eutrophication assessment in coastal waters. *Journal of Marine Systems* 94: S2–S15. DOI: 10.1016/j.jmarsys.2011.11.001

Beaugrand, G., F. Ibanez, J.A. Lindley, and P.C. Reid. 2002. Diversity of calanoid copepods in the North Atlantic and adjacent seas: species associations and biogeography. *Marine Ecology Progress Series*. 232:179–195.

Berkelmans, R. 2002. Time-integrated thermal bleaching thresholds of reefs and their variations on the Great Barrier Reef. *Marine Ecology Progress Series* 229: 73–82.

Blossey, B. 2004. Monitoring in weed biological control programs, pp. 95–105. In: E.M. Coombs, J.K. Clark, G.L. Piper and A.F. Cofrancesco, Jr. (eds.). Biological control of invasive plants in the United States. Oregon State University Press, Corvallis.

Chapron, B., R. Garello, and D.E. Weissman. 2008. Ocean remote sensing: challenges for the future. Technical Society and Oceanic Engineering Society of IEEE, 15–18 September 2008, Quebec City, Quebec, Canada. *Oceans 2008*. pp. 1–7. DOI: 10.1109/OCEANS.2008.5289439

Charles, H. and J.S. Dukes. 2007. Impacts of invasive species on ecosystem services. *Biol Invas* 193:217–237.

Chassot, E., S. Bonhommeau, G. Reygondeau, K. Nieto, J.J. Polovina, M. Huret, N.K. Dulvy, and H. Demarcq 2011. Satellite remote sensing for an ecosystem approach to fisheries management. ICES *Journal of Marine Science*, 68: 651–666.

Chen, M., Q. Zhuang, D.R. Cook, R. Coulter, M. Pekour, R.L. Scott, J.W. Munger, and K. Bible. 2011. Quantification of terrestrial ecosystem carbon dynamics in the conterminous United States combining a process-based biogeochemical model and MODIS and AmeriFlux data, Biogeosciences, 8, 2665-2688, doi:10.5194/bg-8-2665-2011.

Church, J.A. 2010. The changing oceans. *Science* 328: 1453. DOI: 10.1126/science.1192996

Costanza, R, R. d'Arge, R. de Groote, S. Farber, M. Grasso, B. Hannon, K. Limburg, S. Naeem, R.V. O'Neill, J. Paruelo, R.G. Raskin, P. Sutton, and M. van den Belt. 1997. The value of the world's ecosystem services and natural capital. *Nature* 387:253–260.

Daly, K. 2011. Directions in biology for ocean observing. GEOSS Workshop XLII-Oceans, 2011. 17 September 2011, Kona, HI, US. pp. 1–31. DOI: 10.1109/GEOS-XLII.2011.6105432

Diaz, R.J., and R. Rosenberg. 2008. Spreading dead zones and consequences for marine ecosystems. *Science* 321: 926–29. DOI: 10.1126/science1156401

DiMassa, D.D., M.L. Heron, and S.F. Heron. 2011. HF radar: a tool for coral reef planning and management. Marine Technical Society and Oceanic Engineering Society of IEEE, 19–22 September 2011, Waikoloa, HI, US. *Oceans 2011*. pp. 1–5.

Doney, S.C. 2010. The growing human footprint on coastal and open-ocean biogeochemistry. *Science* 328: 1512–1516. DOI: 10.1126/science.1185198

Hoepffner, N. and G. Zibordi. 2009. Remote Sensing of Coastal Waters, *Encyclopedia of Ocean Sciences (Second Edition)*, Pages 732–741, doi:10.1016/B978-012374473-9.00801-8

Hooker, S.B., N.W. Rees, and J. Aiken. 2000. An objective methodology for identifying oceanic provinces. *Progress in Oceanography*. 45:313–338.

Hovel, K.A. and R.N. Lipcius. 2002. Effects of seagrass habitat fragmentation on juvenile blue crab survival and abundance. *J Exp Mar Biol Ecol* 271:75–98.

Howarth, R., F. Chan, D.J. Conley, J. Garnier, S.C. Doney, R. Marino, and G. Billen. 2011. Coupled biogeochemical cycles: eutrophication and hypoxia in temperate estuaries and coastal marine ecosystems. *Frontiers in Ecological Environments* 9(1): 18–26. DOI: 10.1890/100008

International Ocean-Colour Coordinating Group (IOCCG). 2000. Remote sensing of ocean colour in coastal, and other optically-complex waters, S. Sathyendranath (Ed.), Report of the International Ocean-Colour Coordinating Group, No. 3, IOCCG, Dartmouth, NS (2000)

International Ocean-Colour Coordinating Group (IOCCG). 2006. Remote sensing of inherent optical properties: Fundamentals, tests of algorithms and applications, Z.P. Lee (Ed.), Report of the International Ocean-Colour Coordinating Group, No. 5, IOCCG, Dartmouth, NS (2006)

Irlandi, E.A., W.G. Ambrose, Jr. and B.A. Orlando. 1995. Seascape ecology and the marine environment: how spatial configuration of seagrass habitat influences growth and survival of the bay scallop. *Oikos* 72:307–313.

Klemas, V., 2008. Remote Sensing of Coastal Ecosystems, College of Marine and Earth Studies, University of Delaware, IEEE, US/EU-Baltic International Symposium, 2008 IEEE/OES, IEEE Explore Digital Library,

Kenov, I.A., F. Campuzano, G. Franz, R. Fernandes, C. Viegas, J. Sobrinho, H. de Pablo, A. Amaral, L. Pinto, M. Mateus, and R. Neves. 2014. Advances in Modeling of Water Quality in Estuaries, In: Advances in Coastal and Marine Resources, Editors: C.W. Finkl and C. Makowski. ISBN: 978-3-319-06325-6 (Print) 978-3-319-06326-3

Knudby, A., S.J. Pittman, J. Maina, and G. Rowlands. 2014. Remote Sensing and Modeling of Coral Reef Resilience, In: Advances in Coastal and Marine Resources, Editors: C.W. Finkl and C. Makowski. ISBN: 978-3-319-06325-6 (Print) 978-3-319-06326-3

Langeland, K.A. 1996. Hydrilla verticillata (L.F.) Royle (Hydrocharitaceae), The Perfect Aquatic Weed. *Castanea*. 61(3): 293–304.

Larkum, A.W.D., R.J. Orth, and C. Duarte, C. 2006. Seagrasses: *Biology, Ecology and Conservation*. Springer, The Netherlands. 691.

Lehmann, A. and J.B. Lachavanne. 1997. Geographic information systems and remote sensing in aquatic botany. *Aquat. Bot.* 58, 195–207.

Lira, C. and R. Taborda, 2014. Advances in Applied Remote Sensing to Coastal Environments Using Free Satellite Imagery In: Advances in Coastal and Marine Resources, Editors: C.W. Finkl and C. Makowski. ISBN: 978-3-319-06325-6 (Print) 978-3-319-06326-3.

Madsen, J.D. and R.M. Wersal. 2012. A Review of Aquatic Plant Monitoring and Assessment Methods. Aquatic Ecosystem Restoration Foundation.

Maeder, J., S. Narumalani, D.C. Rundquist, R.L. Perk, J. Schalles, K. Hutchins, and J. Keck. 2002. Classifying and mapping general coral-reef structure using IKONOS data. Photogramm. Eng. Remote Sens., 68:1297–1305.

Maina, J., T.R. McClanahan, V. Venus, M. Ateweberhan, and J. Madin. 2011. Global gradients of coral exposure to environmental stresses and implications for local management. *PLoS One* 6(8): e23064.

Mishra, D.R., S. Narumalani, D. Rundquist, and M. P. Lawson. 2006. Benthic habitat mapping in tropical marine environments using QuickBird imagery, Photogramm. Eng. Remote Sens., 72:1037–1048.

Mumby, P.J. and A.J. Edwards. 2002. Mapping marine environments with IKONOS imagery: enhanced spatial resolution does deliver greater thematic accuracy. Remote Sensing of Environment. 82:248–257.

Miller, R.L., R. López, R.P. Mulligan, R.E. Reed, C.-C. Liu, C.J. Buonassissi and M.M. Brown. 2014. Examining Material Transport in Dynamic Coastal Environments: An Integrated Approach Using Field Data, Remote Sensing and Numerical Modeling, In: Advances in Coastal and Marine Resources, Eds: C.W. Finkl and C. Makowski. ISBN: 978-3-319-06325-6 (Print) 978-3-319-06326-3.

NASA Earth Observatory (2010, January 31). Patagonian Bloom. Accessed December 31, 2010.

Nelson, S.A.C., K.S. Cheruvelil, and P.A. Soranno. 2006. Satellite Remote Sensing of Freshwater Macrophytes and the Influence of Water Clarity. *Aquatic Botany* 85. 4, 289–98.

Nichols, C. and B. Shaw. 1986. Ecological life histories of the 3 aquatic nuisance plants, *myriophyllum spicatum, potamogeton crispus* and *elodea canadensis*. *Hydrobiologia* 131(1):3–21.

Painter, S.C. et al. 2010. The COPAS'08 expedition to the Patagonian Shelf: Physical and environmental conditions during the 2008 coccolithophore bloom. *Continental Shelf Research*, 30 (18), 1907–1923.

Pala, C. 2011. Coral reefs: winners and losers. 10 August 2011. *AAAS Science Now*. http://news.sciencemag.org/sciencenow/2011/08/coral-reefs-winners-and-losers.html?ref=em&elq=3a3e6ccfdd8d43f9b7c976bac991b7f0. Accessed 22 August 2011.

Pan, X., A. Mannino, M.E. Russ, S.B. Hooker, and L.W. Harding, Jr. 2010. Remote sensing of phytoplankton pigment distribution in the United States northeast coast. *Remote Sensing of Environment* 114: 2403–2416. DOI: 10.1016/j.rse.2010.05.015

Penuelas, J., J.A. Gamon, K.L. Griffen, and C.B. Field. 1993. Assessing coummunity type, plant biomass, pigment composition, and photosyntheic efficiency of aquatic vegetation on spectral reflectance. *Remote Sens. Environ.* 46, 110–118.

Pimentel, D., R. Zuniga, and D. Morrison. 2005. Update on the environmental and economic costs associated with alien-invasive species in the United States. *Ecol. Econ.* 52: 273–288.

Pu, R., S. Bell, K.H. Levy, and C. Meyer. 2010. Mapping detailed seagrass habitats using satellite imagery. Geoscience and Remote Sensing Symposium (IGARSS), 2010, IEEE International. pp. 1–4. DOI: 10.1109/IGARSS.2010.5651884

Richardson, et al. 2012. Monoecious Hydrilla—A Review of the Literature. Northeast Aquatic Nuisance Species Panel. Accessed Online: http://www.nyis.info/user_uploads/files/Monoecious%20Hydrilla%20Lit%20Review%20-%20Final.pdf

Santoki, M., S. Ratheesh, R. Sharma, K.N. Joshipura, and S. Basu. 2012. Assimilation of drifter data in a circulation model of the Indian Ocean. *IEEE Geoscience Remote Sensing Letters*, **9**, pp. 100–103.

Simpson, J.H. and J.R. Hunter. 1974. Fronts in the Irish Sea. *Nature* 250: 404–406.

Simpson, J.J. 1992. Remote sensing and geographical information systems: Their past, present and future use in global marine fisheries. *Fisheries Oceanography*. 1(3):238–280.

Skirving, W.J., M.L. Heron, and S.F. Heron. 2006. The hydrodynamics of a bleaching event: implications for management and monitoring. Pp. 145–161. In Phinney, J.T. et al. (eds.). *Coral Reefs and Climate Change: Science and Management*. AGU Coastal and Estuarine Series, Vol. 61, 350 p. ISBN 0-87590-359-2

Stockdon H.F., A.H. Sallenger, H.J. List, and R.A. Holman. 2002. Estimation of shoreline position and change using airborne topographic lidar data. *Journal of Coastal Research* 18(3): 502–513.

Valley, R.D., M.T. Drake, and C.S. Anderson. 2005. Evaluation of alternative interpolation techniques for the mapping of remotely-sensed submersed vegetation abundance. *Aquat. Bot. 81*, 13–25.

Waycott, M., C.M. Duarte, T.J. Carruthers, R.J. Orth, W.C. Dennison, S. Olyarnik, A. Calladine, J.W. Fourqurean, K.L. Heck, A.R. Hughes, G.A. Kendrick, W.J. Kenworthy, F.T. Short, and S.L. Williams. 2009. Accelerating loss of seagrasses across the globe threatens coastal ecosystems. Proceedings of the National Academy of Sciences 106, 12377–12381.

Wilde, S.B., T.M. Murphy, C.P. Hope, S.K. Habrun, J. Kempton, A. Birrenkott, F. Wiley, W.W. Bowerman, and A.J. Lewitus. 2005. Avian vacuolar myelinopathy linked to exotic aquatic plants and a novel cyanobacterial species. *Environ. Toxicol.* 20:348–353.

Williams, S.K., J. Kempton, S.B. Wilde, and A. Lewitus. 2007. A novel epiphytic cyanobacterium associated with reservoirs affected by avian vacuolar myelinopathy. *Harmful Algae*. 6:343–353.

Wilson, C.J., P.S. Wilson, C.A. Greene, and K.H. Dunton. 2013. Seagrass meadows provide an acoustic refuge for estuarine fish. *Mar Ecol Prog Ser* 472:117–27.

Wilson, J.C. 2011. Mapping coral reef complexes using airborne LiDAR bathymetry technology. Marine Technical Society and Oceanic Engineering Society of IEEE, 19–22 September 2011, Waikoloa, HI, US. *Oceans 2011*. Pp. 1–5.

Additional Reading

Selman, M., S. Greenhalgh, R. Diaz, and Z. Sugg. 2008. Eutrophication and hypoxia in coastal areas: a global assessment of the state of knowledge. Water Quality: Eutrophication and Hypoxia. WRI Policy Note. World Resources Institute, No. 1. March 2008, Pp. 1–6. http://www.wri.org/sites/default/files/pdf/eutrophication_and_hypoxia_in_coastal_areas.pdf

NOAA. Sustainable fisheries management. Fishwatch U.S. Seafood Facts. Accessed Online: http://www.fishwatch.gov/wild_seafood/managing_fisheries.htm

Simpson, J.J. 1992. Remote sensing and geographical information systems: Their past, present and future use in global marine fisheries. Fisheries Oceanography. 1(3):238–280. http://onlinelibrary.wiley.com/doi/10.1111/j.1365-2419.1992.tb00042.x/epdf

Relevant Websites

NASA's OceanColor Web: http://oceancolor.gsfc.nasa.gov/cms/

NASA Goddard Space Flight Center—Perpetual Ocean: https://www.flickr.com/photos/gsfc/7009056027/

Global Ocean Observing System (GOOS): http://www.ioc-goos.org/

Australia's Integrated Marine Observing System (IMOS): https://imos.aodn.org.au/imos123/

European Marine Observation and Data Network (EMODnet): http://www.emodnet-hydrography.eu/

U.S. Integrated Ocean Observing System (IOOS®): http://www.ioos.noaa.gov/

Pacific Islands Ocean Observing System (PacIOOS): http://oos.soest.hawaii.edu/pacioos/

Chapter 7
The Final Frontier: Building New Knowledge Through Planetary and Extrasolar Observation

7.1 Introduction

In 2012, NASA issued the final version of an "all-sky atlas" that consists of 18,000 image mosaics constructed from 2.7 million images captured by its Wide-Field Infrared Survey Explorer (WISE) satellite (Clavin and Harrington 2012). Figure 7.1 is one of those image mosaics, showing the entire sky in infrared light. The Space Sciences Laboratory at the University of California, Berkeley has identified and labeled a number of noteworthy features in the mosaic image, such as the Andromeda Galaxy, the Tarantula Nebula, and Puppis A, a supernova remnant. The bright strip across the middle of the image is the Milky Way. Among the many discoveries made by WISE is a remote galaxy that is more luminous than 300 trillion suns, making it the most luminous galaxy ever identified by astronomers (Tsai et al. 2015). This luminosity may be related to a supermassive black hole at the galaxy's core.

Like supernovae, gamma-ray bursts are explosive events associated with the deaths of massive stars. Gamma-ray bursts are actually more extreme events, exhibiting luminosity that is a million times greater than the peak luminosity of an exploding-star supernova (Gehrels and Mészáros 2012). Although first discovered in the 1960s, much remained unknown about gamma-ray bursts until the launch of NASA's Swift (i.e., the Swift Gamma-Ray Burst Mission) and Fermi (i.e., the Fermi Gamma-Ray Space Telescope) satellites in 2004 and 2008, respectively. The two satellites carry instruments that have contributed significantly to research regarding the gamma-ray burst phenomenon, which is believed to relate to the gravitational collapse of matter that leads to black hole formation (Gehrels and Mészáros 2012). Figure 7.2 shows the December 2011 explosion of a gamma-ray burst, designated GRB 111209A, which was captured by the Swift satellite. While some gamma-ray bursts produce a signal (including both gamma-rays and X-rays) that lasts only a couple of seconds, GRB 111209A produced a signal that lasted 7 h, making it the longest-duration gamma-ray burst ever recorded (Gendre et al. 2013).

The preceding are just two examples of recent developments in the remote sensing of space. Indeed, many of the types of sensors commonly used for remote sensing of the Earth have similarly been applied to the other planets in our solar system

S. Khorram et al., *Principles of Applied Remote Sensing*,
DOI 10.1007/978-3-319-22560-9_7

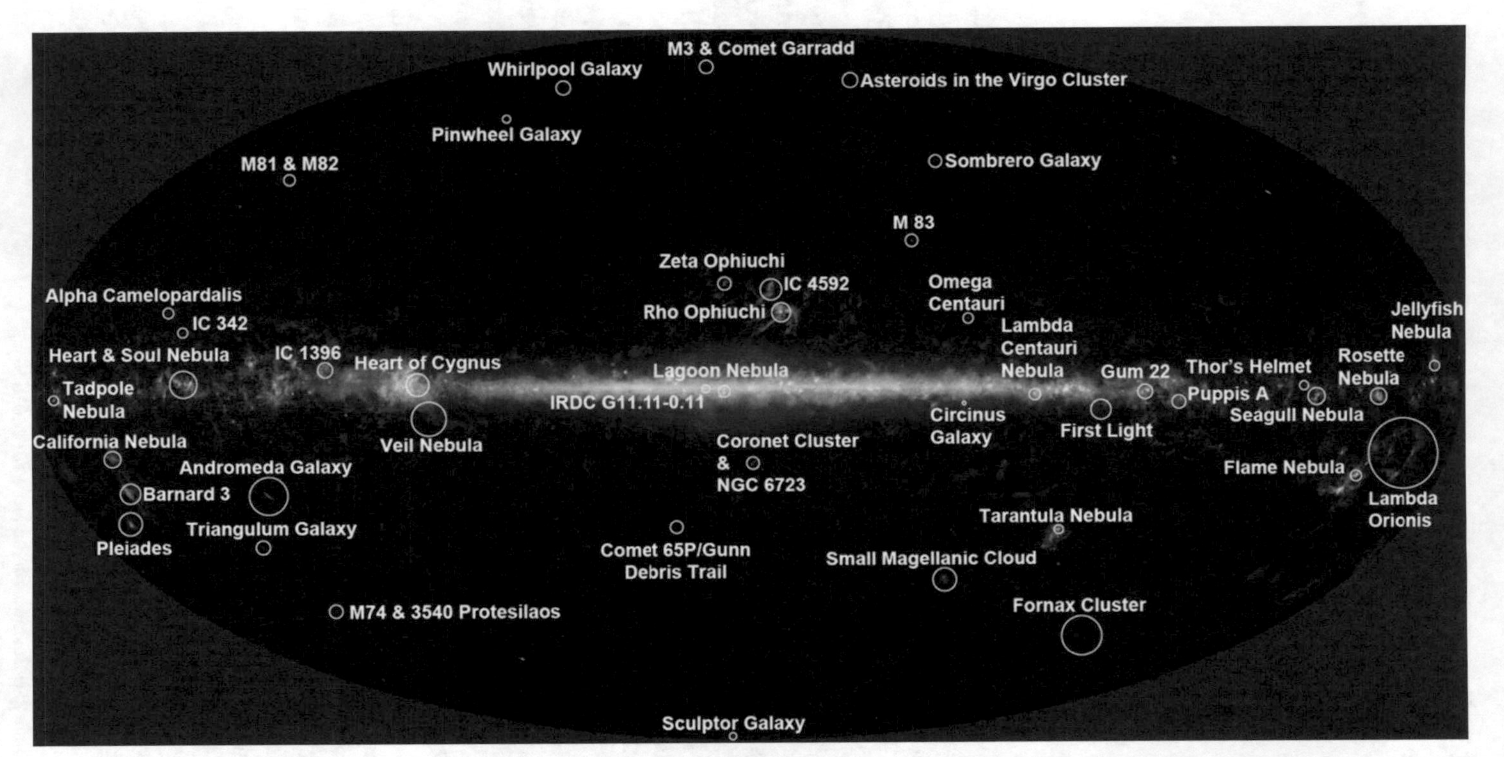

Fig. 7.1 Mosaic image of the entire sky, as captured by NASA's Wide-Field Infrared Survey Explorer *(WISE)* satellite. (Image courtesy of NASA and the Space Sciences Laboratory, University of California, Berkeley)

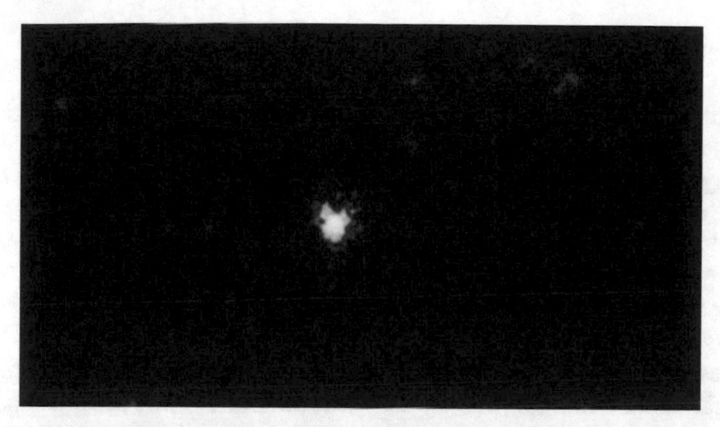

Fig. 7.2 False-color image showing the explosion of GRB 111209A in December 2011. The image was captured by the X-ray telescope aboard the Swift satellite. (Image courtesy of NASA/Swift/B. Gendre)

(Short 2006) or, as illustrated above, for viewing phenomena beyond our system. In fact, many technological developments in remote sensing are at least partially rooted in space exploration efforts; for instance, imaging spectroscopy—hyperspectral imaging—was developed in parallel for terrestrial and planetary applications (Goetz 2009). Of course, what are perhaps the best-known space exploration efforts have actually involved direct (i.e., non-remote-sensing) measurements: the Apollo missions, which landed astronauts on the Earth's Moon during the late 1960s and early 1970s, and more recently, the unmanned Mars Exploration Rovers *Spirit* and *Opportunity*, which began collecting data from the Martian surface in 2004 (*Opportunity* was still active as of July 2015; see http://marsrover.nasa.gov/home/index.html), as well as the rover *Curiosity*, which has been collecting data since 2012 as part of NASA's Mars Science Laboratory mission (http://mars.jpl.nasa.gov/msl/). Notably, all of these surface investigation missions have coincided with remote sensing by orbiting spacecraft. For example, the Apollo Command and Service Modules, which orbited the Moon while their corresponding Lunar Modules were on the surface, were equipped with a variety of remote sensing instruments (Short 2006), and there are currently five different spacecraft orbiting Mars. Ultimately, orbiting or fly-by spacecraft have collected remotely sensed data, in varying amounts, for all of the planets in our solar system and some of their moons (Hanel et al. 2003), as well as asteroids and the Sun (e.g., NASA's Solar Dynamics Observatory). Furthermore, the number of satellite-based instruments targeted at deep space continues to increase. In this chapter, we highlight some prominent historical missions as well as some active efforts.

7.2 Lunar Exploration

The Earth's Moon was the first target of non-terrestrial remote sensing. Prior to the Apollo missions, the former Soviet Union sent a series of spacecraft to the Moon, including Luna 3, which captured the first images of the far side of the Moon in 1959 (Short 2006). NASA's Lunar Orbiter program similarly sent five spacecraft to the Moon between 1966 and 1967, with the primary objective to find candidate landing sites for the Apollo Lunar Modules (Short 2006).

After the final Apollo mission (Apollo 17) in 1972, the next major lunar exploration mission was the 1994 Clementine spacecraft, built and operated by the U.S. Naval Research Laboratory (McEwen and Robinson 1997). The main accomplishment of the Clementine mission was the first acquisition of global-scale compositional and topographic data for the Moon. The instruments on the Clementine spacecraft included ultraviolet-visible, near infrared, long-wave infrared, and high-resolution cameras, a LiDAR system, and a radar-like unit transmitting in the S-band radio frequency (McEwen and Robinson 1997; Short 2006). Specialized products from the Clementine mission included distribution maps of compositional elements determined from recorded spectral reflectance at certain wavelengths. Experiments performed with Clementine's S-band transmitter were also the first to suggest the possibility of water, in the form of ice, in the Moon's polar regions (Nozette et al. 1996). Figure 7.3 is a topographic map of the Moon constructed from Clementine data. The image shows the Moon's most prominent basins, including the impact

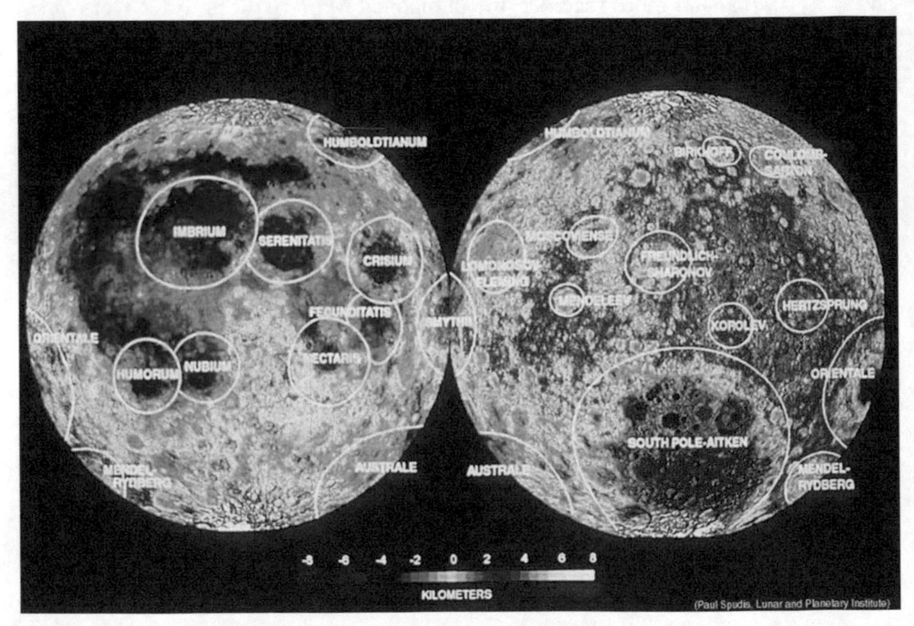

Fig. 7.3 Topographic map of the Moon, created using data from the Clementine mission. (Image courtesy of NASA and Paul Spudis, Lunar and Planetary Institute)

crater known as the South Pole-Aitken basin, which is the largest, oldest, and deepest (Petro and Pieters 2004).

NASA's Lunar Prospector, launched in January 1998 and operated until 1999 (when the spacecraft intentionally impacted the lunar surface), had a goal to further investigate Clementine's apparent finding of polar ice on the Moon; to accomplish this, its instrumentation was designed to obtain high-resolution gravity, magnetic, and compositional data (Binder 1998; Konopliv et al. 2001). The instruments on board the Lunar Prospector included gamma ray and neutron spectrometers to create global maps of the concentrations of elements such as iron, potassium, and hydrogen; a magnetometer/electron reflectometer to measure low-intensity magnetic fields; and a Doppler gravity experiment to derive a lunar gravity field map (Binder 1998).

Although the US and, previously, the Soviet space programs historically dominated lunar exploration, a number of other space agencies have sent spacecraft to the Moon. For example, the European Space Agency (ESA) launched its first lunar mission, SMART-1, in 2003; the spacecraft orbited the Moon from November 2004 to September 2006 (Foing et al. 2006). Other orbiting spacecraft have been sent to the Moon by China (Chang'e-1 in 2007 and Chang'e-2 in 2010), Japan (Hagoromo in 1990 and SELENE/Kaguya in 2007), and India (Chandrayaan-1 in 2008). The Selenological and Engineering Explorer (SELENE), commonly referred to by its Japanese nickname *Kaguya*, carried 13 instruments, and also deployed two micro-satellites, *Okina* and *Ouna*. In particular, Kaguya carried a high-definition television (HDTV) instrument with both wide-angle and telephoto cameras, which provided the first high-resolution videography of the Moon. Figure 7.4 shows a still image extracted from HDTV video recorded by Kaguya in April 2008. In addition, the 10-m resolution terrain camera aboard Kaguya provided the first images from inside the permanently shadowed Shackleton Crater near the lunar south pole (see Fig. 7.5); although it was theorized that the crater contained water-ice deposits, the sensor found no evidence to support the theory (Haruyama et al. 2008).

The Moon exhibits asymmetry between its near (i.e., visible from Earth) and far sides, as demonstrated by very few basaltic "mares," or plains, on the far side, as well as high concentrations of radioactive elements on the near side. One theory proposes that the asymmetry was caused by a large impact early in the Moon's history. Nakamura et al. (2012) used spectral data from two Kaguya instruments, the Spectral Profiler and Multiband Imager, to show the presence of a mineral, low-calcium pyroxene, in the 3000-km diameter Procellarum basin. This discovery suggests that the basin was in fact caused by a massive, ancient impact, thereby supporting the impact origin theory of the Moon's asymmetry.

The Chandrayaan-1 spacecraft, which orbited the Moon from November 2008 until August 2009, had instruments from several nations. In particular, a Brown University and NASA Jet Propulsion Laboratory research team contributed the Moon Mineralogy Mapper, an imaging spectrometer for collecting data at visible and near-infrared wavelengths. Significantly, measurements by the Moon Mineralogy Mapper indicated the presence of very low levels of water over most of the lunar surface (Pieters et al. 2009). Another NASA-commissioned research team

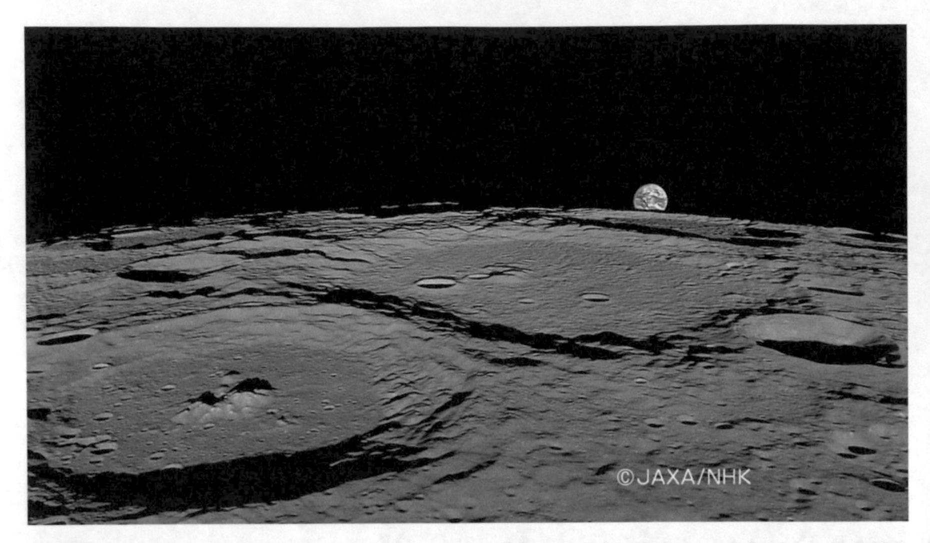

Fig. 7.4 Still image extracted from video recorded by the high-definition television (HDTV) instrument aboard the Kaguya satellite in April 2008. The area in the image is near the Moon's northern pole. In the *left* foreground is the Plaskett crater, and *behind* it is the Rozhdestvenskiy crater. The oblique angle of the instrument allowed for regular capture of "Earth-rise" and "Earth-set" from the perspective of the Moon, as exemplified here. (Image source: © JAXA/NHK)

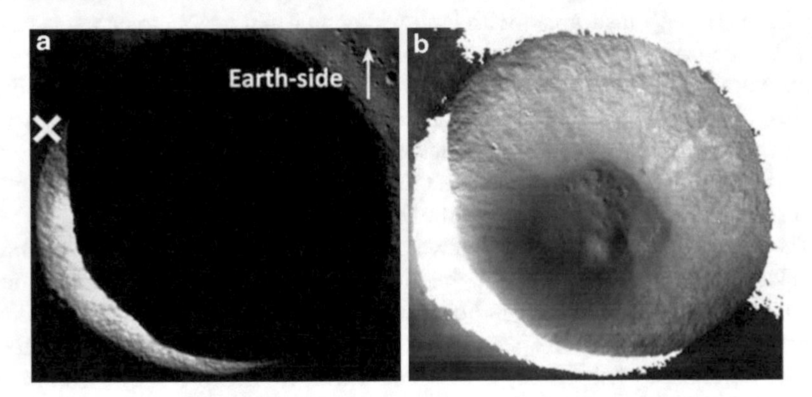

Fig. 7.5 **a** Image of the Shackleton crater recorded by the terrain camera aboard Kaguya. The *cross* indicates the location of the lunar south pole. **b** Enhanced image of the crater, using scattered light from the illuminated portion of the crater's upper wall. (Images excerpted from Haruyama et al. 2008)

designed and constructed an SAR instrument known as the Mini-SAR, with the objective of estimating the water content of the Moon's polar regions. Figure 7.6 is a Mini-SAR map of the lunar north pole, which suggests that more than 600 million m^3 of water may be found in the region's craters.

NASA's Lunar Reconnaissance Orbiter (LRO) was designed to expand upon previous lunar missions by collecting surface morphological data of sufficient

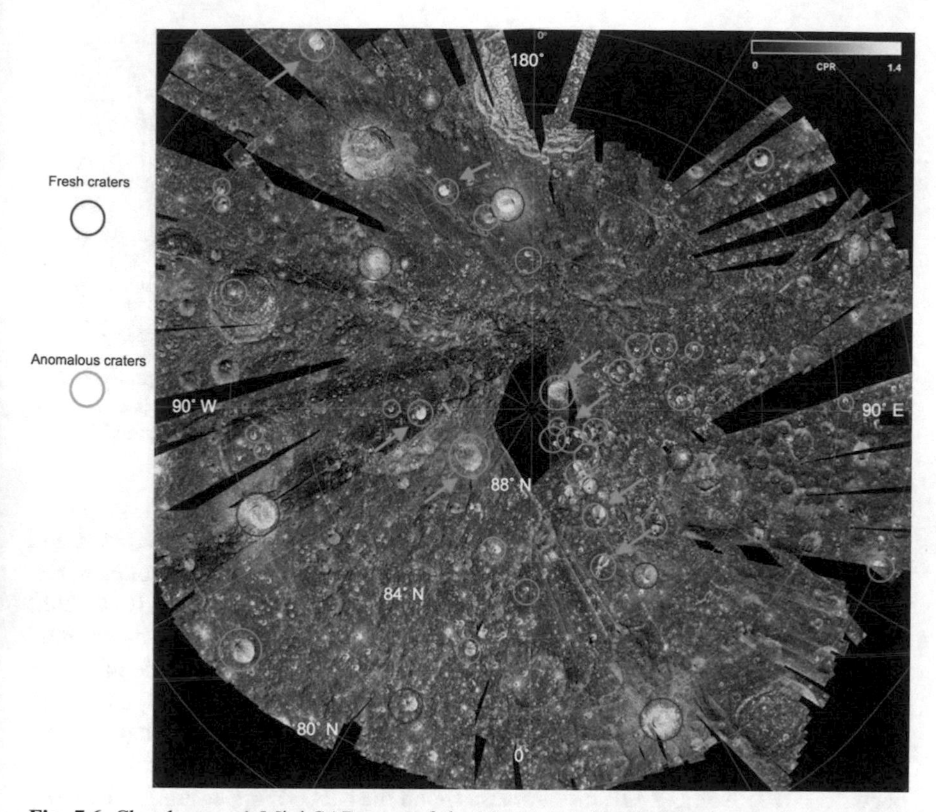

Fig. 7.6 Chandrayaan-1 Mini-SAR map of the Moon's north pole. The image documents the circular polarization ratio (CPR) values of surface features. The *red circles* indicate craters with high CPR values both inside and outside their rims, which is typical of young impact craters. The *green circles* indicate craters that have high CPR values only within their permanently shadowed interiors, which is indicative of water-ice presence. (Image courtesy of NASA)

resolution to support future manned exploration, including the possibility of a permanent outpost on the Moon (Robinson et al. 2010). Launched in 2009 and still operational as of 2015, the LRO contains a variety of instruments similar to those on previous lunar spacecraft; for example, it includes an SAR instrument known as the Mini-RF (Miniature Radio Frequency), which continues the search for polar water-ice deposits initiated by the Mini-SAR instrument aboard Chandrayaan-1 (see Spudis et al. 2013). However, its most distinctive instrument is the Lunar Reconnaissance Orbiter Camera (LROC). The LROC consists of three imaging subsystems: two Narrow Angle Cameras for capturing high-resolution (0.5 m) panchromatic images, and a multispectral Wide Angle Camera, which captures 75-m resolution images in the visible and 384-m images at ultraviolet wavelengths (Robinson et al. 2010). Notably, the LROC has recorded roughly 10,000 before-and-after image pairs that can be used to identify lunar surface changes, including new impact craters. Figure 7.7 shows one such image pair that documents an 18.8-m diameter crater created when a boulder-size object struck the Moon on March 17, 2013.

Fig. 7.7 Before (*left image*, captured in February 2012) and after (*right image*, captured in July 2013) images from the LROC instrument, documenting the formation of a new impact crater in the Moon's Mare Imbrium. (Images courtesy of NASA Goddard Space Flight Center and Arizona State University)

Other recent NASA-sponsored orbital missions include the Gravity Recovery and Interior Laboratory (GRAIL) and the Lunar Atmosphere and Dust Environment Explorer (LADEE). The GRAIL mission consisted of twin satellites, GRAIL-A (Ebb) and GRAIL-B (Flow), that orbited the Moon in 2012. Each satellite was outfitted with a Lunar Gravity Ranging System; working in concert, the two satellites constructed a high-resolution map of the Moon's gravitational field (Fig. 7.8) and found that more than 98 % of the gravitational signature is associated with topographic features, including impact craters, tectonic structures, and volcanic landforms (Zuber et al. 2013). The GRAIL mission also found the Moon's upper crust to be more porous than what was originally thought, emphasizing the influence of large impact events early in the Moon's history (Wieczorek et al. 2013), although widespread magmatic dykes also suggest an early period of expansion, during which time the Moon's radius may have increased by 0.6–4.9 km (Andrews-Hanna et al. 2013).

LADEE was intended to answer key questions about the composition of the lunar atmosphere, including the distribution of dust at high altitudes (Elphic et al. 2014). The spacecraft collected data for 140 days (September 2013–April 2014) from a near-equatorial (±23°) band around the Moon (Benna et al. 2015; Elphic et al. 2014). The neutral mass spectrometer (NMS) instrument aboard LADEE yielded the first global characterization of the dynamics of two prominent atmospheric gases, helium and argon, which had been previously detected by Apollo 17, and discovered that a third gas, neon, also comprised a significant portion of the lunar atmosphere (Benna et al. 2015). The Lunar Dust Experiment (LDEX) instrument uncovered a substantial dust component to the lunar atmosphere, largely concentrated on the morning side of the Moon where meteoroids most commonly impact the lunar surface (Elphic et al. 2014).

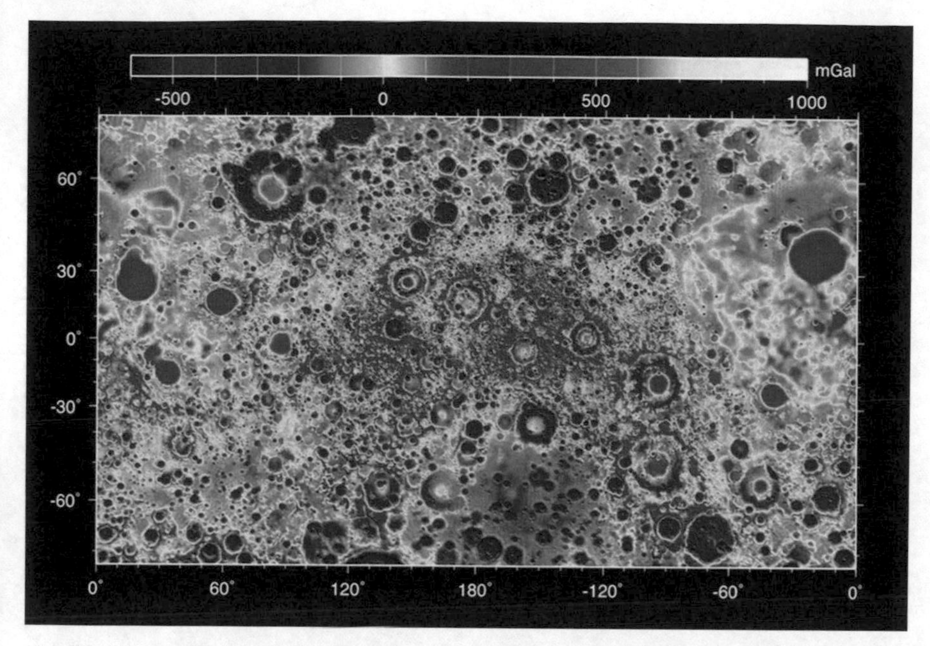

Fig. 7.8 The gravitational field of the Moon as measured by NASA's GRAIL mission. The map, which is in the Mercator projection, shows the far side of the Moon in the center and the near side at the right and left. The units are milliGalileos, where 1 Galileo is 1 cm/s^2. *Red* areas correspond to mass excesses, which create regions of higher local gravity, and *blue* areas correspond to mass deficits, which create regions of lower local gravity. (Image courtesy of NASA Jet Propulsion Laboratory)

7.3 Mercury, Venus, and Mars

Besides the Earth, the other "rocky planets" of our solar system (i.e., Mercury, Venus, and Mars) have been the targets of a number of historical and current remote sensing efforts. NASA's MESSENGER spacecraft was developed to study the geology, gravity field, and chemical composition of Mercury (JHU Applied Physics Laboratory 2015b). Launched in 2004, MESSENGER successfully entered the orbit around the planet in 2011. One of its main objectives was to construct the first global map of Mercury's surface; its predecessor, Mariner 10, had imaged less than half of the planet in the 1970s. To facilitate this, one of its five instruments was the Mercury Dual Imaging System (MDIS), which consisted of both narrow- and wide-angle CCD cameras. The MDIS captured global imagery at 250 m spatial resolution, as well as finer-resolution (20–50 m) imagery of regions of particular interest. Figure 7.9 shows a color-enhanced image of Mercury, constructed from the thousands of images collected during MESSENGER's primary mapping mission. During its more than 4 years in orbit around Mercury (the spacecraft was intentionally crashed into the planet's surface in 2015), MESSENGER made a number of significant discoveries. Similar to the Moon, permanently shadowed craters

Fig. 7.9 Color-enhanced image of Mercury created from images from the primary global mapping mission of the MESSENGER satellite. The colors represent distinct chemical, mineralogical, and physical signatures. The large, *tan-colored* region is the Caloris basin, created by an impact that occurred early in our solar system's formation. The *light blue* and *white* features are comparatively young crater "rays." (Image courtesy of NASA/JHU Applied Physics Laboratory/Carnegie Institution for Science)

near Mercury's north pole were found to contain stable water-ice deposits, but in many cases, these deposits were also found to be covered by a dark insulating layer, believed to be rich in organic material (Neumann et al. 2012; Paige et al. 2012). Scientists postulate that these water-ice and organic materials were introduced to Mercury by impacts from comets or asteroids. MESSENGER also yielded key details about the planet's thin exosphere (i.e., atmosphere), which varies seasonally in terms of its density and chemical composition, and sometime trails behind Mercury as a comet-like tail, as long as 2 million km, due in part to solar wind activity (McClintock et al. 2009).

The planet Venus was visited by 25 spacecraft during the twentieth century (including NASA'S 1990–1994 Magellan mission), but the European Space Agency's Venus Express spacecraft, which orbited the planet from 2006 until 2014, was the first to suggest key details about the planet's turbulent atmosphere, including the occurrence of lightning at a rate at least half that of the Earth (Titov et al. 2009). Figure 7.10 shows a swirling vortex of clouds and gases above the planet's south pole, as captured by the Visible and Infrared Imaging Spectrometer (VIRTIS) on board Venus Express. Such vortices form at both poles of Venus when heated atmospheric gases from the planet's equatorial zone are driven by high winds toward the poles, where they subsequently cool and sink. Overall, the mission's findings suggest that Venus and Earth once had similar surface environments but evolved quite differently. For example, Venus, which is now relatively dry, likely once held much more water but lost it because of high temperatures due to its proximity to the Sun (Titov et al. 2009).

A number of remote sensing efforts have been aimed at finding and characterizing evidence of water on Mars, either currently or historically, as a key condition

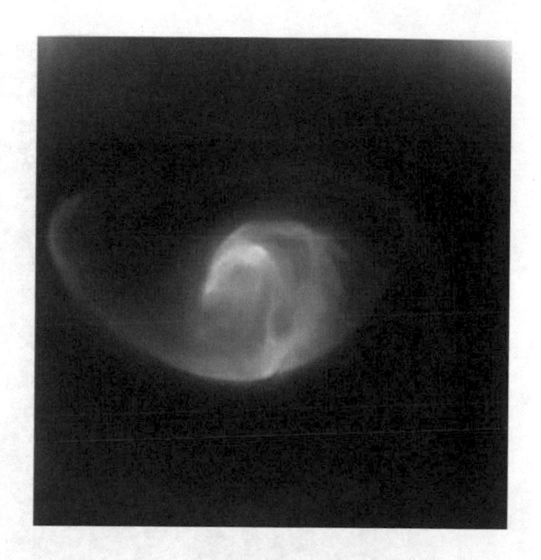

Fig. 7.10 Visible and infrared imaging spectrometer (VIRTIS) image of a swirling vortex over the south pole of Venus. (Image courtesy of the European Space Agency)

for supporting life. This was a major goal of NASA's Mars Global Surveyor, which orbited the planet from 1997 to 2006 with the additional goals of studying its climate and geology and preparing for possible human exploration (NASA 2010). Among the instruments carried by the spacecraft was the Mars Orbiter Camera (MOC), comprising 1.5-m narrow-angle and 240-m wide-angle scanners. Data collected by the MOC suggested the presence of sources of liquid water at shallow surface depths (Malin and Edgett 2000); ultimately, MOC data were used to suggest that liquid water had flowed on Mars as recently as the previous decade (Malin et al. 2006). Figure 7.11 shows August 1999 and September 2005 images of a crater in the Centauri Montes region of Mars, both captured by the Mars Global Surveyor. The light-toned deposit visible near the base of a gully in the later but not the first image suggests the presence of sediment carried by liquid water at some point in the intervening 6 years. Meanwhile, another instrument aboard the spacecraft, the Mars Orbiter Laser Altimeter, facilitated comprehensive topographic mapping of Mars, providing 600 million measurements—with a vertical precision of better than 1 m—of the Martian surface before the sensor failed in 2001 (Kirk 2005; Smith et al. 1999).

NASA built upon the success of the Mars Global Surveyor with the Mars Odyssey spacecraft, which began mapping the planet in 2002; the spacecraft was still active in 2015. One of the primary instruments on Mars Odyssey, the Gamma-Ray Spectrometer, provided the first evidence of large subsurface water-ice deposits near the planet's north and south poles (Boynton et al. 2002). The ESA's Mars Express orbiter, which arrived in December 2003, further expanded on the two prior NASA orbiter missions. In particular, its Observatoire pour la Mine´ralogie, l'Eau, les Glaces, et l'Activite (OMEGA) instrument, which is a hyperspectral visible and near-infrared imaging spectrometer, has identified some surface minerals formed in the presence of water; this suggests that the planet was once much warmer and wetter than it is now (Bibring et al. 2005; Paige 2005). Based on imagery from the

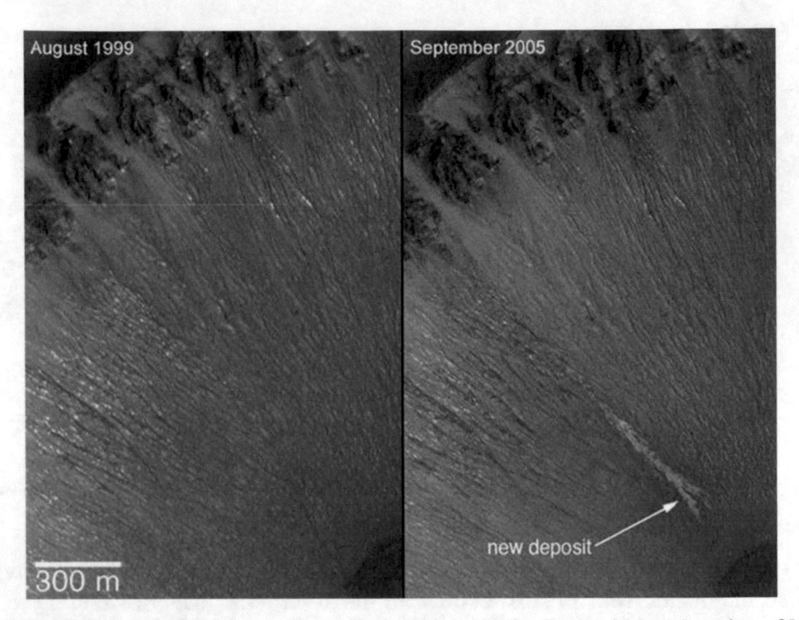

Fig. 7.11 Before-and-after images of a gully in a crater in the Centauri Montes region of Mars, both captured by the Mars Orbiter Camera aboard the Mars Global Surveyor. The *new light-toned deposit* in the 2005 image appears to indicate the movement of sediment by liquid water during the intervening years. (Images courtesy of NASA)

High-Resolution Stereo Camera (HRSC), another instrument on board Mars Express, a team of researchers presented evidence for volcanic eruptions as recently as 2 million years ago on Olympus Mons, Mars' highest volcano, suggesting that the planet may still be volcanically active (Neukum et al. 2004). The HRSC also may have uncovered a possible dust-covered frozen sea near the Martian equator, indicated by an abundance of fractured, plate-like features that resemble pack-ice (Murray et al. 2005). In 2009, Mars Express observed what appears to be the delta of an ancient lake within the Eberswalde crater in the planet's southern highlands (see Fig. 7.12).

NASA's Mars Reconnaissance Orbiter began collecting data from Martian orbit in 2006. One of the mission's primary objectives, as with its predecessors, is to study the historical and current role of water on Mars (Zurek and Smrekar 2007). The spacecraft's High Resolution Imaging Science Experiment (HiRISE) provides images with a spatial resolution as fine as 30 cm. In August 2011, scientists reported that repeat images from the HiRISE sensor reveal narrow, dark markings on steep slopes that appear to grow incrementally during warm seasons and fade in cold seasons; it is possible that these dark streaks represent warm-season flows of briny water (McEwen et al. 2011). Another phenomenon documented by the Mars Reconnaissance Orbiter (and by earlier spacecraft) is the presence of gigantic 'dust devils', often similar to or larger than typical tornadoes on Earth. Most Martian dust devil activity occurs in two narrow bands near 60° north and 60° south, during their

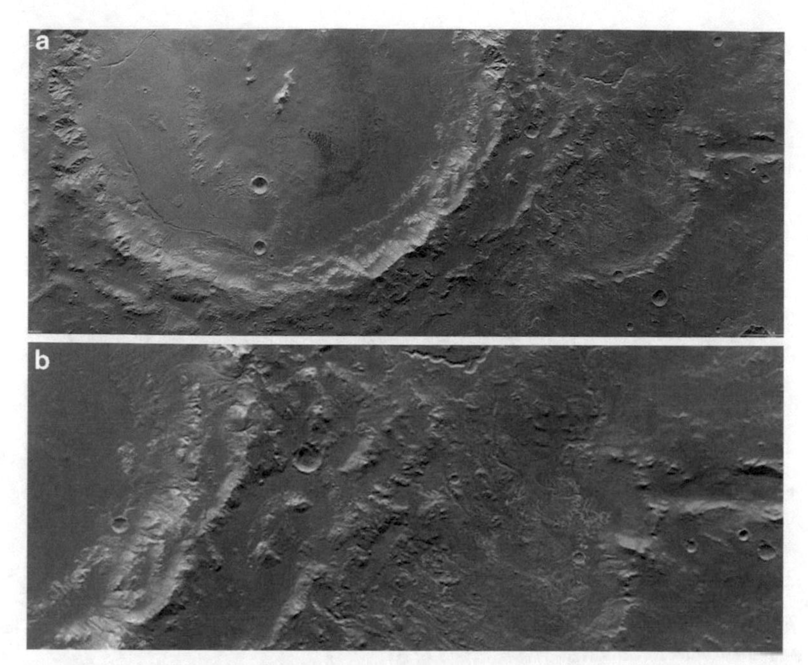

Fig. 7.12 **a** Image of the Holden and Eberswalde craters, captured by the High-Resolution Stereo Camera aboard Mars Express. Holden crater is 140 km across, filling the *left side* of the image, while to the *right* is the remaining part of Eberswalde crater, with a diameter of about 65 km. They are located in the southern highlands of Mars. **b** The Eberswalde crater is partially obscured by debris from the impact that created the Holden crater, but a fan-shaped cluster of feeder channels, typical of a delta, is still evident in this close-up image. (Image courtesy of ESA)

respective spring and summer (Whelley and Greeley 2008). An especially large dust devil in the Amazonia Planitia region of northern Mars was captured in March 2012 by the HiRISE sensor (Fig. 7.13) and the 6-m resolution, panchromatic Context Camera (Fig. 7.14), another sensor aboard the Mars Reconnaissance Orbiter. The long shadow in Fig. 7.14 suggests that the plume reached a height of roughly 20 km, despite being only 140 m in diameter. The soaring heights of Martian dust devils are facilitated by the planet's relatively thin atmosphere, which allows transferred surface heat (i.e., the energy source of dust devils) to reach altitudes higher than those that would be possible in the Earth's atmosphere (Geissler 2012).

Two other spacecraft recently joined Mars Odyssey, Mars Express, and the Mars Reconnaissance Orbiter in orbit around the planet: NASA's Mars Atmosphere and Volatile Evolution (MAVEN) mission and the Mars Orbiter Mission (MOM, also known by its Hindi name, *Mangalyaan*) from the Indian Space Research Organization. Both spacecraft began orbiting Mars in September 2014. MAVEN is focused on analysis of the planet's upper atmosphere, while MOM will explore both the Martian surface and its atmosphere. Nonetheless, the primary objective of MOM is to demonstrate India's technological capabilities for future planetary exploration.

Fig. 7.13 Large dust devil captured in March 2012 by the High Resolution Imaging Science Experiment (HiRISE), an instrument aboard the Mars Reconnaissance Orbiter. (Image courtesy of NASA/Jet Propulsion Laboratory/University of Arizona)

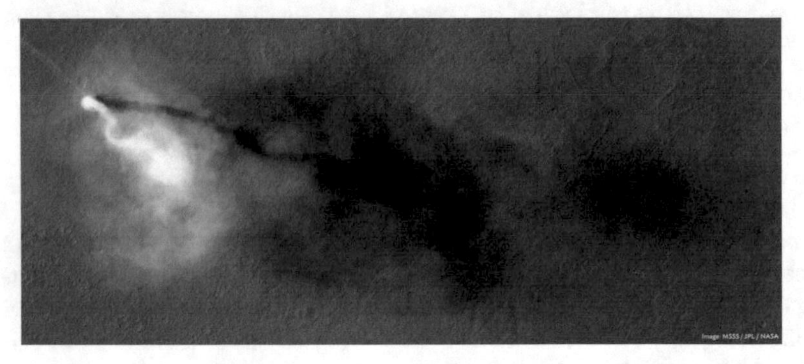

Fig. 7.14 Large dust devil (the same as in Fig. 7.13) captured in March 2012 by the Context Camera aboard the Mars Reconnaissance Orbiter. (Image courtesy of MSSS/JPL/NASA)

7.4 Jupiter, Saturn, Uranus, and Neptune

The "gas giant" planets of our solar system have been the subjects of far fewer exploratory missions than the rocky planets. NASA's Galileo spacecraft, which orbited Jupiter from 1995 to 2003, was the first spacecraft to orbit (rather than fly by, like the earlier Pioneer or Voyager missions) any of the gas giants. Galileo carried a variety of imaging systems, including near-infrared and ultraviolet spectrometers as well as a magnetometer, with which it captured detailed imagery of Jupiter's vertical cloud structure (Banfield et al. 1998) and prominent atmospheric features such as the Great Red Spot (Vasavada et al. 1998). Galileo also observed some of the planet's major moons, recording evidence of a global ocean beneath the icy surface of Europa (and possibly Ganymede and Callisto) as well as a very high level of volcanic activity on Io (McEwen et al. 1998); indeed, Io is considered the most volcanically active body in the solar system. Figure 7.15 shows a Galileo-captured image of an active eruption in the Tvashtar Catena, a sequence of volcanic calderas in the Moon's northern hemisphere. Another NASA spacecraft, Juno, is expected to reach Jupiter in 2016.

On its way to Jupiter, Galileo passed through the asteroid belt that sits between Mars and Jupiter. In August 1993, it flew by and recorded images of the asteroid 243 Ida. Examination of these images revealed that 243 Ida has a small (1.6 km across) satellite, subsequently named Dactyl, which appears to be composed of

Fig. 7.15 An active volcanic eruption in the Tvashtar Catena of Io, as captured by Galileo in February 2000. The image is a mosaic that combines separate images captured with near-infrared, clear, and violet filters. The dark, L-shaped lava (*left* of image center) marks the location of a November 1999 eruption, which was also documented by Galileo. The two small *bright spots* are sites where molten rock is exposed to the surface at the toes of lava flows. The larger *orange* and *yellow* ribbon is a cooling lava flow that is more than 60 km (37 miles) long. (Image courtesy of NASA Planetary Photojournal)

similar materials as 243 Ida (Chapman et al. 1995). Since the 1970s, scientists had suspected that some asteroids have satellites, but Galileo's discovery (see Fig. 7.16) provided the first definitive proof.

NASA's Cassini is the first spacecraft to orbit Saturn. Placed into the orbit in 2004 and still active in 2011, Cassini has provided extensive observations of the planet's atmosphere, rings, and magnetosphere (Gombosi and Ingersoll 2010). Saturn's rings, which are largely composed of water ice, are structurally diverse and dynamic; indeed, certain structural aspects can change within the period of a few days (Cuzzi et al. 2010). The rings are also influenced by at least 150 moons and small (1- to 100-km)-diameter "moonlets" that orbit Saturn. For example, Fig. 7.17 is an image, captured by the Cassini Imaging Science Subsystem (ISS), showing the influence of Prometheus, an oblong-shaped moon, on the planet's outermost

Fig. 7.16 Asteroid 243 Ida and its satellite, Dactyl, as captured by the Galileo spacecraft in August 1993. This color-enhanced image was constructed from a set of images that used violet and infrared filters. (Image courtesy of NASA)

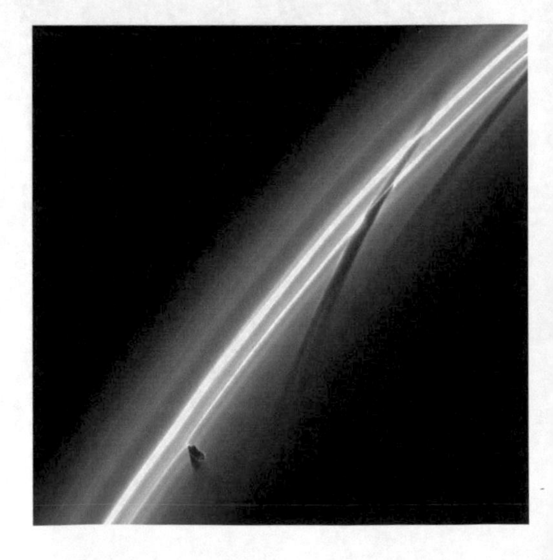

Fig. 7.17 Cassini image, captured in January 2009, showing the creation of light and dark streamers by the periodic penetration of the moon Prometheus into Saturn's outermost F ring. (Image courtesy of Cassini Imaging Team/ISS/JPL/ESA/NASA)

"F" ring. Prometheus orbits Saturn just inside this ring, but penetrates its inner edge approximately every 15 h. The moon's gravity attracts ring particles toward it, creating a streamer effect that is duplicated with every orbital pass into the ring.

Cassini has also made close-scale observations of some of Saturn's moons, particularly Titan. In fact, Cassini served as the launch pad for the ESA's Huygens probe, which landed on the moon's surface in 2005. Scientists believe that Titan, with its dense, hazy atmosphere, strongly resembles a primitive Earth (Kerr 2005). Evidence from Cassini and Huygens suggest that Titan has a methanological cycle that is similar to Earth's hydrologic cycle (Stofan et al. 2007). The Cassini Radar Mapper (an SAR instrument) has captured what are believed to be hydrocarbon (probably methane or ethane) lakes and even seas. Figure 7.18 shows a peculiar change observed in Ligeia Mare, one of the largest of these seas: a presumed coastal land feature that is not visible in April 2007, but which changes its appearance between July 2013 and August 2014. Other SAR observations have suggested that the Ligeia Mare is extremely smooth, with no surface wave activity (Zebker et al. 2014), so scientists have proposed other potential explanations for the recent appearance of the feature, including rising bubbles, floating solids, or solids that are suspended just below the lake's surface. They also suspect that the feature's emergence is related to seasonal changes on Titan (NASA Jet Propulsion Laboratory 2014).

NASA's Voyager 2 remains the only spacecraft to have flown by Uranus (in 1986) and Neptune (in 1989). Data collected by Voyager 2 suggest that Uranus has an ocean of dense, superheated liquid water (plus methane and ammonia) below its thick cloud atmosphere (Moore and Henbest 1986), while Neptune is extremely windy, with clouds of methane ice suspended in a clear atmosphere above a layer of hydrogen sulfide or ammonia ices (Smith et al. 1989). While orbital missions to Uranus and Neptune have been proposed, none has yet advanced to the planning stage.

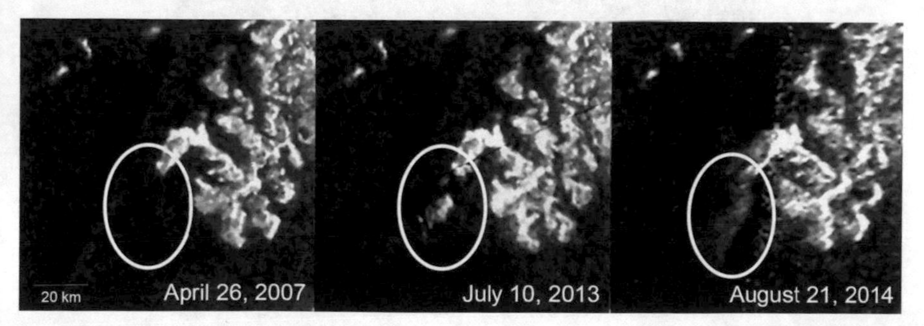

Fig. 7.18 Three images, created from Cassini Radar Mapper data, showing the appearance and subsequent change of a presumed land feature near the coast of Titan's Ligeia Mare. (Image credit: NASA/JPL-Caltech/ASI/Cornell)

7.5 Pluto and the Kuiper Belt

NASA's New Horizons spacecraft is the first mission to study the dwarf planet Pluto and other objects in the solar system's remote Kuiper Belt (Stern 2008). It made its closest approach to Pluto on July 14, 2015. Figure 7.19 is an image of Pluto's near side captured the prior day (July 13), showing a distinctive, heart-shaped area of bright terrain. In terms of the early findings of the New Horizons, mission scientists have confirmed the presence of nitrogen and methane ices in the polar region as well as large amounts of nitrogen leaking from the planet's atmosphere into space. They have also measured Pluto's diameter as 2370 km, larger than some earlier estimates, which reestablishes the planet as the largest object in the Kuiper Belt. A more distant object, Eris, is slightly smaller but greater in mass (Chang 2015; JHU Applied Physics Laboratory 2015a).

New Horizons has also observed noteworthy differences between Pluto and its largest moon, Charon, despite the fact that they were formed from the same material after a collision of two Kuiper Belt objects in the early history of the solar system (Chang 2015). Figure 7.20 was captured by the spacecraft on July 8, 2015, at a distance of 6 million km. The image highlights how large Charon is relative to Pluto: 1208 km in diameter, roughly half the planet's size. Pluto is composed of reddish materials that are apparently absent on Charon. The interior of Pluto is believed to be mostly rock, while that of Charon is believed to be equal parts rock and ice. Charon also lacks a significant atmosphere (JHU Applied Physics Laboratory 2015c).

Fig. 7.19 Image of the near side of Pluto, captured by the Long Range Reconnaissance Imager (LORRI) aboard New Horizons on July 13, 2015. The image has been color enhanced using lower-resolution data from the Ralph instrument aboard New Horizons. A distinctive *heart-shaped* area of bright terrain borders darker equatorial terrains; the smoothness of the heart-shaped area may indicate ongoing geologic processes. (Image courtesy of NASA/JHU Applied Physics Laboratory/ Southwest Research Institute)

Fig. 7.20 Image of Pluto and its largest moon, Charon, captured by the Long Range Reconnaissance Imager (LORRI) aboard New Horizons on July 8, 2015. The image has been color enhanced using lower-resolution data from the Ralph instrument aboard New Horizons. The image was captured at a distance of roughly 6 million km. (Image courtesy of NASA/JHU Applied Physics Laboratory/Southwest Research Institute)

In the fall of 2015, New Horizons will begin operations to travel to other objects in the Kuiper Belt. The Kuiper Belt contains some of the most primitive matter in the solar system, which is also the least affected by the Sun (Jewitt 1999). Thus, the hope is that these additional New Horizons observations, combined with the spacecraft's data on Pluto, will provide clues about the early history and formation of our solar system.

7.6 The Sun

The term "heliophysics" refers to the study of the interactions of our Sun with the Earth and the rest of the solar system (Bentley 2009). Heliophysics is one of the four main research areas within NASA's current science mission, along with earth science, planetary science, and astrophysics. While other space organizations have also invested in satellite-based heliophysics research (e.g., the ESA's Solar Orbiter, expected to launch in 2018), NASA has been especially active in this field; as of July 2015, 17 different NASA-sponsored or jointly sponsored spacecraft are collecting data on various aspects of the Sun–solar system connection.

One of the main concerns of these missions is to provide researchers with a better understanding of "space weather." In addition to solar flares and high-speed solar winds, the Sun periodically emits huge clouds, known as coronal mass ejections, which carry strong magnetic fields and energetic particles (Baker 2002). These emissions cause geomagnetic storms that can disrupt power and communications systems on Earth as well as in orbiting spacecraft (Baker 2002). Data from ongoing heliophysics missions should help scientists to forecast these occurrences. For example, the Solar Terrestrial Relations Observatory (STEREO) consists of two satellites launched in 2006: one positioned in front of Earth in its orbit, the other positioned behind it. Together, the two satellites provide stereoscopic (i.e., three-dimensional) images of the Sun. Davis et al. (2009) presented such an image that depicted an Earth-impacting coronal mass ejection, which was considered a major success for the STEREO mission. Similarly, NASA's Solar Dynamics Observatory

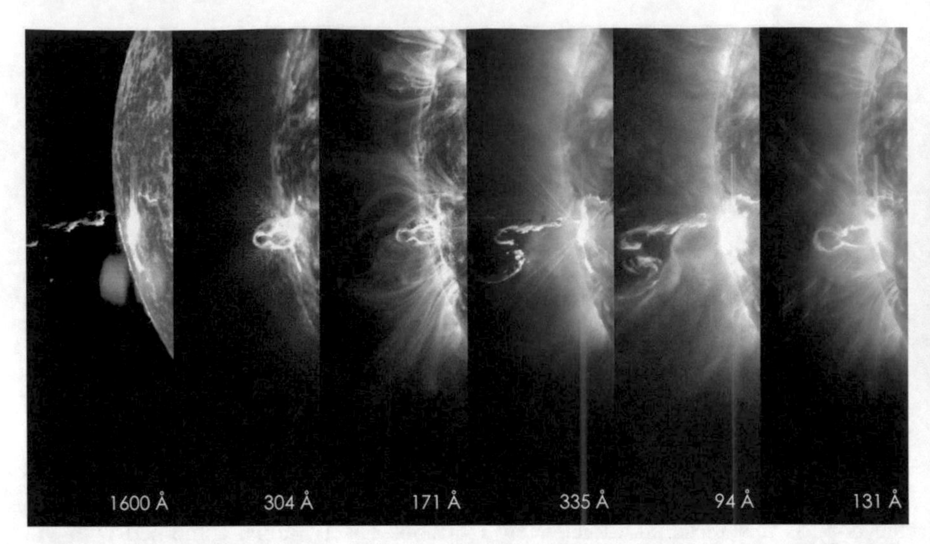

Fig. 7.21 Solar Dynamics Observatory images, in different wavelengths, showing the early moments of an X4.9-class solar flare. The images were captured on February 24, 2014. (Image courtesy of NASA/SDO/Goddard's Scientific Visualization Studio)

(SDO), launched in 2010, was developed to study the Sun's interior and magnetic field, and to learn how stored magnetic energy is released as solar wind and other space-weather-related emissions (Pesnell et al. 2012). Instruments aboard the SDO collect spectral data in a variety of wavelengths, including extreme ultraviolet (10–124 nm) wavelengths. Figure 7.21 is an SDO image, captured in February 2014, of an X4.9-class solar flare. X-class flares are the most intense, while the number further describes its strength, e.g., an X4.9-class flare is 4.9 times as intense as an X1 flare.

7.7 Extrasolar Remote Sensing

The introduction to this chapter presented examples of spacecraft that were developed to observe deep space phenomena. NASA's WISE, Fermi, and Swift satellites are all Earth-orbiting telescopes in the tradition of the Hubble Space Telescope. Launched in 1990, the Hubble Space Telescope is considered one of NASA's four "Great Observatories" along with the Compton Gamma Ray Observatory (1991–2000), the Chandra X-ray Observatory (1999–current), and the Spitzer Space Telescope, an infrared observatory (2003–current). It carries a highly versatile instrument suite that is targeted at near-ultraviolet, visible, and near-infrared wavelengths (115–2500 nm) (STScI 2015). Designed with the expectation that it would be periodically upgraded by astronauts, the Hubble Space Telescope has undergone a series of modifications and updates in its 25-year history, such that it may continue to operate successfully for another two decades, or even longer. It will eventually be replaced by the Webb Space Telescope, which will be as capable of viewing infrared wavelengths as the Hubble is at viewing visible wavelengths (STScI 2015).

The Hubble Space Telescope has made numerous scientific breakthroughs (Klotz 2015), including the most accurate estimate yet of the age of the universe: 13.82 billion years. It also enabled researchers to discover that the universe is expanding at an increasing speed, a still little-understood phenomenon known as "dark energy." Perhaps most notably, the Hubble Space Telescope provided evidence that supermassive black holes can be found at the centers of most galaxies, including the Milky Way (Klotz 2015). For example, Fig. 7.22 shows a disk of dust spiraling around a massive black hole at the center of a galaxy, NGC 4261, which is 100 million light-years from Earth. The mass of this black hole is estimated to be 1.2 billion times the mass of our Sun, but concentrated in an area roughly the size of our solar system (Ferrarese et al. 1996; STScI 1995).

In 2008, Paul Kalas, an astronomer at the University of California, Berkeley, reported the first visible-light images of an exoplanet, captured with the Hubble Space Telescope's Advanced Camera for Surveys (Fig. 7.23). Exoplanets are planets orbiting stars other than our Sun; the planet observed by Kalas and his colleagues was designated Fomalhaut b in reference to the star it orbits, which is located roughly 25 light-years from Earth (Kalas et al. 2008). Fomalhaut b takes 872 years to orbit Fomalhaut at a distance of 119 astronomical units ($\approx$17.8 million km), and is believed to have a mass similar to that of Jupiter (Kalas et al. 2008; Sanders 2008).

Since that first discovery, spacecraft-mounted telescopes such as the Hubble Space Telescope have detected hundreds of exoplanets. As of July 2015, nearly 1600 confirmed exoplanets have been documented in the Exoplanet Orbit Database, which compiles details about exoplanets reported in peer-reviewed literature (Han et al. 2014). A major advantage of space telescopes for exoplanet detection is that they avoid atmospheric turbulence and haze that impede observations by

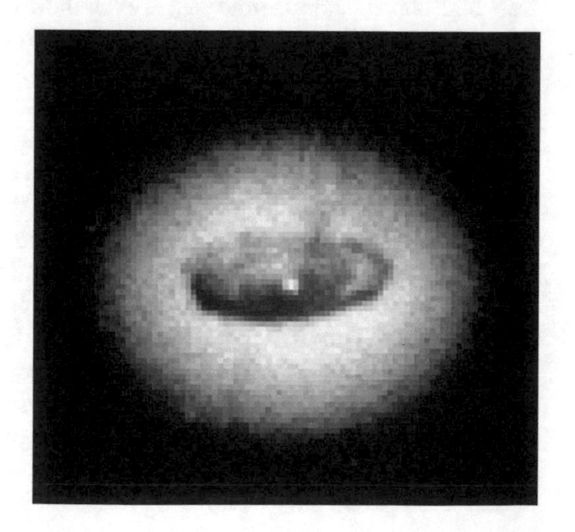

Fig. 7.22 Image captured by the Hubble Space Telescope in 1995, showing a disk of dust spiraling around a supermassive *black hole* at the *center* of galaxy 4261, which is 100 million light-years from Earth. (Image courtesy of L. Ferrarese (Johns Hopkins University) and NASA)

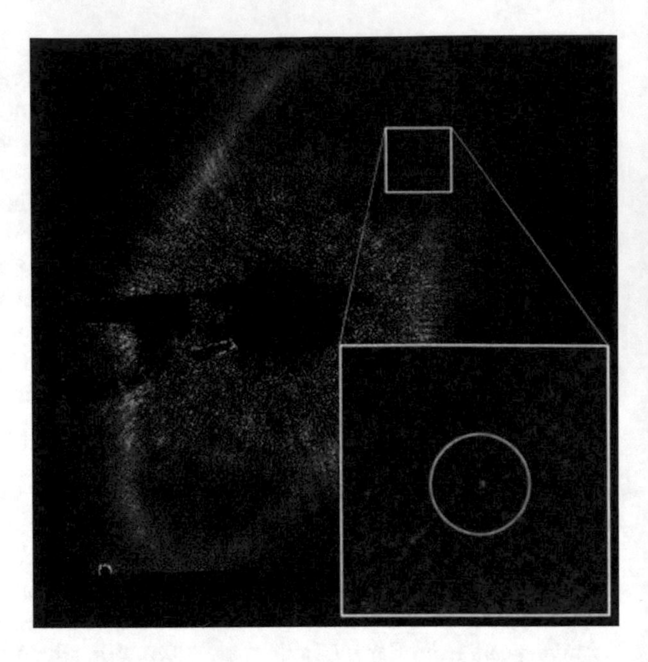

Fig. 7.23 Hubble Space Telescope image, captured in 2006, showing a belt of dust and debris around the star Fomalhaut; the inset image shows a planet, Fomalhaut b, near the outer limits of this belt. Light from the star, which would normally obscure Fomalhaut b, is masked by a coronagraph on the Hubble's Advanced Camera for Surveys. (Image courtesy of Paul Kalas/UC Berkeley, NASA, ESA)

ground-based telescopes. Nevertheless, detection remains challenging because exoplanets are light-years away and orbiting stars that are far larger and brighter than our Sun (Sasselov 2008). The remote sensing techniques involved in these analyses are simple yet provide a great deal of information. One approach is to measure the "wobble" of a star—caused by the gravitational pull of an orbiting planet—using the Doppler Effect (i.e., a shift in the wavelength of the visible light from the star) (Sasselov 2008). Another method, the transiting method, looks at the drop in a star's brightness when an orbiting planet passes in front of it. Through these and other techniques, scientists are able to glean characteristics such as the planet's mass and radius, and in some cases, even more (Sasselov 2008); in 2007, a team of researchers used the Spitzer Space Telescope to create the first-ever map of the surface temperature of the extrasolar planet: HD 189733b, a Jupiter-like gas giant approximately 63 light-years from our solar system (Knutson et al. 2007). Figure 7.24 shows a global temperature map for planet HD 189733b (expressed in terms of infrared brightness). The fact that the brightest (i.e., hottest) spot on the planet's map is offset from the center point, which receives the most light from the star it orbits, suggests the influence of powerful, super-rotating winds (Knutson et al. 2007; Majeau et al. 2012). Subsequent research (Ben-Jaffel and Ballester 2013; Swain et al. 2008) suggests the presence of water vapor, oxygen, and methane in the atmosphere of HD 189733b, and that the planet has a deep blue color (Berdyugina et al. 2011).

Similarly, scientists used the Hubble Space Telescope to construct a temperature map (Stevenson et al. 2014) and to measure water abundance (Kreidberg et al. 2014) in the atmosphere of WASP-43b, a Jupiter-like exoplanet 260 light-years from Earth. Figure 7.25 shows a temperature map of WASP-43b. The planet orbits closely to its parent star, completing one orbit in 19.5 h. The planet is thus very hot,

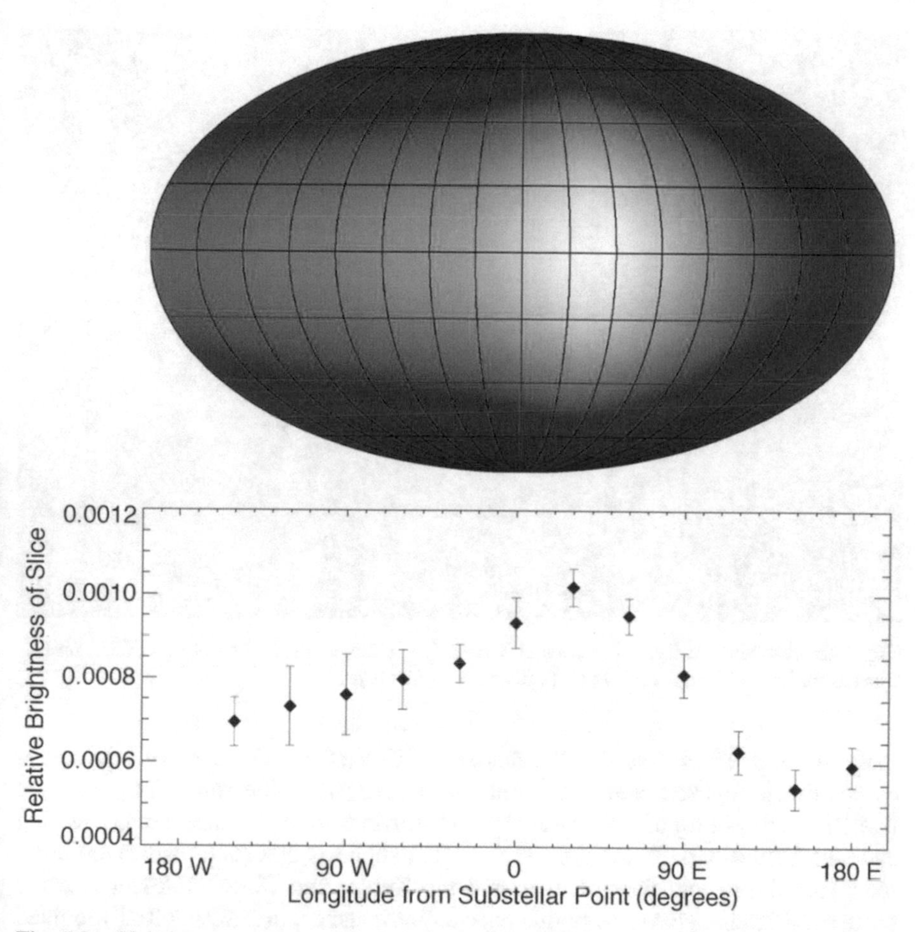

Fig. 7.24 Global temperature map for exoplanet HD 189733b, expressed in terms of infrared brightness. Note the pronounced brightness "hot spot" offset to the *right* of map *center*. (Image courtesy of NASA/JPL-Caltech/H. Knutson (Harvard-Smithsonian CfA)

but unlike other giant planets in its solar system, WASP-43b always has one side of the planet facing its star (i.e., it does rotate on an axis). As a result, there are substantial temperature differences between the two sides: the bright area on the daytime side reaches temperatures above 1500 °C, while nighttime-side temperatures fall below 500 °C. At these temperatures, all of the water in the planet's atmosphere is vaporized, rather than condensed into icy clouds, as seen on Jupiter (Kreidberg et al. 2014; Hubble ESA Information Centre 2014).

The French Space Agency (Centre National D'études Spatiales), in conjunction with the ESA and other partners, launched the COnvection ROtation et Transits planétaires (CoRoT) space telescope in December 2006 (CNES 2011). After a computer failure, CoRoT was decommissioned in 2013. NASA launched the Kepler Space Telescope in 2009 (Batalha et al. 2011). Despite mechanical problems in 2014 that affected its targeting capabilities, the Kepler telescope continues its pri-

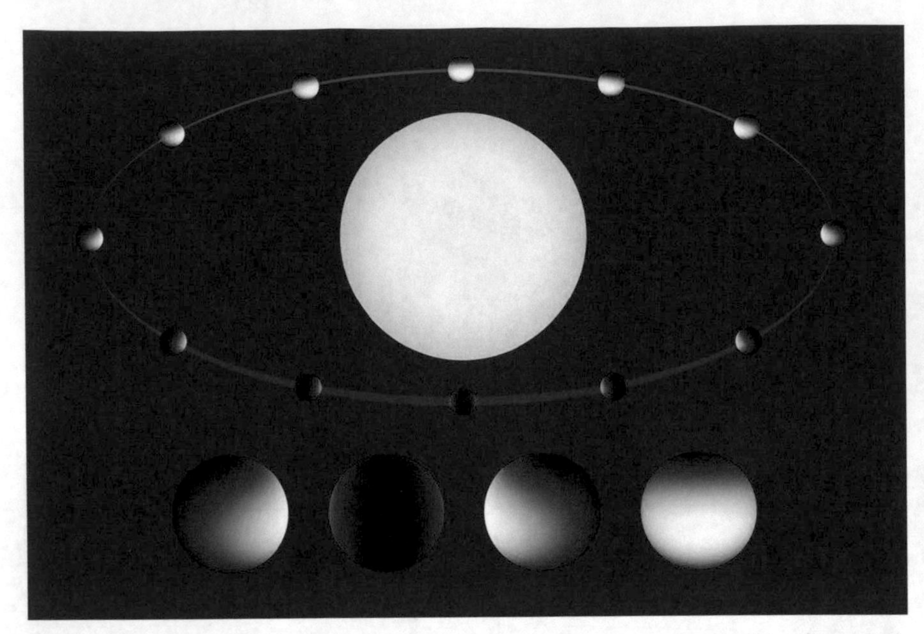

Fig. 7.25 Temperature map of exoplanet WASP-43b. (Image courtesy of NASA, ESA, and K. Stevenson, L. Kreidberg, and J. Bean (University of Chicago)

mary mission, which was also the mission of CoRoT: the detection of exoplanets, especially Earth-sized planets orbiting within the "habitable zone" (i.e., the zone likely to support life) of their parent stars (Borucki et al. 2008). Research (Dressing and Charbonneau 2013; Petigura et al. 2013) suggests that such planets are commonplace throughout the universe, and the Kepler and CoRoT telescopes made some significant discoveries in this regard. For instance, in 2009, CoRoT provided the first characterization of a rocky "super-Earth" planet: CoRoT-7b, which is approximately 1.58 times the size of the Earth (Léger et al. 2011). In 2011, the Kepler telescope discovered its first rocky planet, Kepler-10b, which is only 1.42 times the size of the Earth (Batalha et al. 2011). Two years later, Borucki et al. (2013) reported on Kepler's observation of a five-planet system with two planets that fall within the habitable zone and are 1.41 and 1.61 times the size of the Earth; both planets are believed to be rocky or composed of mostly solid water. Regardless, future research is likely to move beyond the identification of additional rocky exoplanets to more detailed analyses of these planets' atmospheric properties, in an effort to find those with conditions most analogous to Earth (Kaltenegger et al. 2011). Des Marais et al. (2002) argued that remote sensing efforts aimed at terrestrial exoplanets should focus on mid-infrared wavelengths (7–25 μm) and visible to near-infrared wavelengths (0.5 to $\approx$1.1 μm), as these wavelength ranges should provide useful information about planetary properties (e.g., size and mass) as well as potential "biosignatures" (i.e., compounds such as O_2, O_3, H_2O, and CO_2) that may indicate habitability.

7.8 Review Questions

1. Briefly describe three scientific breakthroughs made possible by the Hubble Space Telescope.
2. Which body in our solar system is believed to most resemble a primitive Earth?
3. List two techniques that scientists use to detect exoplanets.
4. What is space weather, and why is it important for scientists to be able to forecast it better?
5. Explain what scientists hope to learn from the New Horizons mission to study Pluto and other Kuiper Belt objects.
6. What conditions on Mars are conducive to the formation of gigantic dust devils?
7. What characteristics of an exoplanet may indicate its habitability?
8. On its way to Jupiter, the Galileo spacecraft found evidence confirming something that had been suspected for at least a couple of decades. What did Galileo discover?
9. What are gamma-ray bursts, and what likely causes them?
10. A common theme of the many missions to the Moon and Mars has been a better understanding of water availability on their surfaces. Why is this so important to scientists?

References

Andrews-Hanna, J.C., S.W. Asmar, J.W. Head III, W.S. Kiefer, A.S. Konopliv, F.G. Lemoine, I. Matsuyama, E. Mazarico, P.J. McGovern, H.J. Melosh, G.A. Neumann, F. Nimmo, R.J. Phillips, D.E. Smith, S.C. Solomon, G.J. Taylor, M.A. Wieczorek, J.G. Williams, and M.T. Zuber. 2013. Ancient igneous intrusions and early expansion of the Moon revealed by GRAIL gravity gradiometry. *Science* 339: 675–678.

Baker, D.N. 2002. How to cope with space weather. *Science* 297: 1486–1487.

Banfield, D., P.J. Gierasch, M. Bell, E. Ustinov, A.P. Ingersoll, A.R. Vasavada, R.A. West, M.J.S. Belton. 1998. Jupiter's cloud structure from Galileo imaging data. *Icarus* 135: 230–235.

Batalha, N.M., W.J. Borucki, S.T. Bryson, L.A. Buchhave, D.A. Caldwell, J. Christensen-Dalsgaard, D. Ciardi, E.W. Dunham, F. Fressin, T.N. Gautier III, R.L. Gilliland, M.R. Haas, S.B. Howell, J.M. Jenkins, H. Kjeldsen, D.G. Koch, D.W. Latham, J.J. Lissauer, G.W. Marcy, J.F. Rowe, D.D. Sasselov, S. Seager, J.H. Steffen, G. Torres, G.S. Basri, T.M. Brown, D. Charbonneau, J. Christiansen, B. Clarke, W.D. Cochran, A. Dupree, D.C. Fabrycky, D. Fischer, E.B. Ford, J. Fortney, F.R. Girouard, M.J. Holman, J. Johnson, H. Isaacson, T.C. Klaus, P. Machalek, A.V. Moorehead, R.C. Morehead, D. Ragozzine, P. Tenenbaum, J. Twicken, S. Quinn, J. VanCleve, L.M. Walkowicz, W.F. Welsh, E. Devore, and A. Gould. 2011. *Kepler*'s first rocky planet: Kepler-10b. *The Astrophysical Journal* 729: 27.

Ben-Jaffel, L. and G.E. Ballester. 2013. Hubble Space Telescope detection of oxygen in the atmosphere of exoplanet HD 189733b. *Astronomy & Astrophysics* 553: A52.

Benna, M., P.R. Mahaffy, J.S. Halekas, R.C. Elphic, and G.T. Delory. 2015. Variability of helium, neon, and argon in the lunar exosphere as observed by the LADEE NMS instrument. *Geophysical Research Letters* 42: 3723–3729.

Bentley, R.D. 2009. Building a virtual observatory for heliophysics. *Earth, Moon, and Planets* 104: 87–91.

Berdyugina, S.V., A.V. Berdyugin, D.M. Fluri, and V. Piirola. 2011. Polarized reflected light from the exoplanet HD189733b: first multicolor observations and confirmation of detection. *The Astrophysical Journal Letters* 728: L6.

Bibring, J.P., Y. Langevin, A. Gendrin, B. Gondet, F. Poulet, M. Berthé, A. Soufflot, R. Arvidson, N. Mangold, J. Mustard, P. Drossart, and the OMEGA team. 2005. Mars surface diversity as revealed by the OMEGA/Mars Express observations. *Science* 307: 1576–1581.

Binder, A.B. 1998. Lunar Prospector: overview. *Science* 281: 1475–1476.

Borucki, W., D. Koch, N. Batalha, D. Caldwell, J. Christensen-Dalsgaard, W.D. Cochran, E. Dunham, T.N. Gautier, J. Geary, R. Gilliland, J. Jenkins, H. Kjeldsen, J.J. Lissauer and J. Rowe. 2008. KEPLER: search for Earth-size planets in the habitable zone. *Proceedings of the International Astronomical Union* 4: 289–299.

Borucki, W.J., E. Agol, F. Fressin, L. Kaltenegger, J. Rowe, H. Isaacson, D. Fischer, N. Batalha, J.J. Lissauer, G.W. Marcy, D. Fabrycky, J.-M. Désert, S.T. Bryson, T. Barclay, F. Bastien, A. Boss, E. Brugamyer, L.A. Buchhave, C. Burke, D.A. Caldwell, J. Carter, D. Charbonneau, J.R. Crepp, J. Christensen-Dalsgaard, J.L. Christiansen, D. Ciardi, W.D. Cochran, E. DeVore, L. Doyle, A.K. Dupree, M. Endl, M.E. Everett, E.B. Ford, J. Fortney, T.N. Gautier III, J.C. Geary, A. Gould, M. Haas, C. Henze, A.W. Howard, S.B. Howell, D. Huber, J.M. Jenkins, H. Kjeldsen, R. Kolbl, J. Kolodziejczak, D.W. Latham, B.L. Lee, E. Lopez, F. Mullally, J.A. Orosz, A. Prsa, E.V. Quintana, R. Sanchis-Ojeda, D. Sasselov, S. Seader, A. Shporer, J.H. Steffen, M. Still, P. Tenenbaum, S.E. Thompson, G. Torres, J.D. Twicken, W.F. Welsh, and J.N. Winn. 2013. Kepler-62: a five-planet system with planets of 1.4 and 1.6 Earth radii in the habitable zone. *Science* 340: 587–590.

Boynton, W.V., W.C. Feldman, S.W. Squyres, T.H. Prettyman, J. Brückner, L.G. Evans, R.C. Reedy, R. Starr, J.R. Arnold, D.M. Drake, P.A.J. Englert, A.E. Metzger, I. Mitrofanov, J.I. Trombka, C. d'Uston, H. Wänke, O. Gasnault, D.K. Hamara, D.M. Janes, R.L. Marcialis, S. Maurice, I. Mikheeva, G.J. Taylor, R. Tokar, and C. Shinohara. 2002. Distribution of hydrogen in the near surface of Mars: evidence for subsurface ice deposits. *Science* 297: 81–85.

Centre national d'études spatiales (CNES). 2011. CoRoT—Astronomy mission. From stars to habitable planets. CNES. http://smsc.cnes.fr/COROT/

Chang, K. 2015. NASA's New Horizons spacecraft zips by Pluto. New York Times, July 14, 2015. http://www.nytimes.com/2015/07/15/science/space/nasa-new-horizons-spacecraft-reaches-pluto.html

Chapman, C.R., J. Veverka, P.C. Thomas, K. Klaasen, M.J.S. Belton, A. Harch, A. McEwen, T.V. Johnson, P. Helfenstein, M.E. Davies, W.J. Merline, and T. Denk. 1995. Discovery and physical properties of Dactyl, a satellite of asteroid 243 Ida. *Nature* 374: 783–785.

Clavin, W. and J.D. Harrington. 2012. NASA releases new WISE mission catalog of entire infrared sky. National Aeronautics and Space Administration (NASA), WISE Mission News. https://www.nasa.gov/mission_pages/WISE/news/wise20120314.html

Cuzzi, J.N., J.A. Burns, S. Charnoz, R.N. Clark, J.E. Colwell, L. Dones, L.W. Esposito, G. Filacchione, R.G. French, M.M. Hedman, S. Kempf, E.A. Marouf, C.D. Murray, P.D. Nicholson, C.C. Porco, J. Schmidt, M.R. Showalter, L.J. Spilker, J.N. Spitale, R. Srama, M. Sremčević, M.S. Tiscareno, and J. Weiss. 2010. An evolving view of Saturn's dynamic rings. *Science* 327: 1470–1475.

Davis, C.J., J.A. Davies, M. Lockwood, A.P. Rouillard, C.J. Eyles, and R.A. Harrison. 2009. Stereoscopic imaging of an Earth-impacting solar coronal mass ejection: A major milestone for the STEREO mission. *Geophysical Research Letters* 36: L08102.

Des Marais, D.J., M.O. Harwit, K.W. Jucks, J.F. Kasting, D.N.C. Lin, J.I. Lunine, J. Schneider, S. Seager, W.A. Traub, and N.J. Woolf. 2002. Remote sensing of planetary properties and biosignatures on extrasolar terrestrial planets. *Astrobiology* 2: 153–181.

Dressing, C.D. and D. Charbonneau. 2013. The occurrence rate of small planets around small stars. *The Astrophysical Journal* 767: 95.

Elphic, R.C., G.T. Delory, B.P. Hine, P.R. Mahaffy, M. Horanyi, A. Colaprete, M. Benna, S.K. Noble, and the LADEE Science Team. 2014. The Lunar Atmosphere and Dust Environment Explorer Mission. *Space Science Reviews* 185: 3–25.

Ferrarese, L., H.C. Ford, and W. Jaffe. 1996. Evidence for a massive black hole in the active galaxy NGC 4261 from Hubble Space Telescope images and spectra. *The Astrophysical Journal* 470: 444–459.

Foing, B.H., G.D. Racca, A. Marini, E. Evrard, L. Stagnaro, M. Almeida, D. Koschny, D. Frew, J. Zender, J. Heather, M. Grande, J. Huovelin, H.U. Keller, A. Nathues, J.L. Josset, A. Malkki, W. Schmidt, G. Noci, R. Birkl, L. Iess, Z. Sodnik, and P. McManamon. 2006. SMART-1 mission to the Moon: Status, first results and goals. *Advances in Space Research* 37: 6–13.

Gehrels, N. and P. Mészáros. 2012. Gamma-ray bursts. *Science* 337: 932–936.

Geissler, P. 2012. A monster dust devil stalks the Martian landscape. High Resolution Imaging Science Experiment, University of Arizona. http://www.uahirise.org/ESP_026394_2160

Gendre, B., G. Stratta, J.L. Atteia, S. Basa, M. Boër, D.M. Coward, S. Cutini, V. D'Elia, E.J. Howell, A. Klotz, and L. Piro. 2013. The ultra-long gamma-ray burst 111209A: the collapse of a blue supergiant? *The Astrophysical Journal* 766: 30.

Goetz, A.F.H. 2009. Three decades of hyperspectral remote sensing of the Earth: A personal view. *Remote Sensing of Environment* 113: S5–S16.

Gombosi, T.I. and A.P. Ingersoll. 2010. Saturn: atmosphere, ionosphere, and magnetosphere. *Science* 327: 1476–1479.

Han, E., S.X. Wang, J.T. Wright, Y.K. Feng, M. Zhao, O. Fakhouri, J.I. Brown, and C. Hancock. 2014. Exoplanet Orbit Database. II. Updates to Exoplanets.org. *Publications of the Astronomical Society of the Pacific* 126: 827–837.

Hanel, R.A., B.J. Conrath, D.E. Jennings, and R.E. Samuelson. 2003. Exploration of the Solar System by Infrared Remote Sensing, 2nd Ed. Cambridge: Cambridge University Press.

Haruyama, J., M. Ohtake, T. Matsunaga, T. Morota, C. Honda, Y. Yokota, C.M. Pieters, S. Hara, K. Hioki, K. Saiki, H. Miyamoto, A. Iwasaki, M. Abe, Y. Ogawa, H. Takeda, M. Shirao, A. Yamaji, and J.-L. Josset. 2008. Lack of exposed ice inside lunar south pole Shackleton Crater. *Science* 322: 938–939.

Hubble European Space Agency (ESA) Information Centre. 2014. Hubble reveals most detailed exoplanet weather map ever. Science Release No. heic1422, October 9, 2014. https://www.spacetelescope.org/news/heic1422/

Jewitt, D. 1999. Kuiper Belt objects. *Annual Review of Earth and Planetary Sciences* 27: 287–312.

Johns Hopkins University (JHU) Applied Physics Laboratory. 2015a. How big is Pluto? New Horizons settles decades-long debate. New Horizons News Center, July 13, 2015. http://pluto.jhuapl.edu/News-Center/News-Article.php?page=20150713

Johns Hopkins University (JHU) Applied Physics Laboratory. 2015b. MESSENGER Website. JHU/APL. http://messenger.jhuapl.edu/index.php

Johns Hopkins University (JHU) Applied Physics Laboratory. 2015c. Pluto and Charon: New Horizons' dynamic duo. New Horizons News Center, July 9, 2015. http://pluto.jhuapl.edu/News-Center/News-Article.php?page=20150709

Kalas, P., J.R. Graham, E. Chiang, M.P. Fitzgerald, M. Clampin, E.S. Kite, K. Stapelfeldt, C. Marois, and J. Krist. 2008. Optical images of an exosolar planet 25 light-years from Earth. *Science* 322: 1345–1348.

Kaltenegger, L., A. Segura, and S. Mohanty. 2011. Model spectra of the first potentially habitable super-Earth—Gl581d. *The Astrophysical Journal* 733: 35.

Kerr, R.A. 2005. Titan, once a world apart, becomes eerily familiar. *Science* 307: 330–331.

Kirk, R.L. 2005. Grids & Datums: Mars. *Photogrammetric Engineering and Remote Sensing* 71(10): 1111–1114.

Klotz, I. 2015. Hubble at 25: space telescope's top science discoveries. Discovery News, April 23, 2015. http://news.discovery.com/space/astronomy/hubble-at-25-science-that-transformed-our-cosmos-150423.htm

Knutson, H.A., D. Charbonneau, L.E. Allen, J.J. Fortney, E. Agol, N.B. Cowan, A.P. Showman, C.S. Cooper, and S.T. Megeath. 2007. A map of the day-night contrast of the extrasolar planet HD 189733b. *Nature* 447: 183–186.

Konopliv, A.S., S.W. Asmar, E. Carranza, W.L. Sjogren, and D.N. Yuan. 2001. Recent gravity models as a result of the Lunar Prospector mission. *Icarus* 150(1): 1–18.

Kreidberg, L., J.L. Bean, J.-M. Désert, M.R. Line, J.J. Fortney, N. Madhusudhan, K.B. Stevenson, A.P. Showman, D. Charbonneau, P.R. McCullough, S. Seager, A. Burrows, G.W. Henry, M. Williamson, T. Kataria, and D. Homeier. 2014. A precise water abundance measurement for the hot Jupiter WASP-43b. *The Astrophysical Journal Letters* 793: L27.

Léger, A., O. Grasset, B. Fegley, F. Codron, A.F. Albarede, P. Barge, R. Barnes, P. Cance, S. Carpy, F. Catalano, C. Cavarroc, O. Demangeon, S. Ferraz-Mello, P. Gabor, J.-M. Grießmeier, J. Leibacher, G. Libourel, A.-S. Maurin, S.N. Raymond, D. Rouan, B. Samuel, L. Schaefer, J. Schneider, P.A. Schuller, F. Selsis, and C. Sotin. 2011. The extreme physical properties of the CoRoT-7b super-Earth. *Icarus* 213: 1–11.

Majeau, C., E. Agol, and N.B. Cowan. 2012. A two-dimensional infrared map of the extrasolar planet HD 189733b. *The Astrophysical Journal Letters* 747: L20.

Malin, M.C. and K.S. Edgett. 2000. Evidence for recent groundwater seepage and surface runoff on Mars. *Science* 288: 2330–2335.

Malin, M.C., K.S. Edgett, L.V. Posiolova, S.M. McColley, and E.Z. Noe Dobrea. 2006. Present-day impact cratering rate and contemporary gully activity on Mars. *Science* 314: 1573–1577.

McClintock, W.E., R.J. Vervack, E.T. Bradley, R.M. Killen, N. Mouawad, A.L. Sprague, M.H. Burger, S.C. Solomon, and N.R. Izenberg. 2009. MESSENGER observations of Mercury's exosphere: detection of magnesium and distribution of constituents. *Science* 324: 610–613.

McEwen, A.S. and M.S. Robinson. 1997. Mapping of the Moon by Clementine. *Advances in Space Research* 19(10): 1523–1533.

McEwen, A.S., L. Keszthelyi, P. Geissler, D.P. Simonelli, M.H. Carr, T.V. Johnson, K.P. Klaasen, H.H. Breneman, T.J. Jones, J.M. Kaufman, K.P. Magee, D.A. Senske, M.J.S. Belton, and G. Schubert. 1998. Active volcanism on Io as seen by Galileo SSI. *Icarus* 135: 181–219.

McEwen, A.S., L. Ojha, C.M. Dundas, S.S. Mattson, S. Byrne, J.J. Wray, S.C. Cull, S.L. Murchie, N. Thomas, V.C. Gulick. 2011. Seasonal flows on warm Martian slopes. *Science* 333: 740–743.

Moore, P. and N. Henbest. 1986. Uranus: the view from Voyager. *Journal of the British Astronomical Association* 96: 131–137.

Murray, J.B., J.-P. Muller, G. Neukum, S.C. Werner, S. van Gasselt, E. Hauber, W.J. Markiewicz, J.W. Head, B.H. Foing, D. Page, K.L. Mitchell, G. Portyankina, and the HRSC Co-Investigator Team. 2005. Evidence from the Mars Express High Resolution Stereo Camera for a frozen sea close to Mars' equator. *Nature* 434: 352–356.

Nakamura, R., S. Yamamoto, T. Matsunaga, Y. Ishihara, T. Morota, T. Hiroi, H. Takeda, Y. Ogawa, Y. Yokota, N. Hirata, M. Ohtake, and K. Saiki. 2012. Compositional evidence for an impact origin of the Moon's Procellarum basin. *Nature Geoscience* 5(11): 775–778.

National Aeronautics and Space Administration (NASA). 2010. Mars Global Surveyor: science summary. NASA. http://mars.jpl.nasa.gov/mgs/science/

National Aeronautics and Space Administration (NASA) Jet Propulsion Laboratory. 2014. Photojournal—PIA18430: mysterious changing feature in Ligeia Mare. NASA Jet Propulsion Laboratory. http://photojournal.jpl.nasa.gov/catalog/PIA18430

Neukum, G., R. Jaumann, H. Hoffmann, E. Hauber, J.W. Head, A.T. Basilevsky, B.A. Ivanov, S.C. Werner, S. van Gasselt, J.B. Murray, T. McCord, and the HRSC Co-Investigator Team. 2004. Recent and episodic volcanic and glacial activity on Mars revealed by the High Resolution Stereo Camera. *Nature* 432: 971–979.

Neumann, G.A., J.F. Cavanaugh, X. Sun, E.M. Mazarico, D.E. Smith, M.T. Zuber, D. Mao, D.A. Paige, S.C. Solomon, and O.S. Barnouin. 2012. Bright and dark polar deposits on Mercury: evidence for surface volatiles. *Science* 339: 296–300.

Nozette, S., C.L. Lichtenberg, P. Spudis, R. Bonner, W. Ort, E. Malaret, M. Robinson, and E.M. Shoemaker. 1996. The Clementine bistatic radar experiment. *Science* 274: 1495–1498.

Paige, D.A. 2005. Ancient Mars: wet in many places. *Science* 307: 1575–1576.

Paige, D.A., M.A. Siegler, J.K. Harmon, G.A. Neumann, E.M. Mazarico, D.E. Smith, M.T. Zuber, E. Harju, M.L. Delitsky, and S.C. Solomon. 2012. Thermal stability of volatiles in the north polar region of Mercury. *Science* 339: 300–303.

Pesnell, W.D., B.J. Thompson, and P.C. Chamberlin. 2012. The *Solar Dynamics Observatory* (SDO). *Solar Physics* 275: 3–15.

Petigura, E.A., G.W. Marcy, and A.W. Howard. 2013. A plateau in the planet population below twice the size of Earth. *The Astrophysical Journal* 770: 69.

Petro, N.E. and C.M. Pieters. 2004. Surviving the heavy bombardment: Ancient material at the surface of South Pole-Aitken Basin. *Journal of Geophysical Research* 109: E06004.

Pieters, C.M., J.N. Goswami, R.N. Clark, M. Annadurai, J. Boardman, B. Buratti, J.-P. Combe, M.D. Dyar, R. Green, J.W. Head, C. Hibbitts, M. Hicks, P. Isaacson, R. Klima, G. Kramer, S. Kumar, E. Livo, S. Lundeen, E. Malaret, T. McCord, J. Mustard, J. Nettles, N. Petro, C. Runyon, M. Staid, J. Sunshine, L.A. Taylor, S. Tompkins, and P. Varanasi. 2009. Character and spatial distribution of OH/H_2O on the surface of the Moon seen by M^3 on Chandrayaan-1. *Science* 326: 568–572.

Robinson, M.S., S.M. Brylow, M. Tschimmel, D. Humm, S.J. Lawrence, P.C. Thomas, B.W. Denevi, E. Bowman-Cisneros, J. Zerr, M.A. Ravine, M.A. Caplinger, F.T. Ghaemi, J.A. Schaffner, M.C. Malin, P. Mahanti, A. Bartels, J. Anderson, T.N. Tran, E.M. Eliason, A.S. McEwen, E. Turtle, B.L. Jolliff, and H. Hiesinger. 2010. Lunar Reconnaissance Orbiter Camera (LROC) instrument overview. *Space Science Reviews* 150: 81–124.

Sanders, R. 2008. Hubble snaps first optical photo of exoplanet. University of California, Berkeley Press Release, November 13, 2008. http://www.berkeley.edu/news/media/releases/2008/11/13_exoplanet.shtml

Sasselov, D.D. 2008. Extrasolar planets. *Nature* 451: 29–31.

Short, N.M. 2006. The Remote Sensing Tutorial [web site]. National Aeronautics and Space Administration (NASA), Goddard Space Flight Center. http://fas.org/irp/imint/docs/rst/

Smith, B.A., L.A. Soderblom, D. Banfield, C. Barnet, A.T. Basilevsky, R.F. Beebe, K. Bollinger, J.M. Boyce, A. Brahic, G.A. Briggs, R.H. Brown, C. Chyba, S.A. Collins, T. Colvin, A.F. Cook II, D. Crisp, S.K. Croft, D. Cruikshank, J.N. Cuzzi, G.E. Danielson, M.E. Davies, E. De Jong, L. Dones, D. Godfrey, J. Goguen, I. Grenier, V.R. Haemmerle, H. Hammel, C.J. Hansen, C.P. Helfenstein, C. Howell, G.E. Hunt, A.P. Ingersoll, T.V. Johnson, J. Kargel, R. Kirk, D.I. Kuehn, S. Limaye, H. Masursky, A. McEwen, D. Morrison, T. Owen, W. Owen, J.B. Pollack, C.C. Porco, K. Rages, P. Rogers, D. Rudy, C. Sagan, J. Schwartz, E.M. Shoemaker, M. Showalter, B. Sicardy, D. Simonelli, J. Spencer, L.A. Sromovsky, C. Stoker, R.G. Strom, V.E. Suomi, S.P. Synott, R.J. Terrile, P. Thomas, W.R. Thompson, A. Verbiscer and J. Veverka. 1989. Voyager 2 at Neptune: imaging science results. *Science* 246: 1422–1449.

Smith, D.E., M.T. Zuber, S.C. Solomon, R.J. Phillips, J.W. Head, J.B. Garvin, W.B. Banerdt, D.O. Muhleman, G.H. Pettengill, G.A. Neumann, F.G. Lemoine, J.B. Abshire, O. Aharonson, C.D. Brown, S.A. Hauck, A.B. Ivanov, P.J. McGovern, H.J. Zwally, and T.C. Duxbury. 1999. The global topography of Mars and implications for surface evolution. *Science* 284: 1495–1503.

Space Telescope Science Institute (STScI). 1995. Hubble finds a new black hole—and unexpected new mysteries. HubbleSite News Release STScI-1995–47. http://hubblesite.org/newscenter/archive/releases/1995/47/image/a/

Space Telescope Science Institute (STScI). 2015. HubbleSite. http://hubblesite.org

Spudis, P.D., D.B.J. Bussey, S.M. Baloga, J.T.S. Cahill, L.S. Glaze, G.W. Patterson, R.K. Raney, T.W. Thompson, B.J. Thomson, and E.A. Ustinov. 2013. Evidence for water ice on the Moon: Results for anomalous polar craters from the LRO Mini-RF imaging radar. *Journal of Geophysical Research* 118: 2016–2029.

Stern, S.A. 2008. The New Horizons Pluto Kuiper Belt Mission: an overview with historical context. *Space Science Reviews* 140: 3–21.

Stofan, E.R., C. Elachi, J.I. Lunine, R.D. Lorenz, B. Stiles, K.L. Mitchell, S. Ostro, L. Soderblom, C. Wood, H. Zebker, S. Wall, M. Janssen, R. Kirk, R. Lopes, F. Paganelli, J. Radebaugh, L. Wye, Y. Anderson, M. Allison, R. Boehmer, P. Callahan, P. Encrenaz, E. Flamini, G. Francescetti, Y. Gim, G. Hamilton, S. Hensley, W.T.K. Johnson, K. Kelleher, D. Muhleman, P. Paillou, G. Picardi, F. Posa, L. Roth, R. Seu, S. Shaffer, S. Vetrella, and R. West. 2007. The lakes of Titan. *Nature* 445: 61–64.

Stevenson, K.B., J.-M. Désert, M.R. Line, J.L. Bean, J.J. Fortney, A.P. Showman, T. Kataria, L. Kreidberg, P.R. McCullough, G.W. Henry, D. Charbonneau, A. Burrows, S. Seager, N. Madhusudhan, M.H. Williamson, and D. Homeier. 2014. Thermal structure of an exoplanet atmosphere from phase-resolved emission spectroscopy. *Science* 346: 838–841.

Swain, M.R., G. Vasisht, and G. Tinetti. 2008. The presence of methane in the atmosphere of an extrasolar planet. *Nature* 452: 329–331.
Titov, D.V., H. Svedhem, F.W. Taylor, S. Barabash, J.-L. Bertaux, P. Drossart, V. Formisano, B. Häusler, O. Korablev, W.J. Markiewicz, D. Nevejans, M. Pätzold, G. Piccioni, J.-A. Sauvaud, T.L. Zhang, O. Witasse, J.-C. Gerard, A. Fedorov, A. Sanchez-Lavega, J. Helbert, and R. Hoofs. 2009. Venus Express: highlights of the nominal mission. *Solar System Research* 43: 185–209.
Tsai, C.-W, P.R.M. Eisenhardt, J. Wu, D. Stern, R.J. Assef, A.W. Blain, C.R. Bridge, D.J. Benford, R.M. Cutri, R.L. Griffith, T.H. Jarrett, C.J. Lonsdale, F.J. Masci, L.A. Moustakas, S.M. Petty, J. Sayers, S.A. Stanford, E.L. Wright, L. Yan, D.T. Leisawitz, F. Liu, A.K. Mainzer, I.S. McLean, D.L. Padgett, M.F. Skrutskie, C.R. Gelino, C.A. Beichman, and S. Juneau. 2015. The most luminous galaxies discovered by WISE. *The Astrophysical Journal* 805: 90.
Vasavada, A.R., A.P. Ingersoll, D. Banfield, M. Bell, P.J. Gierasch, M.J.S. Belton, G.S. Orton, K.P. Klaasen, E. DeJong, H.H. Brenemann, T.J. Jones, J.M. Kaufman, K.P. Magee, D.A. Senske. 1998. Galileo imaging of Jupiter's atmosphere: The Great Red Spot, Equatorial Region, and White Ovals. *Icarus* 135: 265–275.
Whelley, P.L. and R. Greeley. 2008. The distribution of dust devil activity on Mars. *Journal of Geophysical Research* 113: E07002.
Wieczorek, M.A., G.A. Neumann, F. Nimmo, W.S. Kiefer, G.J. Taylor, H.J. Melosh, R.J. Phillips, S.C. Solomon, J.C. Andrews-Hanna, S.W. Asmar, A.S. Konopliv, F.G. Lemoine, D.E. Smith, M.M. Watkins, J.G. Williams, and M.T. Zuber. 2013. The crust of the Moon as seen by GRAIL. *Science* 339: 671–675.
Zebker, H., A. Hayes, M. Janssen, A. Le Gall, R. Lorenz, and L. Wye. 2014. Surface of Ligeia Mare, Titan, from Cassini altimeter and radiometer analysis. *Geophysical Research Letters* 41: 308–313.
Zuber, M.T., D.E. Smith, M.M. Watkins, S.W. Asmar, A.S. Konopliv, F.G. Lemoine, H.J. Melosh, G.A. Neumann, R.J. Phillips, S.C. Solomon, M.A. Wieczorek, J.G. Williams, S.J. Goossens, G. Kruizinga, E. Mazarico, R.S. Park, and D.-N. Yuan. 2013. Gravity field of the Moon from the Gravity Recovery and Interior Laboratory (GRAIL) mission. *Science* 339: 668–671.
Zurek, R.W. and S.E. Smrekar. 2007. An overview of the Mars Reconnaissance Orbiter (MRO) science mission. *Journal of Geophysical Research* 112: E05S01.

Additional Reading

Jin, S., ed. 2015. Planetary Geodesy and Remote Sensing. Boca Raton, FL: CRC Press.
Kitchin, C.R. 2011. Exoplanets: Finding, Exploring, and Understanding Alien Worlds. New York: Springer-Verlag.
Schriver, C.J. and G.L. Siscoe, eds. 2010. Heliophysics: Evolving Solar Activity and the Climates of Space and Earth. Cambridge: Cambridge University Press.

Relevant Websites

Wide-field Infrared Survey Explorer (WISE), NASA: http://www.nasa.gov/mission_pages/WISE/main/index.html
Fermi Gamma-ray Space Telescope, NASA Goddard Space Flight Center: http://fermi.gsfc.nasa.gov/
The Swift Gamma-ray Burst Mission, NASA Goddard Space Flight Center: http://swift.gsfc.nasa.gov/
NASA Lunar Exploration Timeline: http://nssdc.gsfc.nasa.gov/planetary/lunar/lunartimeline.html

Messenger Web Site, Johns Hopkins University Applied Physics Laboratory: http://messenger.jhuapl.edu/

Venus Express, European Space Agency: http://www.esa.int/Our_Activities/Space_Science/Venus_Express

NASA Mars Exploration Timeline: http://nssdc.gsfc.nasa.gov/planetary/chronology_mars.html

NASA Mars Exploration: http://mars.nasa.gov/

Voyager—The Interstellar Mission, NASA Jet Propulsion Laboratory: http://voyager.jpl.nasa.gov/

Galileo Legacy Site, NASA: http://solarsystem.nasa.gov/galileo/

Cassini—Solstice Mission, NASA Jet Propulsion Laboratory: http://saturn.jpl.nasa.gov/

New Horizons—NASA's Mission to Pluto, Johns Hopkins University Applied Physics Laboratory: http://pluto.jhuapl.edu/

Heliophysics Missions, NASA Goddard Space Flight Center: http://sec.gsfc.nasa.gov/sec_missions.htm

SDO—Solar Dynamics Observatory, NASA Goddard Space Flight Center: http://sdo.gsfc.nasa.gov/

HubbleSite, Space Telescope Science Institute: http://hubblesite.org/

Hubble European Space Agency Information Centre: https://www.spacetelescope.org/

Exoplanet Orbit Database/ Exoplanet Data Explorer: http://exoplanets.org/

Kepler—A Search for Habitable Planets, NASA Ames Research Center: http://kepler.nasa.gov/

Chapter 8
International Laws, Charters, and Policies

8.1 Introduction

Remote sensing has been identified as one of the most significant technological achievements of the twentieth century (Gabrynowicz 1993). Earth observation satellites transcend national boundaries and geophysical space, creating transparency into activities and places that were once concealed from foreign states. This raises many issues—the ideals of cooperation, societal openness, and information sharing juxtaposed with the very real fears that spatial information could be used for sparking military conflicts and other malevolent purposes.

Historically—throughout the 1950s and 1960s—remote sensing was developed and implemented predominantly for military applications and weather observation, with nearly 75% of satellites employed for military reconnaissance and surveillance purposes (Keeley and Huebert 2004). The launch of Landsat-1 in 1972, for the purpose of "gathering facts about the natural resources of our planet" (Roccio 2011), signaled the dawn of a new era by enabling access to widespread Earth observation by researchers and the public.

Since then, satellite image technology has rapidly evolved, facilitating the opportunity for virtually anyone with access to a computer to download high-resolution images easily and inexpensively. Almost completely open access and ease of use is now so commonplace that we would have a hard time imagining life without this opportunity; this has opened the gateways to better scientific understanding of a wide variety of phenomena. However, with such easy access to *millions* of images covering the entire Earth's surface (and subsurface), there are understandable concerns about regulating the distribution and use of remotely sensed images and safeguarding against threats to personal and national security.

This chapter covers international laws, agreements, and policies regarding the use of remotely sensed data, the more notable national laws and policies, and some closing thoughts concerning future policy directions.

S. Khorram et al., *Principles of Applied Remote Sensing*,
DOI 10.1007/978-3-319-22560-9_8

8.2 Origin and Focus of International Space Law

It should not be surprising that international committees to govern the use of space and remotely sensed data were organized shortly after the first artificial satellite, Sputnik I (illustrated in Fig. 8.1), was launched by the Soviet Union in 1957.

During the height of the Cold War between the United States and the former Soviet Union, the quest to be the first in space exploration was intense. Although the (spatial, temporal, and spectral) resolutions of early satellite systems were coarse by today's standards, the ability to launch satellites not long after the end of World War II triggered widespread concerns about the ability of these two superpowers to launch nuclear weapons, fears that national defense department reconnaissance missions would be compromised, and geopolitical concerns regarding the photographing of foreign countries without their permission. Consequently, in 1955, (US) President Eisenhower proposed an "Open Skies" policy that would allow both free and reciprocal observations between the US and the Soviet Union at an international conference in Geneva (Leghorn and Herken 2001). The intent of this (failed) proposal was to ease tensions and allow mutual aerial reconnaissance of military resources.

In 1958, the United Nations (UN) General Assembly established the Committee on the Peaceful Uses of Outer Space (COPUOS) as an 18-member ad hoc group to deal with the emerging issue of space exploration governance and the need to facilitate international cooperation (Fig. 8.2). The following year, through UN Resolution 1472 (XIV), COPUOS was formally established as the only international forum for the development of international space law, including satellite remote sensing issues. Since 1958, COPUOS has met annually in Vienna, Austria, to discuss issues relating to current and future activities in space such as maintaining outer space for peaceful purposes, safe operations in orbit, space debris, space weather, the safe use of nuclear power in outer space, global navigation satellite systems, as well as space law and national space legislation.

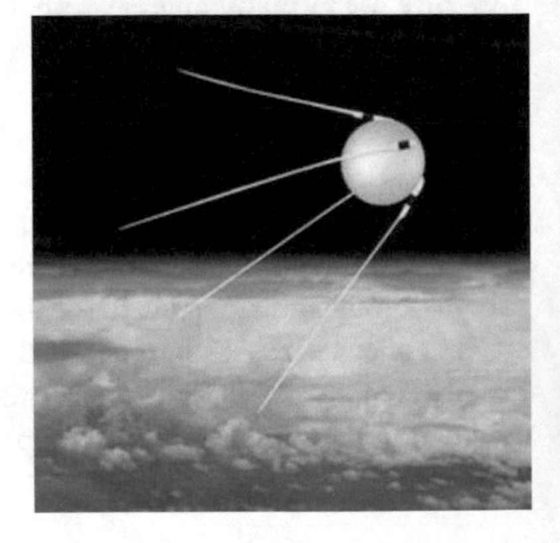

Fig. 8.1 On October 4, 1957, the Soviet Union launched Sputnik I, the first artificial satellite. At 58 cm and weighing 83.6 kg, it took about 98 min to orbit the Earth (http://history.nasa.gov/sputnik/)

Fig. 8.2 United Nations Committee on the Peaceful Uses of Outer Space (COPUOS) 2015 meeting (http://www.oosa.org/oosa/en/ourwork/cupuos/comm-subcomms.html)

Since its inception, COPUOS has concluded five international legal instruments and five sets of legal principles governing space-related activities. The five international treaties are the following:

1. The Treaty on Principles Governing the Activities of States in the Exploration and Use of Outer Space, including the Moon and Other Celestial Bodies (Outer Space Treaty)
2. The Agreement on the Rescue of Astronauts, the Return of Astronauts, and the Return of Objects Launched into Outer Space (Rescue Agreement)
3. The Convention on International Liability for Damage Caused by Space Objects (Liability Convention)
4. The Convention on Registration of Objects Launched into Outer Space (Registration Convention)
5. The Agreement Governing the Activities of States on the Moon and Other Celestial Bodies (Moon Treaty)

With regard to issues dealing specifically with remotely sensed data, international laws and legal principles generally focus on three matters of concern:

1. The *right to acquire* remotely sensed imagery/the right to launch remote sensing satellites
2. The *right to disseminate* remotely sensed imagery without prior consent of the sensed State
3. The *right to obtain* remotely sensed satellite imagery from a particular State

After extensive discussions in the COPUOS, the United Nations General Assembly unanimously adopted Resolution 41/65, *Principles Relating to Remote Sensing of the Earth from Outer Space*,[1] in December 1986. The 15 principles place international customary obligations on states and form the basis for remote sensing

[1] United Nations Resolution (RES) 41/65. Principles Relating to Remote Sensing of the Earth from Outer Space. Adopted without a vote, December 3, 1986. Available online: http://www.unoosa.org/pdf/gares/ARES_41_65E.pdf.

activities globally, regulating and encouraging technical cooperation between sensing and sensed States. Several are particularly noteworthy:

- Remote sensing activities … shall be *carried out for the benefit and in the interests of all countries*, irrespective of their degree of economic or scientific development … Activities shall be conducted on the basis of respect for the principle of full and permanent sovereignty of all States and peoples … in accordance with international law… (Principle IV).
- In order to maximize the availability of benefits from remote sensing activities, States are encouraged, through agreements, to provide for the establishment and operation of data collection, storage, processing and interpretation facilities, in particular within the framework of arrangements wherever feasible (Principle VI).
- Remote sensing shall *promote the protection of mankind from natural disasters*. To this end, States participating in remote sensing activities that have identified processed data and analyzed information in their possession that may be useful to States affected by natural disasters, or likely to be affected by impending natural disasters, shall transmit such data and information to States concerned as promptly as possible (Principle XI).
- As soon as the primary data and the processed data concerning the territory under its jurisdiction are produced, the sensed State shall have access to them on a *non-discriminatory basis and on reasonable cost terms*. The sensed State shall also have access to the available analyzed information concerning the territory under its jurisdiction in the possession of any State participating in remote sensing activities on the same basis and terms, taking particularly into account the needs and interests of the developing countries (Principle XII).

While the principles are not a binding source of international law per se, these watershed resolutions are regarded by most space law scholars and legal researchers as the primary international legal document that addresses issues of remote sensing (Harris 2003; Jakhu 2004; Macauley 2005; Rao and Murthi 2006; Smith and Doldirina 2008). As such, the principles are regarded as a codification of customary law and have acquired the "evidence of a general practice accepted as law" according to Article 38(1)(b), Statute of International Court of Justice (Gabrynowicz 1993; Harris 2003; Williams 2006; Smith and Doldirina 2008).

In addition to the UN Principles, several other factors have contributed to a presumption of open access. Nondiscriminatory access policies have been adopted by major remote sensing nations (e.g., Japan, the US, and Canada), and the data policies of some remote sensing missions (e.g., ENVISAT, RADARSAT) specifically incorporate nondiscriminatory access (Harris 2003). At present, there are at least 65 statutes worldwide that govern access to information, of which at least 50 establish a right of access to information, rather than a mere "limited right of access" to documents (Gabrynowicz 2007; Smith and Doldirina 2008). The right of access to environmental information is, in certain circumstances, guaranteed by the European Convention on Human Rights (Smith and Doldirina 2008).

8.3 The International Charter on Space and Major Disasters

In July 1999, following the UNISPACE III conference held in Vienna, Austria, the European and French space agencies (ESA and CNES) initiated the International Charter on Space and Major Disasters. This policy framework, which was fully operational by November 1, 2000, was the realization of Principle XI of UN Resolution 41/65—that is, to promote international cooperation and support from member space agencies and a pool of national/international space system operators during natural or technological disasters by quickly providing critical information and managing the crisis and subsequent reconstruction operations. The satellite operator pool consists of numerous organizations and companies, including DigitalGlobe, the (US) National Oceanic and Atmospheric Administration (NOAA), the European Organization for the Exploitation of Meteorological Satellites (EUMETSAT), and others. Each brings a complement of satellite to help image the disaster area.

Since 2000, the Charter has been increasingly activated to request satellite imagery for floods, wildfires, earthquakes, oil spills, and other catastrophic events. Authorized users are given a confidential phone number to request the mobilization of satellite remote sensing resources of the member agencies. As depicted in Fig. 8.3, a 24-h on-duty operator takes the call, confirms the identity of the caller, and delivers the information to an emergency on-call officer. The emergency on-call officer assesses the request and the scope of the disaster with the authorized user. The Operational Satellite Applications Program (UNOSAT)—a subbranch of the UN Institute for Training and Research (UNITAR)—is the coordinating agency for all the images collected from the satellites responding to Charter activation. UNOSAT determines which of the members' pool of satellites are immediately available to fly over the affected area. The satellite with the most appropriate sensors is used to collect the data. UNOSAT ensures the images get to the correct organizations and people. Data acquisition and delivery occurs on a rapid basis and a project manager assists the user in providing accurate disaster maps. The Charter is widely seen as an enormously successful example of international cooperation and has been invoked an increasing number of times since it was established. The devastating Haitian earthquake of April 15, 2010 (Fig. 8.4) is one example of how the Charter has been applied. It should be noted that while the Charter operates within the legally binding framework of the Outer Space Treaty and the nonbinding UN Principles, it uses a "best effort" approach with no legal binding; consequently, no liability is assumed for the resulting mapping products (Ito 2005).

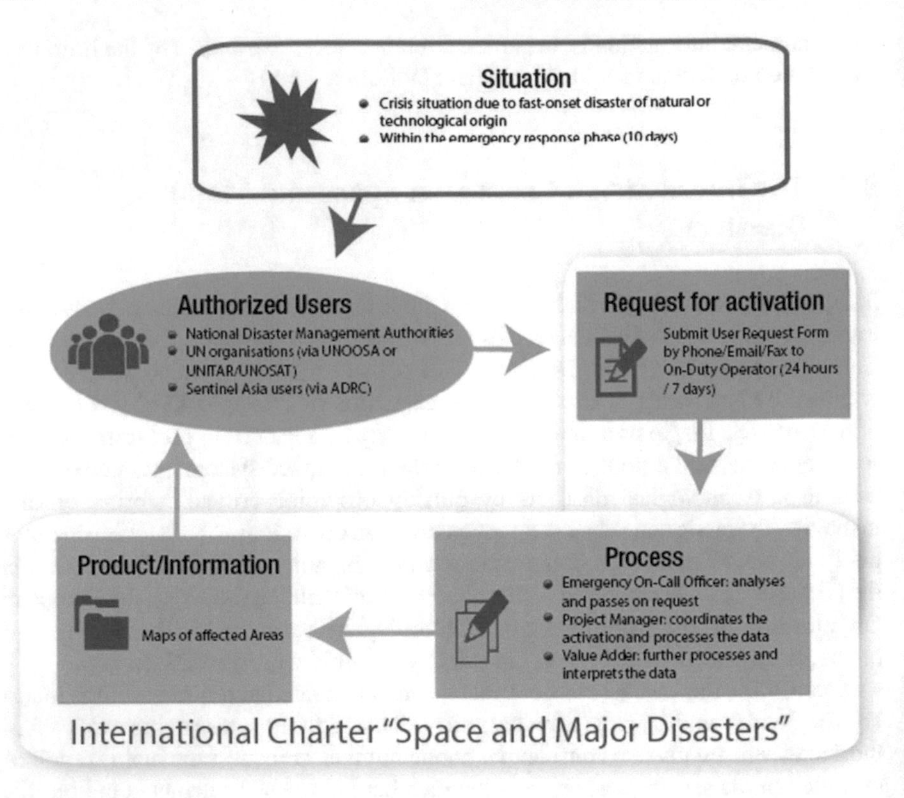

Fig. 8.3 Flow chart of activating the Charter on Space and Major Disasters. Source: http://www.un-spider.org/space-application/emergency-mechanisms/international-charter-space-and-major-disasters

8.4 National Policies Governing Remotely Sensed Data

After the end of the Cold War in the late 1980s, geopolitical thinking faded and has increasingly been replaced with discourse on the collaborative uses of remote sensing data and transnational economic competition. However, the fact that remote sensing is a dual-use technology—that is, it can be used for both military and nonmilitary purposes—creates a tension between commercial industry interests and those of national governments. There are many complex national security concerns and policy issues that have yet to be resolved. Added to the rapid increase in the worldwide availability of high-quality commercial satellite imagery, remote sensing technology and distribution has overwhelmingly outpaced the development of policy solutions.

Currently, there are approximately 30 countries with satellite-based Earth observation capabilities, compared with only 3 in 1980, along with an increasing number of countries that have their own image-receiving stations for remote sensing systems, due to the significant reduction in acquisition costs (de Montluc 2009).

Fig. 8.4 Satellite-provided images of Haiti after a massive earthquake struck on April 15, 2010. Satellite images of adjacent areas, one captured on April 15, 2010 *(left)* and the other on November 6, 2004 *(right),* illustrate the extensive destruction caused by the earthquake. Images courtesy of DigitalGlobe

The US, Canada, China, India, Israel, Japan, Russia, and the European Union—particularly France, Germany, and Italy—are the predominant leaders in geospatial technology. However, Argentina, Australia, Brazil, Malaysia, Nigeria, South Korea, South Africa, and Thailand have been building and launching their own satellite remote sensing systems.

Global leaders in satellite remote sensing, such as the US, have established national laws and legal precedents, while other nations primarily rely on internal or national policies.

8.4.1 Common Themes and Policy Solutions

Overall, national laws and policies are rooted in similar fundamental principles—allow access to remotely sensed imagery for scientific, social, and economic benefit, and confine access to protect national security. Before any commercial firm can design and build a remote sensing satellite, or an entity can gain access to images, they need to obtain government licenses that set or limit imaging capabilities such as panchromatic and multispectral resolutions. In addition, once the satellites are launched, national policies come into play to determine how remote sensing systems are allowed to be operated and the resulting images are distributed. The major differences in various national policies concern high spatial and spectral resolution images (e.g., tighter restrictions or denying access). Another method of limiting access to specific locales during specific times is known as *shutter control*.

Shutter Control

Does Google Inc. or other commercial remote sensing providers need to ask permission to post online satellite images of properties?

Governments exert shutter control over all commercial remote sensing carried out—whether foreign or domestic companies.

Shutter control is a government-authorized mechanism to protect national security, international obligations, and/or foreign policy interests by interrupting, withholding, altering, or blocking access to particular data. Under licensing agreements, a government temporarily limits the imaging operations of commercial remote sensing satellites (e.g., a "24-hour rule"). An example of shutter control is a government order (based on recommendations by national security or state department officials) to commercial remote sensing operators to turn off their satellite imaging sensors when viewing specific areas on the globe where government military operations are occurring, or are about to occur, or instances when national security could be compromised (Thompson 2007). To date, shutter control mechanisms have not been used by the US, but rather pseudo-shutter control, termed "persuasion to buyouts," has been used to purchase all images of a particular location to keep data on US military operations from falling into the wrong hands (Gallucci 1994). Alternatives to physical shutter control include delaying the transmission or distribution of data, restricting the field of view of the system, encryption of the data, or political/economic restrictions through contracts.

8.4.2 US Laws and Policies

We begin with the United States because they were the first to develop a national law concerning remotely sensed data. Nations typically look to other nations for precedents and then model their policies on this framework with differences mainly concerning particular national interests. Thus, other countries with specific remote sensing policies have very strong similarities to US policies.

The political ideals of transparency and societal openness (in sharp contrast to the former Soviet Union's block on access) were promoted through nondiscriminatory policies by installing ground stations in various regions around the world, allowing those nations nondiscriminatory access to the Landsat imagery collected, and allowing the sale of the imagery (Gabrynowicz 1993). The first law concerning satellite images was passed in 1984. The Land Remote Sensing Commercialization Act (15 USC § 4201) recognized that "competitive private sector participation in remote sensing was in the national interest" (Reagan 1984) and was intended to facilitate the commercialization of the Landsat system (Roccio 2011). The primary intention was to shift remote sensing away from being a government-run enterprise to private industry. This initiative was a policy failure for many reasons, including the lack of government subsidies to assist in the transition to a commercial industry, and an insufficient satellite imagery data market.

During the early 1990s, concern about the US's prominence in global remote sensing[2] technology led to the Land Remote Sensing Policy Act of 1992 (15 USC § 5601–5672). This landmark Act was initiated to ensure that the US retained its international leadership role in remote sensing by promoting commercialization of unenhanced data (Peters and Woolley 1999). The aim was to maintain Landsat as "an unclassified program that operates according to the principles of open skies and non-discriminatory access" (15 USC § 5601, 10). The Act transferred management of the Landsat program from the Department of Commerce to an integrated program management by the Department of Defense and the National Aeronautics and Space Administration (NASA) (Roccio 2011).

The US Commercial Remote Sensing Policy, authorized by President George W. Bush on April 25, 2003, substantially relaxed restrictions on the data that commercial remote sensing industries could provide domestically and internationally. The policy strongly encouraged and promoted the commercial remote sensing industry by shifting away from US government satellite systems and to specifically "build and operate commercial remote sensing space systems whose operational capabilities, products, and services are superior to any current or planned foreign commercial systems" (OSTP 2003). The 2003 policy guides the licensing, operation, and distribution of data and requires additional controls to protect national security and foreign policy interests. US policies have created a small but growing commercial remote sensing industry and related domestic and international market for geospatial imagery. Concurrently, with the recent entry of commercial remote sensing firms into the previously military domain of satellite reconnaissance, the US faces some complex policy issues and challenges (Thompson 2007).

[2] France had launched SPOT in 1986 and the resulting images quickly outsold US Landsat images. In 1988, India launched IRS-1A. These events placed pressure on the United States to determine how to promote the commercial remote sensing industry, ultimately leading to the passage of the Remote Sensing Policy Act of 1992. To implement that legislation, the President William J. Clinton Administration issued Presidential Decision Directive (PDD)–23 on March 9, 1994 to support and encourage foreign sales of images while protecting national security (Thompson 2007).

Defense Bill Encourages Commercial Satellite Industry

Tucked within the Defense Authorization Act for FY 2015 (US Senate Bill S2410) is language that encourages the Secretary of Defense to "support relaxation, as soon as practicable, of panchromatic, spectral, and infrared imagery resolution limits so that the United States commercial space imaging industry may promptly begin to:

1) Attract investment in new satellite capabilities; design and build new satellites; and create new processing capabilities, business strategies, and marketing capacity; and
2) The Under Secretary of Defense for Policy should provide a recommendation to Congress by April 1, 2015 on the design and development of a flexible and dynamic capability to control the collection and sale of commercial space imagery to protect national security."

Similarly, the US House of Representatives Bill, HR 4435, advocated a pilot program for acquisition of commercial satellite communication services with the goal of "incentivizing private-sector participation and investment in technologies to meet future requirements of the Department of Defense with respect to commercial satellite services."

8.4.3 Legal Frameworks Within the European Union

One of the main issues confronting the European Space Agency (ESA) and member states of the European Union is the lack of comprehensive policy governing remote sensing data. Most European satellite operators protect data through licensing and copyright laws; however, implementation differs depending on national law. Many individual countries control access through licensing regulations such as the Federal Republic of Germany's Satellite Data Security Act of 2007, which seeks to maximize the public's use of remotely sensed data while protecting national security. With regard to high-resolution data, Germany uses a two-layered approach for first-time dissemination of information: a sensitivity check of the data requested by a customer and a permit/refusal by governmental authority. An algorithm determines sensitivity of the data and potential user on a case-by-case basis using factors such as whether the data are classified, the location, the customer, and other internal procedures (Schneider 2010).

8.4.4 Asian Policies

India developed a less restrictive Remote Sensing Data Policy, released in July 2011, which allows nondiscriminatory access for images up to 1-m resolution. An interagency "High Resolution Image Clearance Committee" screens private and foreign users for access to submeter resolution imagery. The previous policy gave the National Remote Sensing Center (NRSC) a monopoly within India to restrict access to images with less than 5.8-m resolution—those from Indian and foreign satellites.

The Japan Aerospace Exploration Agency (JAXA), an independent administrative public corporation, promotes space development and use, including remote sensing data. Japanese agencies issue "directions," "requests," "warnings," "encouragements," or "suggestions," which are not considered to be legally binding, but rather well respected as administrative guidance (Aoki 2010). The recent guidance concerning remotely sensed data includes JAXA's Basic Plan for Space Policy (June 2, 2009). In principle, all data are available to the public without limitations on spatial resolution. The policy promotes the peaceful uses of Earth observation data; JAXA retains intellectual property rights.

8.4.5 *Australian Remote Sensing Policy*

Australian remote sensing policy generally concerns access control (as is the case with virtually every country's policy). Customers sign a legal agreement when they order data stating that they will be used for noncommercial purposes. There are also restrictions regarding the sharing and redistribution of data. Unprocessed data cannot be transferred to a third party if they retain their original pixel structure and can be converted back to primary data. One may save a copy as a backup, but cannot otherwise duplicate the images (Geoscience Australia 2009).

8.4.6 *Remote Sensing Policies on the African Continent*

The African continent is increasingly involved in Earth observation, both as providers and as users of data. Algeria, Egypt, Nigeria, and South Africa operate remote sensing satellites. The African Resources Management (ARM) Satellite Constellation is a joint program of South Africa, Nigeria, Kenya, and Algeria (and any other interested African country), aimed at fulfilling the need for regular high-resolution data over the continent for resource management applications (Dowman and Kufoniyi 2010). Many African countries are members of the UN COPUOS as well as the International Telecommunication Union (ITU) and INTELSAT, the two intergovernmental organizations, respectively, responsible for the regulation and provision of communication services (Dowman and Kufoniyi 2010). However, there is a lack of national policies governing remotely sensed data. In addition, some African countries see maps and map products as highly sensitive and classified information, and they are frequently controlled by the military. National geospatial information policies serve to increase opportunities for capacity building, economic development, as well as innovative applications of imagery.

8.5 The Future of Remote Sensing Laws and Policy

International and national laws and policies are dynamic and ever changing in response to changes in politics, technologies, as well as to real or perceived risks to national security (Ghosh 2010). While it is difficult to know what new changes will be implemented in the near future, there are some indications of particular policy directions. Internationally, two relatively new major organizations are involved with remote sensing: the *Committee on Earth Observation Satellites* (CEOS) and the voluntary partnership of governmental and intergovernmental organizations, *Group on Earth Observations* (GEO). CEOS works toward coordinating international Earth observation systems and activities to meet the common good of member states, with special attention paid to developing countries. In response to an appeal for more international cooperation and coordination in data sharing regarding atmospheric, land, and water data, GEO—a collaborative of over 130 governments and international organizations—was formally created by resolution at the third Earth Observation Summit (EOS), held in Brussels in February 2005.

In creating GEO on a voluntary and legally nonbinding basis, the founding governments and international organizations resolved that GEO would establish the Global Earth Observation System of Systems (GEOSS) by 2015 using the 10-Year (2005–2015) Implementation Plan (Koike et al. 2010). The anticipated benefits for nine "Societal Benefit Areas[3]" Plan rely on cooperatively sharing Earth observation data by GEO members and participating organizations through the GEOSS. A major initiative of the plan was the establishment of the GEOSS Data Collection of Open Resources for Everyone (Data-CORE). Many challenges remain to be resolved including determining who pays for infrastructure, training, and administration; whether to control data access; how to include the private sector; and whether problems of collective action will continue to hamper the effort.

With regard to national laws and policies, as high-resolution (i.e., submeter resolution) imagery continues to be widely available to the public through a multitude of sources, the divide between open (public) access and restricted (military) access has vanished. As a result, shutter control is not a viable national security policy; there are numerous alternative sources of high-quality imagery. We have moved from an era in which a handful of developed countries had access to high-resolution imagery to one in which virtually everyone has access. Transparency offers both enormous benefits and challenges. Governments throughout the world are not adequately prepared for this new era of access and transparency; policies are predominantly ad hoc, reactive, and not based on a working knowledge of geospatial technology. A harmonized international framework of international legal norms that goes beyond

[3] The GEOSS 10-year Implementation Plan includes nine "Societal Benefit Areas": disasters, protecting health/forecasting disease outbreaks, optimizing energy use, improving climate observations, managing water resources, forecasting weather, mapping and classifying ecosystems, improving agriculture and fisheries management, and protecting biodiversity.

the scope of the UN Remote Sensing Principles will be needed to resolve these challenges.

8.6 Review Questions

1. How often does the Committee on the Peaceful Uses of Outer Space (COPUOS) meet? When did they first meet?
2. What was the intent of the "Open Skies" treaty proposed by US President Eisenhower?
3. How many international treaties are there? Name them.
4. What three issues are at the heart of virtually every national and international law or policy?
5. Is the use of remote sensing an invasion of privacy? Personal? National?
6. Describe one of the reasons Space Law was needed?
7. What would be a reason to not honor a request via the International Charter on Space and Major Disasters?
8. Define "Major Disaster."
9. Compare and contrast two different "Remote Sensing Policies."
10. How should private remote sensing capabilities be considered in light of multinational private companies? Should private company data be private from governmental organizations?
11. What is meant by "transparency" and how can or should it be enforced? Describe a future or law or policy need that should be enacted now.

References

Aoki, S. 2010. Regulation of space activities in Japan. In Jakhu, R.S., ed. *National Regulation of Space Activities*. NY: Springer. 499 p.

de Montluc, B. 2009. The new international political and strategic context for space policies. *Space Policy* 25: 20–28.

Dowman, I. and O. Kufoniyi. 2010. Policies for applying earth observation in Africa: an ISPRS perspective. In: International Archives of the Photogrammetry, Remote Sensing and Spatial Information Science, Volume XXXVIII, Part 8, Kyoto Japan 2010, pp. 1088–1093.

Gabrynowicz, J.I. 1993. The promise and problems of the Land Remote Sensing Policy Act of 1992. *Space Policy* 9: 319–328.

Gabrynowicz, J.I. 2007. The Land Remote Sensing Laws and Policies of National Governments: A Global Survey. National Center for Remote Sensing, Air, and Space Law at the University of Mississippi School of Law. US Department of Commerce, National Oceanic and Atmospheric Administration's Satellite and Information Service Commercial Remote Sensing Licensing Program. p. 42.

Gallucci, R. 1994. Statement in U.S. House of Representatives; Committee on Science, Space, and Technology and Permanent Select Committee on Intelligence. *Commercial Remote Sensing in the Post-Cold War Era*, 73.

Geoscience Australia. 2009. End User License Agreement. Geoscience Australia, Department of Resources, Energy, and Tourism. http://www.ga.gov.au/image_cache/GA18442.pdf. Accessed 3 October 2011.

Ghosh, A. 2010. *Remote Sensing Data Policies*. Geospatial World—Geospatial Communications Network. http://www.geospatialworld.net/index.php?option=com_content&view=article&id=20307&Itemid=1304. Accessed 20 July 2011.

Harris, R. 2003. Current policy issues in remote sensing: report by the International Policy Advisory Committee of ISPRS. *Space Policy* 19: 293–296.

Ito, A. 2005. Issues in the implementation of the international charter on space and major disasters. *Space Policy* 21: 141–149.

Jakhu, R. 2004. International law governing the acquisition and dissemination of satellite imagery. In: *Commercial Satellite Imagery and United Nations Peacekeeping: a view from above*. Burlington, VT: Ashgate Publishing Company, p. 259f.

Leghorn, R.S. and G. Herken. 2001. The origins and evolution of openness in overhead global communications. In: Baker, J.C., K.M. O'Connell, and R.A. Williamson, Eds, *Commercial Observation Satellites: at the leading edge of global transparency* (pp. 17–36). Santa Monica, CA: RAND-ASPRS.

Keeley, J.F. and R.N. Huebert. 2004. *Commercial Satellite Imagery and United Nations Peacekeeping: a view from above*. Burlington, VT: Ashgate Publishing Company.

Koike, T., M. Onoda, D. Cripe, and J. Achache. 2010. The global earth observation system of systems (GEOSS): supporting the needs of decision making in societal benefit areas. In: Kajiwara, K., K. Muramatsu, N. Soyama, T. Endo, A. Ono, and S. Akatsuka, Eds. Networking the world with remote sensing, ISPRS Technical Commission VIII Symposium, 9–12 August 2010, Kyoto, Japan. ISPRS Proceedings, Volume XXXVIII, Part 8. Pp. 6.

Macauley, M.K. 2005. Is the vision of the Earth Observation Summit realizable? *Space Policy* 21: 29–39.

OSTP (U.S. Office of Science Technology and Policy). 2003. *U.S. Commercial Remote Sensing Policy Fact Sheet 25 April 2003*. http://www.au.af.mil/au/awc/awcgate/space/2003remotesensing-ostp.htm. Accessed 11 October 2011.

Peters, G. and J.T. Woolley. 1999. George Bush: Statement on Signing the Land Remote Sensing Policy Act of 1992. 28 October 1992. *The American Presidency Project*. http://www.presidency.ucsb.edu/ws/?pid=21693. Accessed 25 September 2011.

Rao, M. and K.R.S. Murthi. 2006. Keeping up with remote sensing and GI advances—Policy and legal perspectives. *Space Policy* 22: 262–273.

Reagan, R.W. 1984. Statement on signing the Land Remote-Sensing Commercialization Act of 1984. 17 July 1984. *Ronald Reagan Presidential Library Archives*. http://www.reagan.utexas.edu/archives/speeches/1984/71784e.htm. Accessed 25 September 2011.

Roccio, L. 2011. Landsat then and now. NASA. http://landsat.gsfc.nasa.gov/about/. Accessed 21 September 2011.

Schneider, W. 2010. German National Data Security Policy for Space-based Earth Remote Sensing Systems. http://www.oosa.unvienna.org/pdf/pres/lsc2010/tech-02.pdf. Accessed 24 October 2011.

Smith, L.J. and C. Doldirina. 2008. Remote sensing: A case for moving space data towards the public good. *Space Policy* 24: 22–32.

Thompson, K.P. 2007. A Political History of U.S. Commercial Remote Sensing, 1984–2007: Conflict, Collaboration, and the Role of Knowledge in the High-Tech World of Earth Observation Satellites. Doctoral dissertation, Virginia Polytechnic Institute and State University, Alexandria, Virginia, US.

Williams, M. 2006. Legal aspects of the privatization and commercialization of space activities, remote sensing, and national space legislation. 2nd Report. International Law Association, Toronto, Ontario, Canada, p. 2

Suggested Reading

Ito, A. 2011. Legal Aspects of Satellite Remote Sensing. Leiden, Netherlands: Brill. DOI: 10.1163/ej.9789004190320i-354 372 p.
Jakhu, R. ed. 2010. National Regulation of Space Activities. Dordrecht, Netherlands: Springer. DOI: 10.1007/978-90-481-9008-9 500 p.

Relevant Websites

International Space Law Index: http://www.unoosa.org/oosa/en/ourwork/spacelaw/index.html
UN Documents and Resolutions Concerning Outer Space Affairs: http://www.unoosa.org/oosa/documents-and-resolutions/search.jspx?&view=resolutions
International Space Law Committee: http://www.ila-hq.org/en/committees/index.cfm/cid/29
http://www.disasterscharter.org/web/charter/home
United Nations Platform for Space-based Information for Disaster Management and Emergency Response: http://www.un-spider.org/
National Center for Remote Sensing, Air, and Space Law (U.S.): http://www.spacelaw.olemiss.edu/

Chapter 9
Future Trends in Remote Sensing

In the previous chapters, we have discussed how the scientific community, government agencies, nongovernmental organizations (NGOs), private industry, and the general public use the wealth of information provided by airborne and satellite remote sensing data. We have presented specific examples of its cost-effective and timely use in a wide range of disciplines including engineering, forestry, geology, public health, archaeology, humanitarian aid, natural resources, and geography. Finally, we explored the linkages between remote sensing, geographical information systems, and spatial modeling. It is this continued fusion of remote sensing and big data science where the future of remote sensing lies.

Advances in data science and associated fields will initiate and demand advances in all facets of remote sensing and geospatial technologies. We foresee developments in both hardware and software as well as the application of remote sensing technology in new and varied ways.

9.1 Future Advances in Hardware and Software

In the past, space-borne remote sensing instruments were primarily designed around a small, carefully-selected number of wavelength ranges due to resource constraints on system configurations. However, public–private partnerships are becoming the standard for satellite construction and operation. These partnerships are bringing innovation while reducing the cost of satellite systems. The coming decades are likely to see innovation in satellite construction and operation. These may include innovative approaches to refractive and reflective optics; larger antenna and mirror structures; use of compact Fourier transform and grating spectrometers to allow for the reductions in overall size and mass of optical instruments; more efficient cooling systems that will allow the extensions of these size benefits far into the infrared region of the electromagnetic spectrum (Pagano and Kampe 2001); and the development of solar power and solid-state radio frequency amplifiers.

S. Khorram et al., *Principles of Applied Remote Sensing*,
DOI 10.1007/978-3-319-22560-9_9

These technological innovations should improve the spatial resolution of satellites in low-Earth orbit (LEO) and allow for the inclusion of multispectral and hyperspectral, Radio Detection and Ranging (RADAR), Light Detection and Ranging (LiDAR), and Synthetic Aperture Radar (SAR) measurements from geostationary orbits (GEO). They may even allow tunable (i.e., with custom selection of desired wavelengths) imaging systems to become commonplace, allowing for much greater flexibility in spectral and spatial domains (Hartley 2003). An example of this sort of innovation is the hyperspectral data being acquired by the Airborne Visible and Infrared Imaging Spectrometer (AVIRIS). It is expected that these high-spectral-resolution sensors, with improved spatial resolution, will be regularly placed on future Earth-orbiting satellites (U.S. Congress, Office of Technology Assessment (OTA) 1993; Vane and Goetz 1993), building on early successes such as the Hyperion sensor.

These technological advances may also allow for the development of sensors with apertures of several meters from LEO and GEO, allowing for differential absorption LiDAR (DIAL) and laser Doppler wind measurement sensors from space (Hartley 2003). Also, further development of the terrestrial laser scanners (TLS) and airborne laser scanners (ALS) is expected. The development of larger antenna mirror and structures may allow the placement of satellites at what are known as the Lagrange points, L1 and L2, which are located one and a half million kilometers from Earth on the Earth–Sun line (Hartley 2003). Because they will be operating in a nearly disturbance-free environment, sensors placed at the Lagrange points may be able to make highly precise measurements (Hartley 2003).

Some of these advances are already evident in the current generation of commercial satellites such as GeoEye-1 (data of 41 cm GSD), WorldView-2 (providing data of 46 cm GSD), and new satellites from India, Europe, China, Japan, Russia, Brazil, and other countries. In addition, in the coming years, several commercial providers expect to launch satellites with 1 m resolution or better. For example, the Indian Cartosat-3 is planned to collect imagery at 25 cm resolution. These higher resolution data will allow for new applications of remotely sensed data. An example that could impact everyday life for most people is the detection of automobiles with some level of uncertainty (depending on their size) with 1 m resolution imagery, whereas 50 cm resolution allows for the delineation of vehicle windshields. Automobile side mirrors can be detected only with 30 cm imagery, clearing a path for automated computer vision techniques permitting car model identification. It is also worth noting that yellow lines in the parking area appear clearer at 30 cm resolution, while they are barely visible at 1 m resolution. Finally, these newer classes of satellites will have high-performance camera control systems capable of rapid re-targeting, allowing the collection of dozens of images over a single target, each with a unique angular perspective.

Beyond innovation in traditional satellite technology, a range of new platforms for the acquisition of targeted remotely sensed data are being developed. These include nanosatellites, microsatellites, and uninhabited aerial vehicles (UAVs). These miniaturized platforms have three advantages as compared to larger systems. First, manufacturing, launching, and operating costs are less than larger satellites. This facilitates "risk taking" during design and encourages technical innovation. Second, small satellites can be developed in only a few years or months. Shortening the time to launch adds resilience and flexibility to the satellite systems, allowing for nimble response to emerging issues such as the need for a deeper understanding of global climate change. The current systems have a 10-year development cycle from conception to launch. Third, operating only a small number of instruments per satellite allows orbits to be optimized for a particular set of measurements. The growing popularity of UAVs and ongoing commercial exploration of their use by companies such as Amazon, UPS, and Federal Express (FedEx), along with the fact there is currently no need to have a pilot's license to fly one in the US (FAA), suggest continued growth and innovation.

These advances and continued decrease in the size of sensor electronics will come with and likely drive further increases in processing speed for both programmable and general-purpose processors as well as storage density and onboard recording capacity. By the time today's technologies are fully exploited, innovative new technologies, such as *quantum* and *biological computing*, will likely be realized. The ultimate result will be, for all practical purposes, unlimited computing power on orbit within the sensor itself (Chao et al. 2002; Hartley 2003).

Paralleling advances in hardware will be advances in algorithm to process the data. One key area of innovation is the development of parallel processing algorithms for remotely sensed data. Currently, parallel processing strategies for remotely sensed data are in their infancy. Those that have been implemented have assumed homogeneity in the underlying computing architecture (e.g., a dedicated cluster of workstations with identical specifications, connected by a homogeneous communication network). However, today we are beginning to see the emergence of both heterogeneous, distributed computer networks, built simply on local (user) computing resources and innovative heterogeneous parallel algorithms for extracting information from high-dimensional remotely sensed data. Plaza (2006) provides an example of the promise of these algorithms for the processing of hyperspectral imagery.

Overall, advances in remote sensing and associated technologies are already providing more focused agricultural and natural resources inventory and monitoring, quicker and more efficient emergency response, more accurate map products, improved navigation, and better geospatial information and derived products for the general public and professionals in a wide variety of fields (Peri et al. 2001).

9.2 Open, Social, and Timely

The rapidly developing and growing industry of commercial remote sensing along with expanded free access to governmental remotely sensed data has already made high-quality digital imagery widely available, leading to a substantial web-based market potential for remote sensing in media. Remote sensing data have allowed television, Internet, and print news media around the world to deliver more timely information (Lurie 2011). The Internet offers a unique medium to educate potential customers and to sell remote sensing data to them. News agencies, web sites, and a host of other visual media services will continue to benefit from the advances in remote sensing and continue to use remotely sensed data to provide current, relevant, and near-real-time information regarding events around the world (Lurie 1999). The widely-used Google Earth platform is an example of such applications.

Widespread access to remotely sensed data has made it integral and influential in our daily lives. Historically, NGOs and other entities who advocated for policy development (relating to economic, environmental, social, and other issues) have limited access to remotely sensed information because the technical expertise required to process satellite imagery was restricted to government agencies. The abundance of open-access satellite imagery and image processing capacity has resulted in nearly unlimited capacity outside government. This movement is being further stimulated by the availability of web-based information systems, such as Google Earth and Google Maps, which provide users easy access to satellite imagery of your own backyard and beyond. These days, researchers, along with NGOs and the general public are applying remote sensing to mobilize support for situations requiring policy development—a process that has been called *satellite imagery activism* (Baker and Williamson 2006). The rise in satellite imagery activism has created its own set of challenges. These include issues such as who pays for infrastructure, training, and administration; whether to control data access; how to include the private sector; and whether problems of collective action will continue to hamper the effort.

Government agencies have already begun to respond to these challenges by assembling, processing, and providing information at broad scales to a broad community of users. For instance, the US Geological Survey (USGS) realigned its science programs to better address relevant issues, concerns, and challenges to societies. New mission areas include Climate and Land Use Change, which is currently developing a science strategy that will guide USGS prioritization for research in the 2020s, and aid in the USGS developing methods to explain how changes in land use, cover, condition, and management alter climate, impact natural systems, and affect human health and welfare. This change in focus will allow USGS to take a leadership role in advancing science, leading to better understand (1) the interactions between the natural and physical environments (e.g., contaminants and pathogen exposure) and (2) the concept of *One Health*—the cohesive study of wildlife, human, ecosystem, and agricultural health.

For example, remote sensing is currently influencing research regarding climate, earth surface processes, and ecosystems across space and time through its contributions to the strategic goals of global climate change studies, including the US Climate Change Science Program. Remote sensing has guided other scientific research strategies and the linkages to other mission areas, including ecosystems and environmental health.

The application of remote sensing data is suggestive of its real-time nature. Once we begin to think of remote sensing in terms of more real-time (or near real-time) applications, then the scope of activities and applications may change significantly, opening the door to new possibilities and new realities for using spatial information. Data are already turned into information and will increasingly become knowledge. Remote sensing will become more valuable and we will become more reliant upon it (Thurston and Ball 2008).

We are already seeing this transformation with radio-frequency identification (RFID) tags, global positioning system (GPS) trackers, cell phones, and arrays of satellites. For example, DigitalGlobe's constellation of satellites has intraday revisit anywhere across the globe and it is capable of collecting over three million square kilometers of imagery every day. The company's archive has complete coverage of most nations and urban areas have imagery as recent as three months old.

In addition, the integration of mobile technology with real-time remotely sensed geospatial data is already supporting the development of decision support tools for a variety of environmental, disaster warning, and emergency monitoring and management applications such as wildland fire risk and damage assessment, large-scale drought conditions and impacts, irrigation and water allocation systems, flooding risks and damage assessments, earthquake, tsunami, and public health as well as infrastructure development of maintenance. For example, certain health issues are now benefiting from advances in information technology and numerous smart devices can now collect remotely sensed data to characterize the environmental and landscape causes for such problems. Sensors and scanners are being integrated into the mobile devices for continuous health assistance and disease attack prevention. Scientists are developing system framework for modeling and analyzing individual exposure to environmental triggers of asthma attacks (http//www.enviroflash.info/). Finally, air quality indices, for example, are now monitored regionally in real time and available as "apps" for mobile devices such as Android and iPhone.

Beyond the use of real-time data, many of these applications are using remotely sensed crowd-sourced data. Much of these crowd-sourced data are often based on volunteer services that may be coupled with the mobile apps using smart mobile devices for collecting and transferring information. Currently, remote sensed data are being integrated with crowd-sourced data in agriculture to allow for the collection, distribution, and exchanges of data to farmers for drought, fertilizing, irrigation, and disease control (Amy J. Harris, NASA Ames Research Center). We expect this combination will be used more frequently in the future for disaster management practices by the remote sensing practitioners without having to be present in the disaster area for deploying the proper systems and allocating more relevant and timely resources for risk assessment and damage prevention or control.

While promising, the technologies described above are only the tip of the iceberg. If Martin Wikelski has his way, the International Space Station will one day keep a close eye on thousands of our planet's bats, songbirds, dragonflies, rodents, and other small creatures. A specialist in animal movements, the ecologist at the Max Planck Institute for Ornithology in Radolfzell, Germany, has for a decade pursued what a collaborator once dubbed a "ridiculous idea": a new space-based system to track animals too small to be monitored globally with current instruments. Wikelski and his colleagues have begun testing whether a new animal tag can eventually communicate with the International Space Station (Pennisi 2011).

In an era of global land-use, land-cover, and climate change, a space-based way to track animals would revolutionize the study of animal movement and the field of conservation biology. As described by Ropert-Coudert and Wilson (2005), biologging is the use of data-recording units to acquire huge, multivariate datasets about the behavior of animals moving freely in their natural environment. This approach allows scientists to study wild animals in the field, behaving normally, with approximately the same degree of rigor as laboratory studies. A future space-based system along with advances involving higher data resolutions, faster data processing, and free access to data would allow scientists and policy makers to better understand the biotic and abiotic factors affecting behavior of mobile species as well as development of effective management plans.

9.3 Interdisciplinarity and Big Data

Remote sensing is providing a platform for building vital bridges between scientific disciplines, pushing the frontiers of big data, and advancing science diplomacy (Godfray et al. 2010).

Recent work on cropland mapping (Pitman et al. 2010) is an excellent example of the application of ongoing advances in remote sensing to provide accurate and reliable crop maps and statistics at local, regional, and global scales. Given that about one in eight people in the world suffer from chronic hunger, impacting their abilities to have an active and healthy life (FAO 2011), timely and accurate information on crop inventory, health condition, and yield estimation can be invaluable in helping to combat these problems. Remote sensing also provides a means to practice precision agriculture, which optimizes resources both within and across fields. In addition to agriculture, hyperspectral imagery and a thermal infrared scanner are currently being used to monitor a variety of environmental, ecological, and geological features at a wide range of wavelengths (e.g., over 200 bands of data). Specific applications include monitoring deforestation; water resources; disaster management such as hurricanes, tsunamis, earthquakes, urban and wildland fires as well as ecosystem management.

Remote sensed data underlies the rise of "Big Data." Moderate- to high-resolution remote sensing data available from multiple satellite systems such as Landsat, SPOT, MODIS, VIIRS, IRS, ERS, JERS along with airborne and satellite hyperspectral data such as AVIRIS, HyspIRI, Hyperion, PRISM, and HICO, and airborne UAV-based data such as UAVSAR when combined with vast amount of ground-based and crowd-sourced data over large areas have given birth to the explosive growth of remotely sensed information leading to the "Big Data" concept. Factors contributing to this massive growth include open access to land cover data (e.g., Landsat, MODIS, VIIRS); open access to vast and frequently collected meteorological data (e.g., GOES, Meteosat, Meteor); open access to field-based data collected by many government agencies and researchers; open access to other large-scale datasets such as digital elevation models (DEM); and organized approaches to collection and preprocessing of some of these datasets such as NASA NEX.

Despite the enormous computational power, the cluster-based high-performance computing (HPC) systems still remain considerably challenged with remotely sensed "Big Data" issues. These issues include the difficulties in storing massive and complex data, intensive irregular data access patterns, managing remotely sensed "Big Data" on multilevel memory hierarchy, optimal scheduling of large amount of interdependent tasks as well as the efficient programming for these vast databases. There is no doubt that the existing techniques and systems are limited in their ability to solve remotely sensed "Big Data" problems completely (Yan Ma et al. 2015).

Finally in terms of enhancing scientific diplomacy, remote sensing has been at the heart of activities organized by the American Association for the Advancement of Science (AAAS) Center for Science Diplomacy. The Center has been instrumental in advancing scientific engagement as an essential element of foreign policy and of building vital bridges between societies. For example, during December 2014, AAAS organized a delegation to Havana, meeting with Cuban colleagues to discuss opportunities for joint research on coral reefs, hurricane dynamics, and other areas of common concern, and in February 2015, an AAAS-led group traveled to Myanmar, the second time in as many years, to meet with government ministries and scientific NGOs to discuss health science, forestry, education, and the role of science in public policy.

9.4 Concluding Thoughts

Several trends are expected to continue far into the future (Hartley 2003). These include the following:

- Miniaturization and integration of electronics
- Further development of UAV-based data acquisition systems
- Increases in computational power such as heterogeneous parallel computing, cloud computing, and quantum and biological computing
- Progress in large apertures and larger antennas
- Increases in transmitter power for active systems
- Miniaturization of optics
- Increase in storage technology
- Development of small satellites
- Advances in screen technology and mobile computing
- Increases in tunable systems and flexible frequencies
- Advances in techniques for processing "Big Data"

Ultimately, sizes of observing platforms will be dictated solely by the physics of their parts. Despite the established utility of remote sensing technology in a wide variety of applications, the state of the US and the global economy and the burden of an increasing federal deficit in the US and major European countries will force decision makers to seek ways to reduce the costs of remote sensing systems. This may cause challenges for the growth of remote sensing technology. These challenges can best be met through the combined efforts of the international scientific and user communities.

Despite the short-term economic pressures on the development of new sensors, payloads, and platforms, along with the international issues involved in acquisition and distribution of remotely sensed data, we are optimistic about the future of remote sensing. In addition to the rapid growth in the technological aspects of acquiring and processing remotely sensed data, the remote sensing community is a direct beneficiary of an ever-advancing state of the art through broad collaborative efforts between developers and user communities such as biologists, engineers, social scientists, the public health and legal communities, media, urban planners, and environmental resource managers. The opportunities to use remote sensing are limitless.

9.5 Review Questions

1. What are the next expected developments in LiDAR technology?
2. What do you envision to be the most unconventional commercialized impact of remote sensing technology that could impact everyday life of most people?
3. What advantages remote sensing offer to the Internet, television, news, and social media as existing today?
4. What US governmental change policy has been enabled by remote sensing technology that will have a potential major impact on global climate change?
5. What is satellite imagery activism?
6. List a few examples of the integration of remote sensing with other geospatial data for routine environmental and emergency monitoring.
7. What are the three advantages of nanosatellite, microsatellites, and UAVs?
8. What role can the AAAS Center for Science Diplomacy play in the socioeconomic advancement of remote sensing technology?
9. What are some of the remaining challenges in using "Big Data"?
10. List five of the trends given by Hartly concerning the future of remote sensing.

References

Baker, J.C., and R.A. Williamson. 2006. Satellite imagery activism: Sharpening the focus on tropical deforestation. *Singapore Journal of Tropical Geography* 27: 4–14.

Chao, T.-H., H. Zhou, G. Reyes, D. Dragoi, and J. Hanan. 2002. High-density high-speed holographic memory. In: Proceedings of the Earth Science Technology Conference, June 11 & 13, 2002, Pasadena, CA.

Food and Agriculture Organization (FAO). 2011. The State of Food Insecurity in the World: How Does International Price Volatility Affect Domestic Economies and Food Security? Food and Agriculture Organization of the United Nations, Rome.

Godfray, H.C.J., J.R. Beddington, I.R. Crute et al. 2010. Food security: the challenge of feeding 9 billion people. Science (New York, N.Y.), 327, 812–818.

Hartley, J. 2003. Earth remote sensing technologies in the twenty-first century. In: Proceedings of the International Geoscience and Remote Sensing Symposium, July 21–25, 2003, Toulouse, France, Vol. 1, pp. 627–629.

Lurie, I. 1999. The commercial future: making remote sensing a media event. In: Fujisada, H., and J.B. Lurie, eds. SPIE Proceedings Vol. 3870, Sensors, Systems, and Next-Generation Satellites III, pp. 601–610. doi:10.1117/12.373224

Lurie, J. 2011. State-of-the-art satellite remote sensing. *Optics and Photonics News* 22(1): 28–35.

Pennisi, E. 2011. Global tracking of small animals gains momentum. *Science* 334: 1041–42.

Pagano, T. and T. Kampe. 2001. The Spaceborne Infrared Atmospheric Sounder (SIRAS) Instrument Incubator Program demonstration. In: Proceedings of the Earth Science Technology Conference, August 28–30, 2001, College Park, MD.

Peri, Jr., F., Hartley, J.B, and Duda, J.L. 2001. The future of instrument technology for space-based remote sensing for NASA's Earth Science Enterprise. In: Proceedings of the International Geoscience and Remote Sensing Symposium, July 9–13, 2001, Sydney, Australia, Vol. 1, pp. 432–435.

Pittman, K., M.C. Hansen, I. Becker-Reshef, P.V. Potapov, and C.O. Justice. 2010. Estimating global cropland extent with multi-year MODIS data. Remote Sensing, 2, 1844–1863.

Plaza, A.J. 2006. Heterogeneous parallel computing in remote sensing applications: current trends and future perspectives. In: Proceedings of the IEEE International Conference on Cluster Computing, Sept. 25–28, 2006, Barcelona, Spain, pp. 1–10.

Ropert-Coudert, Y. and R.P. Wilson. 2005. Trends and perspectives in animal-attached remote sensing. *Frontiers in Ecology and the Environment* 3: 437–444.

Thurston, J. and M. Ball. 2008. Perspectives: What do you think the future role of remote sensing will be? Vector One Magazine, online article, posted May 2, 2008. http://www.vector1media.com/Dialogue/Perspectives/. Accessed 30 October 2011.

U.S. Congress, Office of Technology Assessment (OTA). 1993. The future of remote sensing from space: civilian satellite systems and applications. OTA-ISC-558. Washington, DC: U.S. Government Printing Office. 213 p.

Vane, G. and A.F.H Goetz. 1993. Terrestrial imaging spectrometry: Current status, future trends. *Remote Sensing of Environment* 44(2/3): 117–126.

Suggested Reading

Lurie, I. 1999. The commercial future: making remote sensing a media event. In: Fujisada, H., and J.B. Lurie, eds. SPIE Proceedings Vol. 3870, Sensors, Systems, and Next-Generation Satellites III, pp. 601–610. www.portent.com/library/comfut.pdf

Lurie, J. 2011. State-of-the-art satellite remote sensing. Optics and Photonics News 22(1): 28–35. http://www.opticsinfobase.org/abstract.cfm?URI=opn-22-1-28

Ropert-Coudert, Y. and R.P. Wilson. 2005. Trends and perspectives in animal-attached remote sensing. *Frontiers in Ecology and the Environment* 3: 437–444. http://www.esajournals.org/doi/abs/10.1890/1540-9295(2005)003[0437:TAPIAR]2.0.CO;2

U.S. Congress, Office of Technology Assessment (OTA). 1993. The future of remote sensing from space: civilian satellite systems and applications. OTA-ISC-558. Washington, DC: U.S. Government Printing Office. www.fas.org/ota/reports/9348.pdf

Yan, M., H. Wu, L. Wang, B. Huang, R. Ranjan, A. Zomaya, and W. Jie. 2015. Future Generation Computer Systems, Elsvier

Appendix 1: Answers to Questions

Chapter 1

Answers to Review Questions for Chapter 1

Question 1 The French inventor Joseph Niépce is generally credited with producing the first permanent photograph in 1827, which depicted the view from his upstairs workroom window.

Question 2 The oldest existing aerial photograph is a view of Boston, taken from a balloon by James Wallace Black in 1860.

Question 3 In 1900, Eastman's company, Kodak, released the Brownie, an inexpensive box camera for rolled film, making photography accessible to a mass audience for the first time.

Question 4 The Wright brothers' first successful flight, in 1903, took place on the shores of the Outer Banks, North Carolina.

Question 5 The first satellite image of the Earth, showing a sunlit portion of the Pacific Ocean and its cloud cover, was captured in August 1959 by Explorer 6.

Question 6 "The Blue Marble," which is often cited as the most widely reproduced image of the Earth, was taken in December 1972 by the crew of Apollo 17.

Question 7 Electromagnetic radiation (EMR) is defined as all energy that moves with the velocity of light in a harmonic wave pattern (i.e., all waves are equally and repetitively spaced in time).

Question 8 The following interactions are possible: absorption, reflection, scattering, or emission of EMR by the matter, or transmission of EMR through the matter.

Question 9 Landsat of the US and Satellite pour l'Observation de la Terre (SPOT) of France.

S. Khorram et al., *Principles of Applied Remote Sensing*,
DOI 10.1007/978-3-319-22560-9

Question 10 Remote sensing provides data with high richness values (high spectral and spatial, spectral, and radiometric resolution) that reach a very broad user community (global). This is an example of disrupting the relationship between the reach and richness.

Question 11 Going from rooftop photography to birds, balloons, helicopters, airplanes, satellites, space shuttle, unmanned aerial vehicles (UAVs), and the International Space Station.

Question 12 Acquiring data over large areas very quickly and very cost-effectively.

Chapter 2

Answers to Review Questions for Chapter 2

Question 1 The primary benefit of higher radiometric resolution is that spectral information is more finely quantified. This may make it possible to distinguish subtle differences between the spectral signatures of target features that might otherwise appear the same at a lower radiometric resolution.

Question 2 One of the main challenges of using high-spatial-resolution satellite image data is that target features (e.g., trees) may be represented by multiple pixels. These pixels can vary considerably in terms of their spectral values, for example because of shadows on a portion of the target feature.

Question 3 The first multispectral environmental remote sensing satellite was Landsat-1, also called ERTS-A and ERTS-1 at early stages in its development. It was launched in July 1972.

Question 4 Both WorldView-2 and the Operational Land Imager (OLI) sensor on Landsat-8 have a specialized band for coastal applications. In both cases, the range of wavelengths captured by the band (0.43–0.45 μm for OLI, 0.4–0.45 μm for WorldView-2) sits at the lower end of the blue portion of the visible electromagnetic range. Other bands in the blue and near-infrared wavelengths may be useful, respectively, for penetrating water bodies and distinguishing between land and water features.

Question 5 Airborne remote sensing is still a viable option because users can tailor a mission to meet their project needs, for example in terms of extent and desired resolution. Now that most aerial photography is digital, many of the automated techniques used for satellite imagery can also be applied for airborne imagery. With respect to LiDAR, no satellite-based systems are currently suitable for applications such as surveying or precise topographic mapping. Ultimately, airborne remote sensing remains more cost-effective for many local- or regional-scale projects.

Question 6 The two primary instruments aboard Landsat-8 are the OLI and the Thermal Infrared Sensor (TIRS). The OLI has eight 30-m resolution multispectral bands, ranging from 0.433 to 1.39 μm as well as a 15-m resolution panchromatic band. The TIS has two 100-m resolution bands, one at 10.6–11.2 μm and the other at 11.5–12.5 μm. They both have 12-bit radiometric resolution, as compared to the 8-bit radiometric resolution of the Enhanced Thematic Mapper Plus (ETM+) sensor. Together, the OLI and TIS have 10 multispectral bands covering a similar range of wavelengths as the eight multispectral bands of ETM+; most notably, the OLI has two specialized bands for coastal applications and cirrus cloud detection that the ETM+sensor does not. However, ETM+offers somewhat better spatial resolution in the thermal infrared range than the TIRS.

Question 7 The Indian Space Research Organization (ISRO) is an especially active agency, launching 18 Earth-observing satellites since 1988.

Question 8 A sun-synchronous orbit means that a satellite passes over a given location on Earth at approximately the same local time each day. A geostationary orbit means that a satellite remains fixed in a particular location above the Earth, typically the equator, and orbits in the direction (and at the speed) of the planet's rotation.

Question 9 The red edge refers to the narrow range of wavelengths where vegetation reflectance transitions from low to high. It sits between the red and near-infrared portions of the electromagnetic spectrum, hence the name red edge. The red edge may provide information about plant health, especially by analyzing differences in red edge values between images of the same location captured at different times.

Question 10 Radar and LiDAR are both active remote sensing systems, meaning that they generate and send their own electromagnetic pulses toward target features and then process the returned signal. Interferometric radar and LiDAR can both be used to map terrain and develop digital elevation models. Radar systems use microwaves, while LiDAR systems use lasers. Radar systems can operate both day and night and in all weather conditions. However, LiDAR systems are sensitive to weather and other conditions that interfere with laser pulses.

Chapter 3

Answers to Review Questions for Chapter 3

Question 1 The processing of multispectral image data is typically divided into three stages: preprocessing, processing, and post-processing.

Question 2 In a false color composite (FCC) where the green band of the input image is displayed using the blue color ramp, the red image band is displayed using the green color ramp, and a near-infrared band from the input image (e.g., Band 4 from a Landsat TM image) is displayed using the red color ramp, this type of FCC display is popular in vegetation studies because it can be used to highlight the presence of healthy green vegetation. Healthy green vegetation may also be used to determine the health of a forest or stand of trees in relation to water stress, drought, insect infestation, etc.

Question 3 A remote sensing analyst may employ these various operations in order to prepare the best possible input data for the actual image processing stage with the objective in mind to minimize distortions and/or errors in an image that could hinder successful classification or to ensure the extraction or enhancement of an image's most critical information, thus making classification more straightforward.

Question 4 The process of image pan-sharpening may be of great value to an image analyst, because by combining a lower resolution image (typically multispectral) with a higher resolution image (typically panchromatic), the resulting image is usually nearly equal to the resolution of the higher panchromatic image. This color, higher resolution image may be useful in aiding the interpreter in distinguishing features on the ground with subtle differences, such as tree canopy heights, vegetation growing season differences, or even built surfaces compared to natural surfaces.

Question 5 A successful image classification, when working with multispectral data, should aim to convert the original spectral data, which are variable and may exhibit complex relationships across several image bands, into a straightforward thematic map (typically a land cover map) that is easily understood by end users.

Question 6 In a supervised classification procedure, the analyst selects multiple training sites during the initial stage of the classification project that serve to establish the relationships between the classes of interest and the image spectral data. These training sites are carefully selected by the analyst based on a priori knowledge (i.e., background knowledge or familiarity) of the study area being classified. Additionally, the analyst will typically rely on ancillary data sources, such as aerial photography, existing GIS coverages, or field visits to identify the training sites, thus providing some valid assumption that training sites are in fact real representations of what is being seen in the image to be classified. This step of training site selection is not conducted in an unsupervised classification, thus actually category classifications may only be validated at the conclusion of the classification process and after an appropriate accuracy assessment has been completed.

Question 7 Image accuracy assessment should be an important part of remotely sensed land use and land cover (LULC) study, because without assessing the accuracy of the classified data the reliability and repeatability of the output products are in question.

Question 8 In developing an LULC change detection study, an appropriate spatial and temporal resolution is necessary to provide the foundation for sound change detection and should be considered prior to starting an image-based change detection procedure or delineation of study area. Spatial considerations should include the assessment of the same or near the same geographic areas for the images being considered. Temporal considerations should include the assessment of the same or near the same anniversary dates for the images being considered. Similar atmospheric conditions should also be considered in the change analysis.

Chapter 4

Answers to Review Questions for Chapter 4

Question 1 Terrestrial applications of remote sensing are very diverse, and remote sensing continues to play such a vital role in large geographic area observations because it may be the only viable mechanism for synoptic and continuous tracking, mapping, and monitoring LULC changes at the subregional, regional, and global scales. As well with programs such as the Landsat program, a valuable historic record of these large areas remains easily assessable for comparative studies.

Question 2 LULC maps, created from imagery and used to categorize natural and human-made features into classes and provide important information to resource managers and researchers, are used to study everything from plant composition to fossil fuel and mineral deposit detection, from regional agriculture production to vegetation and forest health, from human settlement patterns to national security observations in water, air, land use, and political and military operations.

Question 3 The urban sprawl's impacts on urban growth have made research an issue of increased interest as this expansion increases demand on physical area and ecological and social resources. Remote sensing may aid decision-makers at the regional scale to complement traditional census sources by providing an efficient means to monitor the extent of land area expansions in relation to socioeconomic and demographic data provided by the census. This combination of data sources from both remote sensing and census may help to better understand urban change dynamics, including the spatial distribution of the population and the related socio-economic drivers.

Question 4 The limitation usually incurred in traditional field and aerial-based large animal movement studies may be overcome or supplemented by satellite remote sensing, in that satellite remote sensing may be useful for monitoring, over a large area, known migration and habitat utilization patterns that are typically tied to the landscape and vegetation patterns and/or changes in the landscape and vegetation patterns. While satellite remote sensing may be limited in directly identifying the animals, regional-scale remote sensing can acquire data over very large areas that would be cost-, labor-, and physically prohibitive.

Question 5 Remote sensing of water and water-related resources has shown advantages in using spectral remotely sensed data for various applications. However, to gain useful information in these types of studies, spectral remote sensing has to overcome the difficulties inherent in interpreting reflectance values of water, as clear water provides little spectral reflectance, the longer wavelengths are absorbed, and the reflected shorter wavelengths—which are typically the wavelengths sensors rely on for surface feature detection—are subject to higher atmospheric scattering.

Question 6 Remote sensing applications to identify status and patterns of deforestation in the Amazon have been manifold. Remote sensing techniques used to study large area deforestation in this region range from LULC classification approaches, statistical modeling procedures derived from remotely sensed data, to the development of vegetation indices that are related to spectrally reflective values of the forest tree canopy and the intensity of vegetation change.

Question 7 The Normalized Difference Vegetation Index (NDVI) is essentially an index of vegetation "greenness" that results from light interacting with the vegetation canopy. Within the leaf's structure, chlorophyll a pigments found in the leaves of healthy vegetation, that make up the forest canopy, interact with incoming solar radiation by strongly absorbing visible light in the blue and the red range of the electromagnetic spectrum. This absorption, along with light being reflected within the green and the near-infrared ranges of the electromagnetic spectrum, makes up the typical healthy vegetation spectral signature curve. This healthy vegetation curve may be compared to vegetation that is undergoing drought, insect invasions, suffering from exposure wildfires, climate change, or other stressors.

Question 8 Remote sensing has been shown to play a critical role in the study of wildland fires, delineating burned area, deriving indices of burned and non-burned vegetation, recording the frequency at which different vegetation types are affected, prevention of fires, supporting existing fire propagation models, and even developing effective strategies for extinguishing an ongoing fire event through the identification of the dryer, water-stressed areas. Remotely sensed data applied to wildfire studies have a unique advantage in that these data can be used to collect timely measurements over larger fire-prone areas, including burn severity, extent, fuel and vegetation composition, etc. Additionally, remotely sensed data may be used to provide estimates of fire characteristics that are relevant for the ecosystem as well as smoke and fire fuel load parameters, including species composition, biomass estimates, landscape structure, fire history, fuel moisture content, and fuel availability.

Question 9 Airborne Synthetic Aperture Radar (SAR) or LiDAR data as active systems, and multispectral data such as Landsat, SPOT, or even hyperspectral data such as AVIRIS as passive systems.

Question 10 No remote sensing technology directly measures biomass or forest carbon. Rather, they all measure a variety of structural parameters of vegetation that can then be used with field-derived allometric equations to estimate vegetation volume, biomass, and carbon. Allometric relationships or allometry are mathematical equations describing the relationship between one or more parameters of an object and its shape. Chave et al. (2004) reviewed the literature on the use of allometric equations to estimate forest biomass and carbon. They found four types of uncertainty that could affect biomass estimates: (i) errors related to tree parameter estimates; (ii) errors related to the selection of allometric equations; (iii) sampling error due to sample plot size; and (iv) how representative the study plots were of the entire ecosystem. Overall, they concluded that the choice of allometric equations was the most important source of uncertainty. These findings point to the care that must be taken when using remote-sensed data to estimate forest biomass and carbon.

Question 11 The most common landscape parameters used in smoke emission modeling are vegetation cover, fuel load, fuel moisture, fuel consumption rate, and fire boundary.

Question 12 Figure 4.15 shows that leaf area index (LAI) is strong correlated with field biomass measures (R^2=0.68) while fCover (R^2=0.50) and NDVI (R^2=0.30) are every bit as strongly correlated. This suggests from the perspective of estimating biomass that using technologies that directly measure the amount of vegetation present is important for accurately estimating biomass.

Question 13 The challenge of effectively tracking the movements of refugees and internally displaced peoples in many developing nations may be aided by the use of very high spatial resolution (VHSR) remotely sensed imagery. This high-resolution imagery may be useful in identifying past and current human activities across a large region, such as verifying burned and razed villages, documenting the existence of mass graves, identifying food, grazing, and water sources, as well as identifying the extent of violent conflicts within an area that may arise as a result or resulted from the movement of the refugees or internally displaced peoples.

Question 14 The evolution of satellite image analysis has transformed the field of archaeology, allowing researchers to exploit an enormous wealth of data of the earth's surface and subsurface contained in various types of satellite images along with aerial photography. Archaeologists can examine a broad spectrum of reflectivity signatures and bands within the remotely sensed data to focus in on prospective archaeological sites to determine if there are evident soil or vegetation disturbances in the surface structure of the landscape, the potential existence of subsurface structures, or even the proximity to possible sources of building materials or other historic human activities.

Chapter 5

Answers to Review Questions for Chapter 5

Question 1 Extreme weather is defined as weather at the extreme of its historical distribution beyond the usual range that has been seen in the past. Extreme weather includes unusual, severe, or unseasonal weather. It is typically based on a particular location's recorded weather history and defined as lying in the most unusual 10%.

Question 2 Remote sensing aids in the precision of weather forecasting by sensing minute details such as the phase changes of water within the clouds of a storm system. Weather forecasting accuracy is improved by remote sensing by using each variable detected by sensors to develop better predictive models.

Question 3 Pollution is defined as an excess of naturally occurring substances or chemical compounds that cause harm to human and natural ecosystem health. Pollution can take the form of chemical substances or energy (e.g., heat, light, or noise). Pollution can be localized or widespread (e.g., air pollution over a region).

Question 4 Algorithms and models can be developed to harmonize the variety of data collected by satellite remote sensing.

Question 5 Remote sensing can predict episodic events such as volcanoes by measuring thermal changes. Interferometric synthetic aperture radar (InSAR) uses the phase component of radar images to determine the position of the Earth's surface. Digital elevation models (DEMs)—crucial in predicting pyroclastic flows and lahars—with centimeter-scale accuracy, are produced from simultaneously recorded images from different radars. Deformation is measured by using time-separated images. Satellite data provide a global perspective, mapping tectonic strain across continents.

Using visible through short-wave infrared (Vis-NIR-SWIR) optical spectra data recorded in the 900–2500 nm wavelength range by the ARTEMIS sensor (flown on the TacSat-3 spacecraft), the temperature and heat flux of an active lava lake within a crater can be estimated. Elevated radiance in the NIR-SWIR wavelength regions recorded a portion of the blackbody radiation function from small, hot areas of the lava lake, which were inverted to determine the temperature and power output of the crater.

Question 6 Satellites do not measure temperature. The intensity of upwelling microwave radiation from atmospheric oxygen is measured by microwave sounding units (MSUs), which must then be mathematically inverted to obtain indirect inferences of temperature. Sensors deteriorate over time and corrections are needed for orbital drift and decay. An understanding of this is critical because the amount of deterioration can affect the accuracy of the data used in creating models.

Question 7 No right or wrong answers. This is a thought question.

Chapter 6

Answers to Review Questions for Chapter 6

Question 1 Sensors such as scatterometers and SeaWIFS. Ocean Surface Currents Analyses Real-time (OSCAR) is one such product.

Question 2 Monitoring hypoxia and dead zones; evaluating the status of an ecosystem; identifying locations of phytoplankton (useful in commercial fishing).

Question 3 Color scanners map chlorophyll and suspended solids. Hyperspectral data is used for water quality and water pollution detection including nitrogen and phosphorous.

Question 4 The most important variable in assessing coral reef health is determining the conditions that promote coral bleaching. This is evaluated by measuring the relative magnitude of ocean currents necessary to produce sufficient tidal mixing. Where surface currents do not meet the threshold for sufficient mixing of the water column, coral bleaching is more likely to occur. Coral bleaching is caused by high ocean temperatures, pollution (e.g., oil spills), and excessive algal levels. Coral die-off has consequences for species that depend on them and increases hypoxic zones.

Question 5 Sustainable fisheries techniques largely function by developing management plans that map out critical fisheries habitats and nursery grounds, as well as limit the number of species caught in a particular area to prevent overfishing.

Question 6 Chlorophyll, sea surface temperature, salinity, mesoscale ocean structures, pollution levels, presence of oil slicks, etc. One could argue that a national government should provide this free of charge. Alternatively, this could be provided by a private service that a commercial fisherman subscribes to on an annual basis.

Question 7

- Color scanners: used for ocean color mapping (chlorophyll and suspended sediments, diffused attenuation coefficients, etc.)
- Multispectral data: used for water quality studies (chlorophyll a, suspended solids, and turbidity)
- Infrared Radiometer data: used for sea surface temperature and currents mapping
- Synthetic Aperture Radar: used for surface waves, swells, internal waves, oil slicks, etc.
- Hyperspectral data: used for water quality and pollution detection studies
- Altimeter data: used for sea surface topography, currents, and surface roughness
- Scatterometer data: used for amplitude of short surface waves (surface wind velocity, roughness)
- Microwave radiometer data: used for microwave brightness temperature (salinity, surface temperature, water vapor)

Question 8 Case 1 waters are characterized by a ratio in which the concentration of phytoplankton (chlorophyll a) is higher compared to other dissolved inorganic particles in the water column. In Case 2 waters, the dissolved inorganic particles are higher compared to the level of the chlorophyll a concentration in the ratio. Thus, optical properties in Case 1 waters are determined primarily by phytoplankton and related colored dissolved organic matter (CDOM), and Case 2 waters optical properties are largely influenced by constituents other than phytoplankton concentration.

Question 9 Eutrophication is the biological process by which aquatic primary production is augmented through an increase in the rate of organic matter and nutrients (e.g., nitrogen and phosphorus from fertilizers, sewage effluent, and other pollutants) delivered to surface water (e.g., river, lake, estuary, and coastal waters). Eutrophication promotes excessive algae and plankton growth and can cause a severe reduction in water quality.

Question 10 Synthetic Aperture Radar (SAR) in various forms.

Question 11 The rate of marine primary production is determined by temperature, light (strongly influenced by surface turbulent mixing depths), and limiting nutrients—particularly nitrogen, phosphorus, iron, and silicon for some plankton (Doney 2010). Thermal bands and multispectral and hyperspectral bands in remotely sensed data can be used for quantifying marine primary production.

Question 12 We still have trouble interpreting the reflectance values of water. For example, clear water provides little atmospheric reflectance, with the majority of the shortwave radiation scattered in the atmosphere and longer wavelengths absorbed within the few millimeters of the water's surface. In addition, most sensors were designed to for detecting land features and not for use in aquatic environments.

Question 13 The US Integrated Ocean Observing System (IOOS®) is a national–regional partnership created to ensure the sustained observation of US coastal areas, oceans, and the Great Lakes, and to develop near real-time and retrospective information products from those observations to assist people in their lives and livelihoods. The primary focus is to provide timely information through national and regional collaboration.

The Pacific Islands Ocean Observing System (PacIOOS) is one of 11 regional associations in the IOOS. PacIOOS addresses familiar question such as: where are the fish today? Which beaches are safe to visit today? Can I bring my vessel in to the harbor safely? Is my home going to be inundated?

Question 14 LiDAR data once integrated with global positioning systems can be used in obtaining accurate topographic and bathymetric maps, including shoreline positions. LiDAR surveys can produce a 10 cm vertical accuracy at spatial densities greater than one elevation measurement per square meter.

Chapter 7

Answers to Review Questions for Chapter 7

Question 1 The Hubble Space Telescope enabled scientists to make the most accurate estimate to date of the age of the universe (13.82 billion years). It also allowed researchers to discover that the universe is expanding at an increasing speed, a phenomenon referred to as "dark energy." In addition, the Hubble Space Telescope provided evidence of supermassive black holes at the center of most galaxies, and yielded the first visible-light images of an exoplanet.

Question 2 Saturn's moon Titan is believed to most resemble a primitive Earth. Titan has a dense, hazy atmosphere as well as extensive hydrocarbon lakes and seas.

Question 3 One approach scientists use to detect exoplanets is to measure the "wobble" of a star, which is caused by the gravitational pull of an orbiting planet, using the Doppler Effect; in short, they are looking for a shift in the wavelength of the star's visible light. A second approach, the transiting method, looks for a characteristic decrease in a star's brightness when an orbiting planet passes in front of it.

Question 4 Space weather describes solar flares, solar winds, and coronal mass ejections emitted by the Sun. These emissions cause geomagnetic storms that can disrupt satellites, communications, and power systems. Better forecasting of solar weather will allow scientists to better anticipate and prepare for these potential disruptions.

Question 5 The Kuiper belt contains some of the most primitive and least thermally affected matter in the solar system. Scientists hope that data from Pluto and other Kuiper belt objects collected by the New Horizons mission will provide insights about the early history of our solar system, including its formation.

Question 6 Dust devils arise when surface heat is re-radiated to near-surface "air" (i.e., atmospheric gases). The heated air rises into cooler air above it, which can cause it to rotate. Because Mars has a comparatively thin atmosphere, the heated air is able to rise to much higher altitudes than would be possible on Earth.

Question 7 The size and orbit of an exoplanet are the primary indicators of potential habitability: scientists typically focus on planets that are roughly Earth-sized and orbiting reasonably close to their parent star. In addition, observations in certain spectral wavelengths may provide information about an exoplanet's temperature and composition, including whether it is likely to be rocky or contains certain indicator compounds like oxygen (O_2), ozone, water, and carbon dioxide.

Question 8 In August 1993, Galileo passed through the asteroid belt between Mars and Jupiter, capturing images of the asteroid 243 Ida. Those images revealed that 243 Ida has a satellite, later named Dactyl. Scientists had previously suggested that some asteroids probably had satellites, but this was the first confirmatory evidence.

Question 9 Gamma-ray bursts are extreme explosive events with luminosity a million times greater than the luminosity of an exploding-star supernova. They are associated with the deaths of massive stars, and are likely caused by the gravitational collapse of matter that results in black holes.

Question 10 Water is considered essential to support life, so the availability of water (even in the form of ice) on the Moon or Mars has important implications for possible future human exploration and/or habitation. In the case of Mars, evidence of once-extensive water bodies suggests that the planet may have once supported microbial life, although no data have been found to substantiate this.

Chapter 8

Answers to Review Questions for Chapter 8

Question 1 Every year (in Vienna, Austria). 1958.

Question 2 Ease tensions between the US and the former Soviet Union as both were becoming superpowers and both countries were flying reconnaissance missions over military installations. President Eisenhower understood that preventing the acquisition of data by another entity would be nearly impossible.

Question 3 Five. These include:

1. The *Treaty on Principles Governing the Activities of States in the Exploration and Use of Outer Space, including the Moon and Other Celestial Bodies* (Outer Space Treaty)
2. The *Agreement on the Rescue of Astronauts, the Return of Astronauts and the Return of Objects Launched into Outer Space* (Rescue Agreement)
3. The *Convention on International Liability for Damage Caused by Space Objects* (Liability Convention)
4. The *Convention on Registration of Objects Launched into Outer Space* (Registration Convention)
5. The *Agreement Governing the Activities of States on the Moon and Other Celestial Bodies* (Moon Treaty)

Question 4 The three basic issues that virtually every national and international law address include:

The *right to acquire* remotely sensed imagery/the right to launch remote sensing satellites
The *right to disseminate* remotely sensed imagery without prior consent of the sensed state
The *right to obtain* remotely sensed satellite imagery from a particular state

Question 5 This is a thought question that requires critical thinking; there is no right or wrong answer. One could say that it is an invasion of privacy (particularly on a national scale), but no more so than security cameras installed in urban areas and that it provides more benefits than harm.

Question 6 Space law was needed to ensure governance of this common space along with peaceful cooperation between countries in outer space. This includes harmonization of global navigation systems, the use of nuclear power in space, weather monitoring, maintaining safe operations in orbit, space debris issues, etc.

Question 7 A determination that the request is a hoax or that it is a means for a country to obtain information by false pretenses. Assistance is provided regardless of the political status of a country. Other reasons?

Question 8 A crisis situation that impairs a substantial land area or affects a significant population of a country due to human or natural causes (e.g., the Haitian earthquake, the Great Sendai Earthquake, the Indian Ocean earthquake and tsunami, the Black Saturday Bushfires in Australia, etc.).

Question 9 Pick any two discussed in the chapter.

Question 10 Thought question.

Question 11 With respect to remote sensing, transparency is open access without restrictions. Second part of the question is a thought question.

Question 12 Thought question; no right/wrong answer.

Chapter 9

Answers to Review Questions for Chapter 9

Question 1 Placement of LiDAR data acquisition devices on satellites, further development of Terrestrial Laser Scanners (TLS) and Airborne Laser Scanners (ALS), differential absorption LiDAR (DIAL) and laser Doppler.

Question 2 Detection of cars via very high resolution image acquisition at 50 cm or higher resolution enabling automated computer vision techniques leading to car model identification and the associated parameters.

Question 3 News agencies, Web sites, and a host of other visual media services will continue to benefit from the advances in remote sensing and continue to use remotely sensed data to provide current, relevant, and near-real-time geospatial information regarding events around the world. Google Earth is a good example.

Question 4 A change in the mission of the US Geological Survey (USGS) includes climate and land use change, which is currently developing a science strategy that will guide USGS prioritization for research in the 2020s, and aid in the USGS developing methods to explain how changes in land use, cover, condition, and management alter climate, impact natural systems, and affect human health and welfare.

Question 5 The use of remote sensing by researchers, along with NGOs and ordinary citizens to mobilize support for situations requiring policy development, that is, the crisis in Darfur.

Question 6 The integration of mobile technology with real-time remotely sensed geospatial data is already supporting the development of decision support tools for a variety of applications such as wildland fire risk and damage assessment, large-scale drought conditions and impacts, irrigation and water allocation systems, flooding risks and damage assessments, and public health as well as infrastructure development of maintenance. In addition, “apps” for mobile devices such as Android and iPhone are now being developed for routine applications and monitoring for disaster warning and emergency management.

Question 7 First, manufacturing, launching, and operating cost less than larger satellites. This facilitates "risk taking" during design and encourages technical innovation. Second, small satellites can be developed in only a few years or less. Shortening the time to launch adds resilience and flexibility to the satellite systems, allowing for nimble response to emerging issues such as the need for a deeper understanding of global climate change. Third, operating only a small number of instruments per satellite allows orbits to be optimized for a particular set of measurements.

Question 8 The AAAS Center for Science Diplomacy has been instrumental in advancing scientific engagement as an essential element of foreign policy and of building vital bridges between societies. Examples include involvement of Cuban and Myanmar government ministries and scientific NGOs to discuss health science, forestry, education, and the role of science in public policy.

Question 9 Challenges include the difficulties in storing massive and complex data, intensive irregular data access patterns, managing remotely sensed "Big Data" on multilevel memory hierarchy, optimal scheduling of a large amount of interdependent tasks as well as the efficient programming for these vast databases.

Question 10

- miniaturization and integration of electronics
- further development of UAV-based data acquisition systems
- increases in computational power such as heterogeneous parallel computing, cloud computing, and quantum and biological computing
- progress in large apertures and larger antennas
- increases in transmitter power for active systems
- miniaturization of optics
- increase in storage technology
- development of small satellites
- advances in screen technology and mobile computing
- increases in tunable systems and flexible frequencies
- advances in techniques for processing "Big Data"

Index

A
Aaronoff, S.G., 100
Accuracy assessment, 15, 99, 100, 102, 127
Ackleson, S.G., 135
Aerial photography, 1, 3, 4, 5, 6, 14, 89, 107, 164
  oblique, 4, 233
Airborne platforms, 1
Al Fugara, A.M., 99
Alparone, L., 79
Al-Saadi, J., 182
Anderson, J.R., 85, 86
Anderson, L.W.J., 213
Andréfouët, S., 215
Andrews-Hannan, J.C., 236
Aoki, S., 271
Armstrong, R., 135
Arnold, C., 107
Arroyo, L.A., 151, 154
Artificial intelligence, 95
Artificial neural network *See* Atmospheric correction, 105
Asner, G.P., 136, 137
Atmospheric correction, 71, 73, 203
Aumann, H.H., 194
Awrangjeb, M., 115

B
Bahro, B., 146
Baker, D.N., 247
Baker, J.C., 280
Ball, M., 280
Band, 75, 77, 107, 110, 148, 158, 165, 202, 204, 207, 236, 282
Banfield, D., 243
Banks, A.C., 211
Barber, K., 146
Barrachina, M., 145
Batalha, N.M., 251, 252
Beaugrand, G., 222
Ben-Jaffel, L., 250
Benna, M., 236
Bentley, R.D., 247
Berdyugina, S.V., 250
Bergen, K.M, 149
Berkelman, R., 217
Bernstein, A., 163
Bibring, J.P., 239
Binder, A.B., 233
Bjorgo, E., 163
Blaisdell, E.A., 91
Blaschke, T., 98
Blauth, D.A., 158
Blossey, B., 214
Bongiovanni, R., 157
Borucki, W., 252
Bowman, D.M.J.S., 150
Boynton, W.V., 239
Brando, V.E., 111
Brightness, 76, 87, 99, 111, 202, 250
Bristow, C.S., 188
Brivio, P.A., 135
Brodley, C.E., 110

C
Cablk, M.E., 113
Cakir, H.I., 79, 81, 82, 102
Carleer, A.P., 107, 112, 113
Carpenter, S.R., 132
Cecchini, F., 158
Celik, T., 88
Cemin, G., 158
Centeno, J.A., 187, 188, 189
Change detection, 74, 82, 94, 99, 103, 105, 106, 107, 113, 114, 125, 129, 195
Chang, K., 246

S. Khorram et al., *Principles of Applied Remote Sensing*,
DOI 10.1007/978-3-319-22560-9

Chapman, C.R., 244
Charbonneau, D., 252
Charles, H., 213
Chassot, E., 221
Chave, J., 147, 168
Chaves, M.M., 159
Chavez, P.S., 81
Chen, D.M., 112, 113
Chen, Q., 114, 115
Chung, P., 95
Church, J.A., 193, 222
Chuvieco, E., 151, 154
Cipar, J.J., 184, 185
Clark, D.B., 114
Classification, 71, 79, 84, 85, 113, 114, 126, 127, 129, 163, 165, 211, 215
  fuzzy logic, 93, 94
  neural network, 95
  object-oriented, 79, 98, 99
  supervised, 84, 87, 89, 92
  unsupervised, 84, 88, 89
Clavin, W., 229
Cloude, S.R., 149
Cloud, J., 6, 7
Clustering, 85, 88, 89, 96, 97
Cocks, T., 145
Cohen, J.A., 102
Color composite, 69, 79, 89, 91, 178
Comiso, J.C., 189
Congalton, R.G., 15, 100, 154
Costanza, R., 213
Cowardin, L.M., 85
Cronin, F., 165
Curran, L.M., 136
Cuzzi, J.N., 244

D
Dahl, T.E., 132
Dai, X.L., 95, 113
Dare, P.M., 113
Data acquisition, 2, 16, 73, 75, 135, 283
Davis, C.H., 98
Davis, C.J., 247
Dekker, A.G., 111, 134, 135
DeLaune, R.D., 132
De Montluc, B., 266
Deshayes, M., 154
Des Marais, D.J., 252
Diaz, R.J., 210
Digital elevation model (DEM), 14, 114
Digital numbers, 76
DiMassa, D.D., 217, 218
Dobson, M.C., 149
Doney, S.C., 205
Donoghue, D.N.M, 148
Dowman, I., 115, 271
Dressing, C.D., 252
Ducati, J.R., 158
Dukes, J.S., 213
Dusseux, P., 143

E
Earth observing system (EOS), 7, 135
Edgett, K.S., 239
Ehrlich, D., 103
Electromagnetic spectrum, 1, 70, 110
  electromagnetic radiation, 11
  infrared, near infrared, 6, 141
  short-wave infrared, 184
  thermal infrared, 185, 282
Ellis, E.C., 114
Elphic, R.C., 236
Elvidge, C.D., 111
Englhart, S., 149

F
False color composite *See* Color composite, 70
Feature space plots, 87
Feierabend, J.S., 132
Feliciano, E., 147
Ferrarese, L., 249
Filters, 99, 100, 115
Fisher, J., 113
Fisher, P.F., 84, 94
Flusser, J., 73
Foing, B.H., 233
Fonseca, L.M.G., 73
Foody, G.M., 100
Forster, B.C., 73
Franklin, S.E., 15, 132
Fregoni, M., 158
Frequency, 11, 85, 99, 154, 218, 232, 277
Friedl, M.A., 109, 110
Fusion, data, 71, 78, 79, 81, 82, 115
Fuzzy logic *See* Classification, 94

G
Gabrynowicz, J.I., 261, 264, 269
Galvão, L.S., 111
Gao, B.C., 74
Gates, D.M., 6
Gehrels, N., 229
Geissler, P., 241
Gendre, B., 229
Geographic information systems/Geographic information science (GIS), 10, 109
Geometric correction, 71, 72

Ghosh, A., 272
Gibbons, C., 107
Giles, K.A., 189
Gitas, I.Z., 99
Goetz, A.F.H., 231, 278
Goetz, S., 113, 114
Gombosi, T.I., 244
Gong, P., 127, 128
Gonzalez, R.C., 88
Goodale, C.L., 146
Goodchild, M.F., 15, 100
Goodenough, D.G., 111
Good, S.A., 195
Greeley, R., 241
Green, K., 15
Gross, J.E., 103
Ground control points, 74
Gupta, V., 163
Gyan, K., 188

H

Haack, B.N., 103
Hagen, A., 84, 95
Hall, A., 158
Han, E., 249
Hanel, R.A., 231
Hanna, E., 191
Hansen, M.C., 136
Hargrove, W.W., 140, 141
Harrington, J.D., 229
Harris, R., 264
Hartley, J., 278, 279, 283
Haruyama, J., 233
Hashim, M., 146
Haykin, S., 95
Henbest, N., 245
Henebry, G.M., 106
Herken, G., 262
Hernandez-Leal, P.A., 152, 154
Herold, M., 99, 111
Hester, D.B., 15, 84, 89, 91, 94, 102, 129
Hirsch, R., 2
Histogram, 74, 81
Hodgson, M.E., 115
Holmgren, J., 115
Homer, C., 113
Hooker, S.B., 221
Hord, R.M., 91
Houghton, R.A., 143, 146
Hovel, K.A., 213
Howarth, R., 210
Howat, I.M., 192
Huebert, R.N., 261
Huete, A., 77
Hunter, J.R., 218
Hyyppä, J., 148

I

Illera, P., 154
Image analysis, 131, 164
Image classification *See* Classification, 83
Image data acquisition *See* Data acquisition, 145
Image processing, 1, 8, 14, 15, 16, 100, 109, 112, 114, 280
  post-processing, 69, 99, 100, 114
  preprocessing, 71, 73, 75, 83
Image registration, 73, 100, 113
Image segmentation, 79, 98, 99, 113
Infrared *See* electromagnetic spectrum, 74
Ingersoll, A.P., 244
Ioannis, M., 127
Irlandi, E.A., 213
Ito, A., 265

J

Jain, A.K., 88
Jaiswal, R.K, 154
Jakhu, R., 264
Jensen, J.R., 10, 13, 74, 85, 87, 88, 89, 91, 94, 95, 96, 103, 106, 109, 111, 135, 136
Jewitt, D., 247
Jiang, Z., 77, 78
Jickells, T.D., 188
Johansen, K., 113
Johnson, L., 157, 158
Johnson, R.D., 163
Jolly, R., 132

K

Kalas, P., 249
Kaltenegger, L., 252
Kampe, T., 277
Keane, R.E., 151
Keeley, J.F., 261
Kennedy, R.E., 103
Kerr, J.T., 10, 15, 132
Kerr, R.A., 245
Keywood, M., 150
Khorram, S., 15, 71, 73, 75, 77, 79, 81, 82, 89, 93, 95, 97, 100, 106, 109, 113, 127, 135, 150, 151
Kirk, R.L., 239
Klemas, V., 135, 210
Klotz, I., 249
Knudby, A., 218, 219
Knutson, H.A., 250
Koh, L.P., 136, 137

Kohorram, S., 100, 106
Koike, T., 272
Konecny, G., 15
Konopliv, A.S., 233
Koren, I., 188
Kreidberg, L., 250, 251
Kreuger, A.J., 185
Kuemmerle, T., 136, 137
Kufoniyi, O., 271
Kurvonen, L., 149
Kutser, T., 111
Kwarteng, A.Y., 81

L
Lacar, F.M., 158
Lachavanne, J.B., 135, 214
Lamb, D.W., 158
Lambin, E.F., 103
Lam, N.S.N., 106
Land use/land cover classification, 92, 96, 99, 102, 103, 113, 125, 132
Langeland, K.A., 213
Langmann, B., 150
Laporte, N.T., 136, 137
Larkum, A.W.D., 213
Lathrop, R.G., 134, 135
Latifi, H., 145
Lee, D.H., 79
Lee, J.K., 87
Léger, A.O., 252
Leghorn, R.S., 262
Lehmann, A., 135, 214
Le Toan, T., 149
LiDAR, 69, 114, 115, 145, 147, 148, 201, 216, 232, 278
Lillesand, T., 91
Lillesand, T.M., 134, 135
Lindell, T., 114
Lindsey, G., 107
Linkie, M., 136, 137
Lipcius, R.N., 213
Liu, Y., 182
Loehr, R.C., 132
Longley, P.A., 10
Lowenberg-DeBoer, J., 157
Lucas, L.A., 149
Lucas, R.M., 148
Luckman, A., 149
Lu, D., 103, 105, 106
Lunetta, R.L., 100
Lurie, I., 280
Lurie, J., 279
Luz, N.B., 158

M
Macauley, M.K., 264
Madden, C.J., 132
Madsen, J.D., 213
Mahowald, N.M., 188
Maina, J., 220
Majeau, C., 250
Majumdar, T.J., 95
Malin, M.C., 239
Manjunath, B.S., 73
Masogo, R., 131
Mather, P.M., 110
Mattison, D., 2, 3
Mausel, P.W., 87
McCarthy, J.J., 7
McCauley, S., 113
McClintock, W.E., 238
McEwen, A.S., 232, 240, 243
McIver, D.K., 109
Meng, X., 115
Menzel, W.P., 151
Mészáros, P., 229
Miller, J.E., 99, 115
Minor, T.B., 113
Misbari, S., 146
Mitchard, E.T.A, 149
Mitri, G.H., 99
Modeling, 99, 108, 113, 114, 115, 125, 132, 134, 146, 151, 154, 177, 189, 202, 214, 218, 224, 277, 281
Mohanty, K.K., 95
Moore, G.K., 135
Moore, P., 245
Moran, E.F., 137
Moran, M.S., 157
Morisette, J.T., 100
Muchoney, D.M., 103
Multispectral, 13, 14, 69, 79, 81, 82, 113, 146, 151, 159, 186, 221, 235, 268, 278
Mumford, G., 165
Munyati, C., 135
Murray, J.B., 240
Murthi, K.R.S., 264

N
Nakamura, R., 233
Narumalani, S., 135
Nelson, S.A.C., 82, 102, 105, 114, 132, 133, 134, 135, 214
Nemani, R.R., 157
Neukum, G.R., 240
Neumann, G.A., 238
Neural network *See* Classification, 95

Newhall, B., 2
Nichols, C., 213
Normalized Difference Vegetation Index (NDVI), 76, 131, 140, 152, 158
Nowak, P., 155, 158
Nozette, S., 232
Núñez, J., 82

O
Object-oriented classification *See* Classification, 79
Oblique aerial photography *See* Aerial photography, 2
Olivier, J.G., 150
Orbit, 7, 8, 9, 112
  Geostationary, 278
Ostrovsky, M., 10, 15, 132

P
Pagano, T., 277
Paige, D.A., 238, 239
Pala, C., 218
Pal, M., 110
Panchromatic, 69, 112
Pandey, U., 149
Pan, X., 211
Papathanassiou, K., 149
Parcak, S., 165
Pathirana, S., 84, 94
Pattern recognition, 15, 95
Payloads, 2, 284
Pedlowski, M.A., 137
Pennisi, E., 281
Penuelas, J., 214, 135
Peri, F., Jr., 279
Pesnell, W.D., 248
Petersen, G.W., 132
Peters, G., 269
Peters, S.W.M., 134, 135
Petigura, E.A., 252
Petro, C.M., 233
Philipson, P., 114
Phinn, S., 113
Pierce, F.J., 155, 158
Pieters, C.M., 233
Pixel, 13, 72, 73, 79
Platforms, 1, 2, 14, 131
Plaza, A.J., 279
Pohl, C., 79, 83
Post-processing *See* Image procssing, 99
Prata, A.J., 185, 186
Preprocessing *See* Image processing, 282
Principal components analysis (PCA), 81
Pringle, H., 165
Processing *See* Image processing, 95
Prospero, J.M., 188, 189

Q
Qiu, F., 95

R
Raber, G., 115
Rack, W., 191
Radar, 115, 177, 180, 184, 185, 201, 232
  radar interferometry, 115
  synthetic aperture radar (SAR), 79, 147, 148, 191, 217
Radiometric correction, 71, 73, 74, 75
Rahman, H., 135
Ramsey, E., 111
Rao, M., 264
Read, J.M., 106, 113
Reagan, R.W., 269
Rees, G., 4, 5, 6
Registration *See* Image registration, 100
Resolution
  radiometric, 111
  spatial, 107, 111, 112, 135
  spectral, 133, 215, 262
  temporal, 112, 132, 154
Richardson, 213
Ricketts, P.J., 135
Rix, M., 185
Robert, P.C., 158
Robinson, M.S., 232
Roccio, L., 269
Rocha, A.V., 78
Rogan, J., 112, 113
Ropert-Coudert, Y., 281
Rosenberg, R., 210
Rottensteiner, F., 115
Rott, H., 191

S
Sabins, M.J., 88
Samuel, H., 159
Sanders, R., 249
Santoki, M., 212
Santoro, M., 149
Sasselov, D.D., 250
Sasser, C.E., 132
Saugier, B., 146
Sawaya, K.L., 114
Schneider, W., 270
Schnur, M.T., 77
Schott, J.R., 74

Seelan, S.K., 114
Seguin, G., 158
Serrano, L., 159
Shackelford, A.K., 98
Shaver, G.R., 78
Shaw, B., 213
Shearman, P.L., 138, 139, 140
Shinn, E.A., 188
Short, N.M., 6, 9, 10, 14, 231, 232
Shutter control, 268, 272
Sidle, J.G., 114
Simone, G., 79
Simpson, J.H., 218
Simpson, J.J., 221
Sithole, G., 115
Skirving, W.J., 217
Smith, B.A., 239, 245
Smith, L.J., 264
Smrekar, S.E., 240
Sohn, G., 115
Solberg, A.H.S., 79
Song, C., 135
South, S., 91
Sparks, R.S.J., 184
Spectral response curves, 70
Spencer, M.M., 73
Sprigg, W., 187, 189
Spruce, J., 140, 141, 142
Spudis, P.D., 235
Stafford, J.V., 158
Stefanov, W.L., 109
Stehman, S.V., 100
Steig, E.J., 191
Stephens, B.B., 146
Stern, S.A., 246
Stevenson, K.B., 250
Stofan, E.R., 245
Stone, B., 107
Stratton, R.D., 151
Stroppiana, D., 150
Struck, D., 189
Supervised classification *See* Classification, 88
Swain, M.R., 250
Swap, R., 188

T

Thomas, N., 99, 112, 113
Thompson, K.P., 269
Thurston, J., 280
Tim, U.S., 132
Titov, D.V., 238
Tou, J.T., 88
Townsend, P.A., 103
Training sites *See* Classification, 89
Trigg, S.N., 136
True color composite (TCC), 70, 79, 178
True color composite *See* Color composite, 91
Tsai, C., 229
Turner, R.E., 73
Turner, W., 132

U

UNHCR, 163
Unsupervised classification *See* Classification, 98
Utterback, J.M., 4

V

Valley, R.D., 214
van Aard, J.A.N., 85
van Donkelaar, A., 182
Vane, G., 278
Van Genderen, J.L., 79, 83
van Leeuwan, C., 158
Vasavada, A.R., 243
Vaughan, D.G., 193
Velicogna, I., 191
Verbyla, D.L., 132
Verlinden, A., 131
Vertical aerial photography *See* Aerial photography, 2
Vosselman, G., 115

W

Wald, L., 79
Wang, H.Q., 114
Wardlow, B.D., 77
Watt, P.J., 148
Wavelength, 11, 70, 86, 184
Waycott, M., 213
Wear, D.N., 132
Weise, D.R., 150
Wersal, R.M., 213
Whelley, P.L., 241
White, N.J., 193
Wickham, J., 127
Wieczorek, M.A., 236
Wilde, S.B., 213
Williams, M., 264
Williams, S.K., 213
Wilson, J., 107
Wilson, R.P., 281
Winker, D.M., 114
Witze, A., 7

Wolff, E., 112
Wood, R.D., 3
Woolley, J.T., 269
Wright, C.S., 150
Wulder, M.A., 15

X
Xiao, X., 77

Y
Yang, C., 95
Yu, Q., 98

Z
Zadeh, L.A., 94
Zebker, H., 245
Zelazny, J.M., 132
Zhang, K., 106, 114, 115
Zitová, B., 73
Zuber, M.T., 236
Zurek, R.W., 240

Printed in the United States
By Bookmasters